PIMLICO

387

THE CHARGE

Mark Adkin was commissioned into the Bedfordshire and Hertfordshire Regiment in 1956 and served with it and the Royal Anglian Regiment in Malaya, Mauritius and Aden. On leaving the British Army he transferred to the Gilbert and Ellice Islands and was one of the last British District officers anywhere in the world.

His final overseas posting was as a contract officer for five years with the Barbados Defence Force, and it was as the Caribbean Operations Staff Officer that he participated in the US invasion of Grenada in 1983. Since 1987 he has lived with his family in Bedford and has written seven books on military subjects, including *Urgent Fury*, *Goose Green* and *Prisoner of the Turnip Heads*.

THE CHARGE

The Real Reason Why the
Light Brigade Was Lost

─────────

MARK ADKIN

PIMLICO

Published by Pimlico 2000

2 4 6 8 10 9 7 5 3 1

First published in Great Britain by
Leo Cooper, an imprint of Pen & Sword Books Ltd, 1996

Pimlico
Random House, 20 Vauxhall Bridge Road,
London SW1V 2SA

Random House Australia (Pty) Limited
20 Alfred Street, Milsons Point, Sydney,
New South Wales 2061, Australia

Random House New Zealand Limited
18 Poland Road, Glenfield,
Auckland 10, New Zealand

Random House South Africa (Pty) Limited
Endulini, 5A Jubilee Road, Parktown 2193, South Africa

Random House UK Limited Reg. No. 954009

A CIP catalogue record for this book
is available from the British Library

ISBN 0-7126-6461-0

Papers used by Random House UK Limited are natural,
recyclable products made from wood grown in sustainable forests.
The manufacturing processes conform to the environmental
regulations of the country of origin

Printed and bound in Great Britain by
Biddles Ltd, Guildford

To my mother and father, God bless them.

Contents

List of Maps and Sketches   ix

Symbols and Abbreviations used on Maps   x

Introduction and Acknowledgements   xi

PART ONE – THE LIGHT BRIGADE

1   'The Light Brigade will advance . . .'   1
2   The Four Horsemen of Calamity   21
3   To War   41

PART TWO – THE ORDERS

4   Orders for the Infantry   67
5   Two Orders for the Cavalry   90
6   Two More Orders for the Cavalry   115

PART THREE – THE CHARGE

7   Running the Gauntlet (1)   139
8   Running the Gauntlet (2)   152
9   Through the Guns   175
10 Back up the Valley   194

PART FOUR – THE CONSEQUENCES

11 The Reckoning   215
12 The Recriminations   230
    Epilogue   247
    Appendix 1: The Light Brigade   254
    Appendix 2: The Cavalry Division   255
    Appendix 3: The Allied Order of Battle   256
    Appendix 4: The Russian Order of Battle   257
    Appendix 5: The Last Survivors   258
    Appendix 6: A Gun Accident, 1811   260

Sources   261

Notes   264

Index   275

Maps and Sketches

MAPS

1.	The Turkish Empire 1854	22
2.	Operations in Turkey	45
2a.	Cardigan's Sore-Back Reconnaissance	45
3.	The Allied March to Balaclava	55
4.	The First Clash – on the Bulganek	57
5.	The Strategic Situation – 25 October, 1854	69
6.	Balaclava Battlefield – The Ground	73
7.	Tactical Situation about 0730 25th October	76
8.	Tactical Situation about 0800 25th October	92
9.	Tactical Situation about 0830 25th October	95
10.	Tactical Situation 0900–0915 (Heavy Brigade attack)	107
11.	Tactical Situation 0930 (Raglan issues 3rd Order)	116
12.	Tactical Situation 1055 (Raglan issues 4th Order)	122
13.	Running the Gauntlet	141
14.	Beyond the Guns	180
15.	Tactical Situation 1200 (Raglan decides not to attack)	211

SKETCHES

1.	The Light Brigade on the Start Line	13
2.	Raglan's view at 0730 – infantry ordered up	66
3.	Raglan's view at 0800 – 1st Order to the Cavalry	93
4.	Raglan's view at 0830 – 2nd Order to the Cavalry	98
5.	Raglan's view at 0930 – 3rd Order to the Cavalry	123
6.	Raglan's view at 1055 – 4th Order to the Cavalry	128
7.	The Charge – viewed from the Fedioukine Heights	169

Symbols and Abbreviations Used on Maps

	British	French	Turkish	Russian
Infantry				
Company	▣		▣	▢
Battalion	▣		▣	▢
Regiment				⌣
Brigade	▣ (x)	▭ (x)		
Division	▣ (xx)		▣ (xx)	
Cavalry				
Squadron	◿	⧅		◻
Regiment	◿			◻
Brigade	◿ (x)	⧅ (x)		◻ (x)
Artillery				
Battery	‖—6—‖			‖—8—‖
Other				
Camp	△△△			
Vineyard	▭V			

Abbreviations

• C	Cardigan, Earl of	• M	Mayow, Lt-Col.
C	C Troop RHA	• N	Nolan, Capt.
E	Espinasse	• R	Raglan, Lord
HB	Heavy Brigade	RHA	Royal Horse Artillery
93H	93rd Highlanders	RM	Royal Marines
I	I Troop RHA	T	Turks
• L	Lucan, Lord	V	Vinoy
LB	Light Brigade	W	W Battery RA

Introduction and Acknowledgements

Today the Crimea belongs to Ukraine, not Russia. Sevastopol, until recently, was the main base for the Soviet Black Sea Fleet. After the break-up of the Soviet Union the Fleet idled, while politicians squabbled fiercely over ownership. The Fleet was to be divided; officers and crews could opt for which navy they wished to belong to; meanwhile both countries declined to pay the sailors. Money for fuel, money for spares, money for maintenance, money for everything, disappeared. A navy was dying at anchor.

Five years on not much has changed. Sevastopol has a seedy, neglected look. It, and Balaclava, belong geographically to Ukraine yet it is the Russians who have military control of the district. The naval facilities and dry dock are seldom used. At Inkerman a large store is heavily guarded while decisions are made as to if, how and when the nuclear missiles inside should be destroyed. Balaclava harbour is home to a few Tartar fishing boats and the dark hulks of nuclear submarines. If you have enough western cigarettes you can transform the sullen, suspicious sailors into friendly, smiling individuals eager to please. It is possible to enter the restricted areas such as that on the eastern promontory at the entrance to the harbour. From here there is an excellent view of the submarines moored along the western side, and of the cavernous hole in the side of the mountain opposite. Submarines sail – vanish – inside the mountain. It provides the perfect shelter, the perfect hiding place for these redundant, rusting, nuclear monsters. Two packets of Marlboro are sometimes sufficient to get invited on board one of these atomic submarines. Inside the stench is awesome, the decay obvious. One gets the impression that not too many dollars would be sufficient to buy the boat and sneak away.

The area to the north was fought over bitterly during the Great Patriotic War, as the Russians call the Second World War. The Sapoune escarpment, from where Raglan watched events unfold, was defended by the Germans in 1944 against Russians advancing on Sevastopol from the north and east. It was a perfect natural obstacle to defend, just as it had been 90 years before. Colonel-General Jaenecke, the commander of the German 17th Army, had virtually the same view as the Allied Crimean generals when he watched Russian guns deploy once again into the North Valley, tanks instead of horsemen roar across the plain, and Russian infantry debouch from the Fedioukine hills. On 8 May, 1944, the vicious struggle for the Sapoune Heights began in earnest. This time the Russians swept into Balaclava, up the escarpment and on to victory.

The Balaclava plain continues to change as more roads are built and new villages spring up and old ones expand. The floor of the North Valley, down which the Earl of Cardigan led the Light Brigade, is now one large vineyard with some dreadful three-story modern buildings under construction in the middle of it. These eyesores are almost precisely at the mid-point of the charge. It is therefore no longer possible to stand where the Light Brigade was drawn up and look down to where the Russian battery was deployed. Fortunately, however, the contours of the land, the hills and the undulations remain much as they were. From near (it is now not possible to stand on the exact spot) Raglan's old viewpoint on the Sapoune it is not hard to reconstruct the changing situation as he saw it below him over 140 years ago. That is what this book sets out to do.

Much has been written about the Battle of Balaclava and the charge of the Light Brigade. It has a special niche in our history books, alongside Hastings, Agincourt and Waterloo. Because of this, and because it was such a controversial event, such an obvious blunder, and so many recriminations and accusations flew around for years afterwards, it still intrigues historians, writers and the public alike. This book attempts to sift the fact from the fiction. As the reader will discover, if he perseveres, the story of the charge itself, gripping and dramatic as it is, owes much of its fascination to having to decide who was to blame for launching it. Four men were particularly involved in its conception and execution, and historians have argued vehemently as to how the responsibility should be shared between them. The man who got the sack for it was the Cavalry Divisional Commander, Lord Lucan – the great-great grandfather of the Lord Lucan who disappeared over twenty years ago. Long and loud were Lucan's complaints of injustice. Was he the scapegoat? If so, who was at fault?

The object of this book is to put the reader as nearly as possible in the saddles of those responsible for issuing the orders that set the charge in motion, and of the participants themselves. To make an informed judgement on an order it is essential to see the situation as the person giving it saw it at the time. Knowledge gained through hindsight should be ignored. This is a far from easy task. It involves knowing the personalities involved intimately, knowing their feelings towards others and their mood, as well as the tactical circumstances.

Perhaps the most neglected aspect of all studies of the Light Brigade's charge has been the lack of a proper understanding of the ground. It is the ground that shapes a battle. The ground dictates what is possible tactically, and what can or cannot be seen from any given point. Military orders should always be given with the ground in mind. Nevertheless, no book on the subject has a contoured map of the battlefield. Without one, supplemented by a visit to the actual site, it is impossible to be sure who could see what, and from where. On that bright morning in October, 1854, this was a crucial factor in sending the Light Brigade down the North (wrong) Valley.

At Balaclava Raglan issued four orders, the last of which launched the Light Brigade. With the aid of maps and sketches I attempt to put the reader alongside Raglan when he made his decisions. Similarly, he will sit beside those who received the orders and acted upon them. While there can be no guarantee that all the maps are absolutely accurate, they have been compiled from the study of existing maps, a contoured map, the accounts of participants and visits to the battlefield. Where doubts still exist a decision has been made, on troop locations for example, based on time and distance and military common sense.

The battlefield sketches are an attempt to see things, at least approximately, as they were on the day at certain key stages of the battle. Most of them show the scene from the viewpoint of the Commander-in-Chief high above the plain when he had decisions to make, orders to give. All except two have been drawn from a unique photograph. It was taken from Raglan's actual viewpoint only a year after the charge by the well-known Crimean photographer, Roger Fenton. The original is in the Royal Archives in Windsor Castle, and consists of five pictures that join to give a panoramic view from Balaclava on the right to the Chernaya River on the left. In the centre is the North Valley, the Causeway Heights, and much of the South Valley. The excavations of several of the redoubts can be clearly seen. By modern standards the picture quality is poor, but it has proved invaluable for the artist in producing realistic sketches of the terrain features on which to superimpose troop dispositions.

Almost all previous accounts conflict over the important details such as the starting-point of the charge, the position of the battery being attacked or the number of guns in it. No author, to my knowledge, has assessed the number of guns actually able to bring fire to bear on the chargers, the time they were in the guns' arcs of fire, and the likely number of rounds they could fire in that time. Nobody seems to have studied the actual casualties to estimate how many men reached the guns – important in assessing the success or otherwise of the charge itself. This book tackles these issues, as well as those of blame, misinterpretation, sloppy staff-work, personality clashes and the friction of war, in an effort to find, as conclusively as possible, the real reason why the Light Brigade was lost.

Many people have helped and encouraged me in this project, and I take this opportunity of thanking them all most sincerely. A book of this nature would have been impossible without them. At this point I must make it clear that all the individuals I mention belong to The Crimean War Research Society. I count my own joining of the Society over a year ago as probably the most worthwhile decision I made before starting to write this book. The Society is a goldmine of expertise on the Crimean War, with its members enthusiastic and eager to help.

Firstly, 'Rifleman' Richard Moore of the 95th Rifles. Richard has recently spent months 'campaigning' in the Crimea as the military adviser to Sharpe's Film Company. The filming of the Peninsular War adventures of Sharpe, in Bernard Cornwell's immensely successful series of books, took place not far from Balaclava. Richard's frequent and lengthy wanderings all over the battle-field have made him an expert on the ground. He has put his knowledge and his photographs unstintingly at my disposal and I am much in his debt.

Next there is Major Colin Robins OBE, an ex-Gunner and editor of the

Society's journal, *The War Correspondent*. He bravely undertook to read, correct, criticize and comment on the completed manuscript. He did so with considerable skill and tact, pointing out a number of pitfalls into which I was about to tumble. His professional knowledge of guns and gunnery was indispensable as the Light Brigade's story is almost entirely concerned with the capabilities of cavalry and artillery, the one against the other.

Mr David Cliff, the secretary and treasurer, was my contact point for seeking information. If I needed an answer on any aspect of the battle, uniform details, or personalities David would invariably know where to find it – or who would know, if he did not. I much appreciate the time and trouble he took to send me information and photocopies from books and articles to which he had access.

The staff artist of the Society, Mr Eric Kemp, deserves a special mention. What I hope sets this book apart from others on the subject is the use of battlefield sketches which depict the situation at critical moments. Eric's drawings, I believe, do this magnificently, and bring the battle to life in a way that more words could never accomplish.

I would like to express my gratitude to Mr Edward Brudenell who most kindly, and at short notice, allowed me to come on a private visit to his lovely home in Deene Park. It was a fascinating experience to be able to explore the place where the Earl of Cardigan lived and died, and to see and photograph so many mementoes of that time, including the head of Cardigan's horse, 'Ronald'.

Finally, I wish to thank Mr Garry Farmer, Mr Norman Gordon and Mr Tony Kitchen. Each of them has visited the battlefield, each of them has given me photographs taken there with permission to use them, and each has taken great pains to assist in interpreting how and where things happened.

I hope they will all feel the book has done justice to an enthralling subject. If it has not, then the fault is mine not theirs.

I

THE LIGHT BRIGADE

CHAPTER ONE
"The Brigade will advance . . ."

'The charge will always remain the thing in which it
will be the cavalryman's pride to die sword in hand'
The Cavalry Journal (1909)

The Light Brigade was having a late breakfast in the field. It was just before
eleven o'clock and the officers and troopers were mostly sitting on the ground
in the pale sunshine. There was much puffing of pipes, almost a picnic atmosphere, although horses remained saddled and bridled: all this with the enemy
in sight, and to the left front almost within effective artillery range. To an uninformed observer the lack of activity, the lack of firing, apart from an
occasional cannon shot away to the south-east, and the stillness that had
encompassed the battlefield for the past hour would surely have indicated that
fighting was at an end.

Colonel Lord George Paget, a son of Wellington's cavalry commander at
Waterloo, was the commanding officer of the 4th Light Dragoons. He had just
resolved an unfortunate difference with his second-in-command, Major John
Halkett. They had not been on speaking terms for some time. Paget heard
Halkett, who was officially on the sick list and as such need not have been on
parade, ask 'one of the group of which I formed part for some rum, and on his
replying that he had not any, I said, "Major, I can give you some", . . . he
accordingly profited by this offer, and thanked me for it.'[1] Paget was never to
forget this small incident as within half an hour Halkett was dead. A few
minutes later Paget lit a cigar.

Barely had he lit it when the familiar notes of a bugle sounding off came
from the direction of the Causeway. As the last echoes died away the Brigade
orderly trumpeter, 'Billy' Britten of the 17th Lancers, repeated the call –
'Mount'. All was bustle and movement as pipes were knocked out, uneaten
food stuffed into pockets or pouches and nearly 700 men settled themselves in
their saddles. Although Britten was a trumpeter he and the others carried and
used a bugle when mounted. All cavalry and horse artillery units had trumpets

1

for use when dismounted – in barracks or in the field. For convenience, however, the bugle, being smaller, lighter and easier to handle, was used when on mounted operations or field training.

Like the rest, Paget sat waiting for further orders. Some five minutes or more elapsed before he saw the brigade commander cantering towards him. Cardigan's instructions to his fellow peer were short and succinct.

"Lord George, we are ordered to make an attack to the front. You will take command of the second line, and I expect your best support, mind, your best support."[2]

Paget was niggled by Cardigan's heavy stress on the words 'best support', and the way he repeated this phrase, giving the distinct impression of a lack of trust. He deliberately responded with equal emphasis:

"Of course, my Lord, you shall have my best support."[3]

Cardigan wheeled his horse and spurred back to the front of his brigade, leaving Paget not much wiser as to what was required. If that was all a colonel commanding a regiment and the supports for a brigade attack was told it is not to be wondered at that little could be passed to the troopers, although realization of what was in store was to come quick enough.

Officers and senior NCOs took up their positions in the front rear of their squadrons, while the troopers sought to dress ranks in accordance with the shouted instructions of the sergeant-majors. While this process was being completed the officers in front sat patiently facing the leading ranks. When the lines were straight the officers turned about to face down a valley.

Five regiments of blue-uniformed horseman had been drawn up in two unequal lines, five regiments of light cavalry, with a combined strength of some 664 all ranks.[4] The Brigade was a sickly shadow of its former self, each regiment parading with less than half the numbers that had embarked some five months earlier. This so-called brigade, which was commanded by a major-general (Cardigan had recently been promoted), was about equal in numbers to a single regiment at a proper war establishment. Cholera, not casualties, had decimated the Allied armies. Cholera killed quickly. Men apparently healthy in the afternoon were often dead by the following morning. It could be even quicker. The first victim of the 4th Light Dragoons to die of it was 'taken ill at 8, died at 3, and buried at 7, simply wrapped in a blanket, thrown into a hole, and field officer read the burial service'.[5] Copious vomiting, violent diarrhoea, agonizing cramps in the stomach, legs and feet totally incapacitated the victim. Within hours the body became blueish, cold to touch, with the features pinched and eyes sunken. With an attack of this severity death was the only outcome. There was not a man in the brigade who had not lost a friend to this dreaded disease. August had been an awful month. So many had died that regiments had to borrow trumpeters to sound the Last Post at gravesides. The

2

commanding officer of the 11th Hussars felt the whole business so bad for morale that he had forbidden funeral music altogether.

The riders were now sitting in their saddles facing east. Visibility was good so the front ranks of the three regiments in the first line could see easily down the gently sloping, wide valley, with low hills on either side. The view was unobstructed. From where they were the valley, never less than 800 metres across, stretched for some 3,000 metres before twisting slightly to the left and disappearing down to the River Chernaya. The 'going', as the cavalry called the state of the ground, was good, except where grass gave way to plough. Here, perhaps, the soft, soggy soil might slow progress. Horse artillery would certainly wish to avoid it.

About 2,000 metres from where they sat they could see a Russian artillery battery of eight guns deployed facing up the valley. Although at that distance it was difficult for individuals to count the cannons, there was an occasional glint of sunlight off a brass barrel and an impression of movement in a dark mass beyond the guns, probably enemy cavalry, which were discernible. More obvious were enemy positions on the higher ground on either side of the valley. Artillery and infantry were evident half left, on the northern slopes, which although steep in places, never exceeded a height of 200 feet above the valley floor. Such hills hardly merited their title of the Fedioukine Heights. These troops were the closest, possibly just within effective roundshot range.[6] Only a slight turning of the head to the right revealed yet more Russian positions on what was known as the Causeway Heights - another misnomer, as the summits of the knolls that were interspersed along a low east-west ridge only slightly exceeded the Fedioukine Heights. Here further guns and infantry were visible, but at least 1500 metres away.

For the moment the tight-packed squadrons were reasonably safe. An advance of 200 metres or more was necessary before it would be worthwhile for the guns on the northern hills to open fire. Distance dictated silence. There was no noise except for the jangle of bits and harnesses as horses shook their heads, or saddle leather creaked as a horseman shifted his weight.

The men knew what the next order would be; they understood what they had to do and realized that in the next few minutes luck was going to decide who lived or died. Before mounting they had watched as one of Brigadier-General Airey's ADC's (Airey was the Quartermaster-General at Lord Raglan's headquarters), Captain Louis Nolan of the 15th Hussars, had galloped past some five minutes earlier with a message for the cavalry division commander, Lieutenant-General Lord Lucan. Nolan had passed through the interval between the 13th Light Dragoons and the 17th Lancers; as he did so Private James Wightman, who thought him, 'an excitable man in an excited state', heard him shout to his commanding officer, Captain Morris,

"Where is Lord Lucan?"

"There", replied Morris, pointing, "there on the right front!" He then added,

"What's it to be Nolan? - are we going to charge?" To which Nolan yelled over his shoulder,

3

"You'll see! You'll see!"[7]

A minute or so later they had seen Lucan, with his little knot of staff officers (including Nolan), trot over to confer briefly with their own commander, seen the look on Cardigan's face, seen the exchange of words and formal salutes, seen Lucan then go across to have a word with Lieutenant-Colonel John Douglas, commanding officer of the 11th Hussars, on the left of the line before riding back towards the Heavy Brigade. They noticed that Nolan remained briefly with the brigade commander. As Cardigan moved off to seek out Paget Nolan rode towards the 17th Lancers.

Wightman on the extreme right of the Lancers was later to write:

'I distinctly remember that Nolan returned to the Brigade and his having a mere momentary talk with Cardigan, at the close of which he [Nolan] drew his sword with a flourish, as if greatly excited . . . then fell back a little way into [sic] Cardigan's left rear, somewhat in front of and to the right of Captain [William] Morris.'[8]

Nolan had taken his position alongside his friend. The youngest trooper did not need an explanation of the immediate future.

So was the Light Brigade about to undertake the impossible? Was it being sent by blunder, stupidity or incompetence on a frontal attack on an artillery battery obviously prepared to receive it? Or was it just a monumental misunderstanding? By the end of this book the reader will have his own views on the answers to these questions. For the present the situation as it must have appeared to the soldiers at the moment will be reconstructed. What did the soldiers see, what did they know, what would they surmise? They had no way of knowing what was in any orders because nobody bothered to tell them; it was not the practice in 1854. They merely reacted to the calls of the trumpeter's bugle or the shouted commands of their officers. As Tennyson aptly put it:

'Theirs not to reason why;
'Theirs but to do or die.'

A cavalry charge was the principal reason cavalry existed, the others being all related to scouting. Cavalry was an offensive arm; cavalry attacked, never defended. A man on a horse armed with a sword, a lance, and either a pistol or short-range carbine was in no position to defend anywhere or anything except by attacking, in order to get to grips with his foe at close quarters. To do that successfully he needed the momentum provided by his horse. He needed to charge. The bulk of his training had this as the ultimate culmination of years of effort. It was the wildly exhilarating highlight of a lifetime; a few thrilling moments that the survivors would never forget and would relive countless times into old age. Once launched, nothing short of the death of horse or rider could halt the frenzied gallop - wounds were seldom enough. Men and animals were swept inexorably forward. As somebody once said, 'It is difficult to be a coward in a cavalry charge.' It was an electrifying adventure that set them apart from comrades who had not participated.

About an hour earlier that morning many in the Light Brigade had seen their comrades in the Heavy Brigade rout an enemy cavalry force twice its own size, while they had been compelled to remain as idle spectators. Some had over-heard Cardigan's comment, "Damn those Heavies! They have the laugh of us this day!" now it seemed his brigade was to have its chance.

But the Heavy Brigade had attacked cavalry, and almost stationary cavalry, at a distance of a mere 300 metres, while the Light Brigade looked as though it was to advance 2,000 metres down a gradual slope with guns as its objective, and with guns and infantry firing from either flank.

Cavalry had certainly charged guns frontally before, and captured them. In the Peninsular War forty-five years earlier a squadron of the regiment that was not on the right of the front line, the 13th Light Dragoons, under a Captain Bowers, was ordered to attack a battery in front and had done so in a most determined manner. In 1808 at the Pass of Somosierra Napoleon's Polish Cavalry of the Guard had charged up a causeway into the jaws of a Spanish battery in a supposedly impregnable position. Only eight years previously in India, at the Battle of Aliwal, the 16th Lancers had also successfully charged guns and infantry. As a young lieutenant of 26, Morris, now commanding the 17th Lancers, was present at that action where he had been wounded attacking Sikh infantry. He well understood cavalry's capabilities.[9]

The training manuals recognized artillery as a possible, if risky, target for cavalry. They accepted that on occasion, when other tactics such as taking them in the flank or while on the move were impossible, guns could be attacked frontally. Ideally, the technique was to advance, in conjunction with horse artillery, with suitably extended files with one metre gaps between riders, then, once the battery had fired its deadly volley of canister (hopefully, at between 200-300 metres), the men would close in, charge and hit the gunners before they could reload. A support line following up, 400 metres or so behind, would deal with other troops nearby or to the rear of the cannons. Yet a third line should be positioned well back as a reserve. Line is not quite accurate, for this last force was expected to be a double column of squadrons. That was the theory.

The Light Brigade was certainly not in open files – it was mounted with its riders formed up knee to knee – but a second line was positioned to offer support. The lines were drawn up with the heads of the horses in the rear ranks almost touching the hindquarters of those in front. The gap between the lines was short, 100 metres or less, so the first line would have to be allowed to forge ahead at the start.

The frontage of the leading line was approximately 188 files (a file being two horsemen, one behind the other). Thus, in the present formation, the width of the advance would be around 200 metres until it started to suffer casualties. A battery of eight guns, deployed at 20-metre intervals would cover 150 metres, so if the charge hit the battery square on there was a reasonable chance that,

5

allowing for losses, most, if not all, the guns would be attacked simultaneously.

There appeared to be no specific reserve, although some officers were re-assured that the brigade was not alone. A glance to the rear, over the left shoulder, revealed a weak French cavalry brigade made up of two squadrons each of the 1st and 4th Chasseurs d'Afrique drawn up within easy supporting distance under the Sapoune Heights. Then, as all were aware, the Light Brigade was a part of the Cavalry Division. To their immediate right rear, just over the ridge, sat the Heavy Brigade and, out of sight perhaps, but equally a part of the division, was I Troop of the Royal Horse Artillery (RHA). There was no reason to suppose that, whatever hazardous undertaking was about to begin, it was to be a solo effort by the Light Brigade. Any advance would surely be supported.

With the assumed objective so far away everybody anticipated the advance would start at a trot (8 mph) and continue at that pace (as in training) until within 300 metres of the guns, then the gallop (12 mph) would be sounded before, finally, at 50 metres the 'charge' which, according to the manuals, meant 'the utmost speed of the slowest horses'. A horse, moving at that pace and bursting through the battery, would be as potent a weapon against the gunners as a sword or a lance. The signal that would initiate and control the advance would normally be from the trumpeter. In this case most expected Cardigan's orderly trumpeter, Britten, to sound 'Walk!', 'Trot!', then at the appropriate time, on the order of the brigade commander, 'Gallop!' and finally the rapid, magically stirring notes of the 'Charge!'. To do this while control-ling a horse at speed required considerable practice, but Britten was a seasoned soldier, having already served for twelve years.

The problem with a charge, as all knew, was control. A line of galloping horsemen is impossible to direct. At that speed no bugle, no shouted word of command, can turn them, slow them or halt them; they are like a bullet, knowing only one course until it hits something or slowly loses its momentum through lack of energy. The principle was to keep them in hand at a slower pace (the trot), for as long as possible, particularly if riding knee to knee. Moving fast, with little interval between squadrons, with the riders and their mounts almost touching, pressure inevitably developed, sometimes running like a wave from flank to flank or into the centre. Nolan had experienced this when serving with Austrian cavalry. He later wrote that, 'The pressure of the horses was often so great as to lift me, with my horse, off the ground, occasioning great pain and making one and all quite helpless.'[10] This was on manoeuvres; it remained to be seen if it was still a problem with gaps being torn in the ranks by gunfire.

Some officers and men undoubtedly made mental calculations of the likely time they would take to reach the enemy. The distance was certainly more than a mile and, assuming they would try to stay at a trot until close to the objec-tive, seven or eight minutes seemed the likely time they would be a target for the Russian batteries. In training cavalry covered about 300 metres in a minute trotting and 400 plus in a minute galloping.

The advance would probably start with swords drawn and lances at the 'carry' (vertical, with the butt in the leather bucket on the stirrup, and supported on the arm by a short sling). In theory swords would be lowered to the 'right engage' and lances to the 'engage' at the time the squadrons were clearly visible to those about to receive the attack (usually as the charge was sounded). The moral effect was considerable – it also gave the rider's adrenalin an additional boost at a crucial moment.

The sword had been the cavalryman's principal weapon since warriors first rode to war. Homer said of it, 'The blade itself incites to violence.' Its shape, size, length and weight had changed hundreds of times as swordsmiths strove to produce the ideal blade and balance. What they sought to reconcile was the problem that the type best suited to cutting was seldom the best for pointing (thrusting), and vice versa. For slashing blows a sabre, whose blade was noticeably curved, broad, and with the point of balance (weight) further towards the point, was the most effective. If razor sharp a powerful swing with a sword of this type could sever an arm or a head as easily as a knife slices salami. Perhaps the finest sword ever used by British cavalry was the light cavalryman's sword first produced in 1796. It was 33 inches long with a curved blade. It was much feared by the French in the Peninsula and at Waterloo. By the time of the Crimean War it had long since been superseded, but by an inferior weapon.

To cut deeply, to cause crippling wounds, required a combination of weight and a finely honed edge. Even without body armour it was not always easy to cut through equipment, helmets, packs or even thick greatcoats. With a blunt blade your opponent was likely to escape an encounter bruised rather than bloody. British swords were often blunt, despite the efforts of all regiments to sharpen them on mobilization. The 11th Hussars took theirs by wagon to Dublin where the work was supervised by 'men from the Tower of London . . . When they were reissued, an order was given that they were not again to be drawn till required, when in the presence of the enemy.'[1] Troop Sergeant-Major (TSM) Loy Smith does not tell us whether this instruction was obeyed. Even if it had been the blade was still likely to have lost its keeness. The British scabbards were made of steel and not only did the swords rattle, thus giving away a person's position, but they rusted easily. Corrosion, and the constant knocking against the inside of the scabbard and drawing for drill purposes soon destroyed the sharpness of the edge. Attempts were made to solve the problem by stuffing straw down the scabbards, but the real answer to the rust and rattle was wooden (or leather) scabbards - a solution fervently advocated by Nolan who had seen its effectiveness in India. This view was strongly endorsed by a sergeant of dragoons when he wrote:

'Two things are much wanting by our cavalry in the East [Crimea], viz – lighter carbines, and curved – not "cut and thrust" swords. The men complain sadly of their weapons . . . also the blunt – almost useless – swords now in use, which are too curved to make good weapons for the

7

"point", but too straight to cut with effectually. The steel scabbards, also, cause the swords to get blunt long before they otherwise would.'[12]

The controversy of whether to cut or thrust was a longstanding one. The British cavalry in the Crimea were taught to prefer the point, the reason being that the point easily penetrates deeply, causing mortal wounds. Nevertheless, in the excitement of the mêlée there was likely to be a lot of hacking as well as thrusting, so the sword carried that day was, as the dragoon sergeant has described with feeling, something of a compromise. The troopers carried the 1829 pattern weapon which had exchanged the broad, double spear-point blade which had so frightened the French for a straighter, longer version. Part of the problem was that, while the point was effective against infantry as the lunge was downwards and had weight, if your opponent was mounted it was far harder to get in a powerful thrust. In a cavalry mêlée the cut was more common.

Nolan, self-styled expert and author of a book on cavalry, was sitting impatiently in front of a regiment armed with a weapon he derided as much inferior to the sword – the lance. Cardigan had deliberately placed the 17th Lancers in the centre of the front line. To face a line of lancers pounding down on you with points levelled was a terrifying experience not easily forgotten by survivors. The lance was supreme in a charge – or in a pursuit. In these circumstances the combination of the point, the reach and the momentum were normally unbeatable. By lying down an infantryman had a good chance of avoiding the stab of a cavalryman's sword, but with a lance he was likely to be skewered to the ground. A well-aimed lance would transfix an opponent, sometimes to the downfall of the rider who could not withdraw his weapon and was thus dragged from his horse. Cardigan had aligned the lancers with the centre of his objective. Of the five regiments under his command they were the best suited to smash through the battery.

A stationary lancer, however, was virtually defenceless. In close combat the lance was cumbersome, if not impossible, to use. How was a man armed with a nine-foot lance to protect himself from a sword cut to the head with his adversary only two or three feet away? The answer was supposed to be the 'St George'. This was the lancer's equivalent to the swordsman's 'head protect' where the sword is held horizontally above the head, and involved twirling the lance above his head rather like the rotor blades of a helicopter. Not surprisingly it seldom worked. Soldiers were taught to attack a lancer from his right rear, or hindquarters. Against this he could do nothing other than turn his horse quickly to try to face his assailant. For these reasons the 17th Lancers (and all Lancer Regiments) also carried swords. It was not uncommon for men to discard their lances just prior to hand-to-hand fighting and resort to their swords.[13]

The lance was also unsuitable for sentry duty or scouting (the pennons were highly visible and the poles snagged easily when moving through scrub or trees). The troopers complained about the sling becoming entangled at the crucial moment with the sword hook, sword belt, saddle cantle or gauntlets. Typically, quartermasters complained that the slings wore out the chevrons and cloth on the arm of the jackets! The lance was not deemed a suitable weapon for officers; none ever carried one in action.

The Light Brigade were also carrying firearms that morning. Carbines were essential weapons for cavalry engaged in picketing or skirmishing duties; pistols could be useful in a mêlée. All cavalry regiments, except for Lancers, carried percussion carbines. They were universally condemned. They were 'far too heavy', according to our dragoon sergeant informant, and the method of carrying them on the off-side (right) by means of a leather bucket in which the muzzle rested, and a swivel hook attached to the belt pouch, caused serious problems. This arrangement was not only inconvenient but

'actually hurtful to the man, by knocking about when he is moving at a rapid rate . . . much time is lost when unstrapping it and springing it for use . . . when the horse is killed the man is without his carbine – for in action there would seldom be time to unstrap the carbine . . . if the horse fell on the off-side it would be utterly impossible.'[14]

The carriage and use of carbines was impractical for lancers, so they were all issued with a clumsy, inaccurate, single-shot, nine-inch-barrel, musket-lock pistol. These weapons were also carried in other regiments by regimental and troop sergeant-majors (RSMs and TSMs) and trumpeters. Again the method of carrying them was flawed. In order to get at it the rider had to remove his right glove, push forward his cloak, or draw back the sheepskin and shabracque, and reach for the butt of the pistol, which was in the holster on the near side, hidden under his left hand – hardly conducive to speed of reaction in a life-threatening situation. Little wonder most officers purchased six-shot revolvers privately, usually the navy Colt or Adams.

*　　*　　*

Before following the fortunes of the Light Brigade during the preceding months, before recounting how it came to be in its present predicament, and before we accompany it down the valley it is important to understand its composition; to be introduced, albeit briefly, to some of the officers and men and to be able to visualize the scene at the outset. To assist with this Sketch 1 shows the Brigade in semi-diagrammatic form moments before the order to advance was given.

Firstly, why the 'Light' Brigade? The Cavalry Division in the Crimea consisted of the Heavy and Light Brigades, but there was little real difference in arms or duties between the two. With minor exceptions the main difference was in the colour of their uniform jackets – blue for the 'Lights' and red for the 'Heavies'. Their titles were inherited from their past, when the roles of cavalry had been more sharply divided. Originally, big horses carrying big

9

men, often protected by cuirasses, were deemed 'heavy' cavalry whose function was solely shock action – the charge, followed by a mêlée in which size and weight were the important factors. These horseman regarded themselves as an élite who did not involve themselves with the more mundane duties of scouting, foraging and picketing. They were perfect on parade, gleaming and magnificent to watch, but because the frequency of charges, even on campaign, was so small they tended to become ceremonial soldiers first and fighters second.

Conversely, light cavalry was the Army's work horse, seeking information, giving early warning of enemy approach, skirmishing, foraging, protecting flanks or rear, escorting the baggage and suchlike less than glorious responsibilities. They were not considered suitable for shock action, although men and mounts were usually far fitter, leaner and hardier than their 'heavy' cousins. Since the end of the eighteenth century these distinctions had blurred so that by the time of the Crimean War only the titles had been retained; training, tactics and weapons were virtually identical.

Within the light cavalry there were supposedly three types – light dragoons, hussars and lancers. The Light Brigade had two regiments of light dragoons (4th and 13th), two of hussars (8th and 11th) and one of lancers (17th). Apart from the obvious differences with the lancers, light dragoons and hussars were now indistinguishable in duties and arms – all light dragoon regiments changed titles to hussars a few years after the war. Dragoons were the original mounted infantry, soldiers who rode to war but fought on foot. By 1854 they had been absorbed into the cavalry, with the traditional cavalry tasks.

In early March, 1854, the regiments that were to form the Cavalry Division received their mobilization orders from the Horse Guards (the 19th century equivalent of the Ministry of Defence). These caused consternation. There had been no major war for 45 years; the Army had been allowed to shrink to a fraction of its former self, and not one of the ten cavalry regiments earmarked for overseas service was up to peace, let alone war, establishment. In terms of numbers the cavalry had shed fifty per cent of its strength since Waterloo, dropping from 22,000 to under 11,000 (of which 3,000 were in India). The shortfall was in both men and horses. To form two understrength cavalry brigades would normally have required three regiments of three squadrons each, but for the Crimea five regiments of two squadrons were needed to make up minimal numbers. A frantic period of drafting, voluntary transfers and cross-posting was put in hand. Regiments remaining at home were milked of men and horses; bands were broken up with bandsmen taking their places in the squadrons.[15] Cardigan's former regiment, the 11th Hussars was, typical:

'The following day the order arrived from the Horse Guards, with a detail of all the arrangements, viz: we were to take out four troops [two squadrons] of 74 men and 62 horses each, and four women per troop. If deficient of men or horses we were to be made up from 16th Lancers,

they being the nearest light cavalry regiment. . . . By breaking up the band
. . . we had men sufficient, but we required about 30 horses to complete.'[16]

The embarkation strength of the Light Brigade was approximately 1570 men,
giving an average regimental strength of some 314 including officers. When
the Brigade formed up that morning there was barely 42 per cent of the
original number left. Even more than the infantry, sickness, not battle, had
crippled the cavalry. As far as can be ascertained the 13th Light Dragoons had
126 on parade, the 17th Lancers 147, the 11th Hussars 142, the 8th Hussars
115, and the 4th Light Dragoons, 126. Active service in the field had also weakened the horses, the supply of food and forage had been erratic and often of
questionable quality. The intense heat of the summer, the bitter winds and
nights of autumn, sleeping rough in the same clothes for weeks on end coupled
with lack of sleep, recent alerts as a result of numerous false alarms, had eroded
the physical stamina of all. Morale had not suffered as severely as might have
been expected as evidenced by the number of men who refused to shelter
behind sickness or other duties which might have excused them from parading.

Uniforms, once bright and colourful, were now worn and shabby. Plumes
had been discarded; jackets and trousers were patched, dirty and discoloured;
cuffs and collars, once white, were stained and muddied, and gold lace dull and
tarnished. Rust was an ever-present enemy.

Although numbers had diminished so dramatically, each regiment strove to
keep its internal structures intact. A headquarters had two field officers – the
lieutenant-colonel commanding, assisted by a major as second-in-command.
There were also several staff officers. The adjutant was often a cornet (these
days a 2nd lieutenant) who had come up from the ranks, such as Cornet John
Yates of the 11th Hussars who had risen to quartermaster-sergeant in the 17th
Lancers and had only been commissioned a month previously. An adjutant's
duties involved overseeing drill, discipline and personnel matters. Another ex-
ranker officer was the quartermaster, responsible for supplies (rations,
ammunition, equipment and clothing). There was also a paymaster, like
Captain Henry Duberly of the 8th Hussars, whose wife watched the charge,
and whose journal was to make her famous in later years. Then there was a
surgeon, an assistant surgeon and a veterinary surgeon. Of these junior officers
only the adjutant would be expected to ride in a charge, the others' duties
normally keeping them in camp.

Assisting with the administration in the headquarters were a number of
NCOs, the most senior of whom was the RSM, a man in line for promotion
to commissioned rank as an adjutant or quartermaster. In addition there was
a quartermaster-sergeant, a trumpet-major, a farrier-major and six sergeants
with special responsibilities (paymaster, armourer, saddler, schoolmaster,
hospital and regimental clerk). Again one might expect to find only the RSM,
trumpet and farrier-majors on parade that morning.

The basic administrative unit within the Regiment was the troop of some
60–70 men, although not one regiment in the Light Brigade could muster much

more than 30–34 now. Each troop was supposedly commanded by a captain assisted by two subalterns (lieutenants or cornets) but, as we shall see, this was often impossible. In the 11th Hussars one troop was commanded by a troop sergeant-major. Within each troop was an establishment for a TSM, four sergeants, three corporals, some 65 troopers, a trumpeter and a farrier.

The tactical unit was the squadron of two troops paraded in two ranks. Because troops were frequently of unequal numbers there was a lot of shuffling around from one to the other to get a reasonably balanced squadron for a battle. This often resulted in comrades being separated and soldiers following a comparatively strange officer – something regarded as unacceptable today. Out in front, in the centre of the Regiment, was the commanding officer with his duty trumpeter behind him. Squadron and troop officers were lined up in front of their sub-units, with the adjutant on the extreme right of the line. The sergeants and corporals were usually posted on the flanks of the troops with the sergeants in the front rank and the corporals behind as 'coverers'. Any supernumerary officers or NCOs, and if on parade the farriers, were distributed in a line behind the rear rank. Also in this line, behind each flank of the squadron, was a trumpeter.

On the right of the first line was the 13th Light Dragoons, like other dragoon regiments distinguishable at a distance by their tall, top-hat style, black beaver-skin shakos, whose gilt and silver cross plate and regimental badges were now concealed under dark waterproof covers. They wore a double-breasted blue jacket with white, officially buff, facings – (cuffs and collars) plus grey overalls (trousers) with a double white stripe down the sides.

The commanding officer, Lieutenant-Colonel Charles Doherty, was sick. Likewise absent was Major William Gore who had been invalided back to the base hospital at Scutari, so command had devolved onto the shoulders of a young captain in his thirties who had spent nearly half of his service as an infantry officer. Captain John Oldham had exchanged into the Regiment in 1847. He almost certainly counted himself fortunate to be sitting in his saddle as acting commanding officer as only a few days earlier he had been under arrest. He had been in charge of an advanced picket when Lucan had arrived and ordered a close reconnaissance of the Chernaya River. He had briefed Oldham personally, specifically emphasizing that nobody should cross the river. Oldham disobeyed, and a sergeant was captured – hence Lucan's wrath and Oldham's arrest (from which he was soon released). Leading the two squadrons were Captains Thomas Goad and Soame Jenyns. Goad, on the right, was worried for his younger brother George, a cornet, who had been severely injured by a shell fragment earlier in the morning. Ironically George recovered but Goad was to die at the guns.

The officer on the extreme right of the line was the acting adjutant, Lieutenant Percy Smith. He was something of a character who enhanced his reputation by parading, and charging, virtually unarmed. Some years previously he had maimed his right hand in a shooting accident, so his brother officers presented him with an iron guard which enabled him to grip a sword.

12

FORMING UP

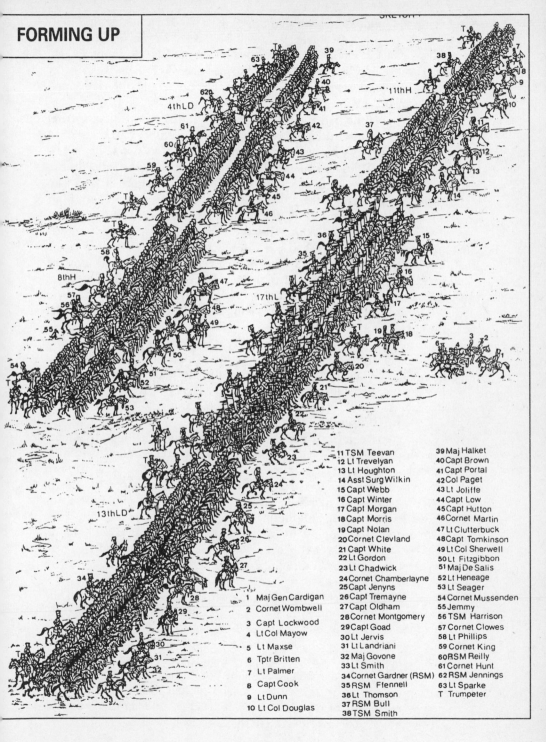

4th LD
8th H
11th H
13th LD
17th L

11 TSM Teevan	39 Maj Halket
12 Lt Trevelyan	40 Capt Brown
13 Lt Houghton	41 Capt Portal
14 Asst Surg Wilkin	42 Col Paget
15 Capt Webb	43 Lt Joliffe
16 Capt Winter	44 Capt Low
17 Capt Morgan	45 Capt Hutton
18 Capt Morris	46 Cornet Martin
19 Capt Nolan	47 Lt Clutterbuck
20 Cornet Clevland	48 Capt Tomkinson
21 Capt White	49 Lt Col Sherwell
22 Lt Gordon	50 Lt Fitzgibbon
23 Lt Chadwick	51 Maj De Salis
24 Cornet Chamberlayne	52 Lt Heneage
25 Capt Jenyns	53 Lt Seager
26 Capt Tremayne	54 Cornet Mussenden
27 Capt Oldham	55 Jemmy
28 Cornet Montgomery	56 TSM Harrison
29 Capt Goad	57 Cornet Clowes
30 Lt Jervis	58 Lt Phillips
31 Lt Landriani	59 Cornet King
32 Maj Govone	60 RSM Reilly
33 Lt Smith	61 Cornet Hunt
34 Cornet Gardner (RSM)	62 RSM Jennings
35 RSM Ffennell	63 Lt Sparke
36 Lt Thomson	T Trumpeter
37 RSM Bull	
38 TSM Smith	

1 Maj Gen Cardigan	
2 Cornet Wombwell	
3 Capt Lockwood	
4 Lt Col Mayow	
5 Lt Maxse	
6 Tptr Britten	
7 Lt Palmer	
8 Capt Cook	
9 Lt Dunn	
10 Lt Col Douglas	

13

When he turned out before dawn that morning he had been unable to find it. Perhaps his best known escapade so far had been when he was on a patrol as part of Cardigan's infamous 'sore back reconnaissance' (see Chapter 3). He had been 'captured' by Turks who were convinced he was a 'Russ' because of the shape of his shako. He had been thrown to the ground and almost excuted by a swarthy swordsman.

All eight officers were in front of the first line with the exception of Cornet George Gardner who was doing the duties of RSM. Gardner had been RSM until a month previously but as his replacement was not available he undertook his old duties. With him behind the rear rank were TSMs John Weston, George Smith and John Linkon. There were, however, two more officers who sat in their saddles close to the adjutant conversing quietly in Italian. Staff Major Giuseppe Govone and Lieutenant Giuseppe Landriani were Sardinians who were attached to the Allied Armies as observers (Sardinia/Piedmont was later to enter the war on the Allied side). Both happened to be attached to the Light Brigade on 25 October and were intent on charging with it. Among the soldiers awaiting the next order were Corporal Joseph Malone, the old Indian campaign veteran John Brooks, and the young ex-shoemaker Private Edwin Hughes who thought the chances of surviving the next few minutes remote, let alone the next 73 years!

In the centre of the first line were the 17th Lancers – Lucan's former Regiment, – instantly recognizable by their lances with red over white swallow-tail pennons and tall, distinctive, square-topped lancer caps with their Death's Head badge concealed under oilskin covers. Like the Light Dragoons they all wore blue jackets, but their trousers were also blue, not grey, with two stripes of gold lace down the seams.

It was another Regiment commanded by a captain. Lieutenant-Colonel John Lawrenson had been invalided home sick, while Major Augustus Willett had died only three days before, after the cavalry division had been stood-to for fourteen hours on the report of a large enemy force marching on Balaclava. Willett was a tyrant who had been detested by the soldiers. His demise was officially caused by cholera, but rumour had it that he froze to death as a result of his own order forbidding the wearing of cloaks on this bitterly cold night. If this was correct he had been the first victim of his own cruel stupidity. There was no regret in Private Wightman's comment, 'He was a corpse before sundown of the following day.'

Captain William Morris was a complete stranger to his men. When the Regiment paraded that morning some were heard to ask, "Who is he?" From a more knowledgeable NCO came the answer, "That's Slacks." – a nickname he had been given by his previous Regiment. He sat in front of his Regiment in the blue frock coat and gold-peaked forage cap of a staff officer, which until the previous day he had been (at Lord Lucan's Cavalry Division headquarters). Recently weakened by cholera, he had faced a dilemma the previous night of whether to stay on the staff or, as the senior captain, to claim the right to command his Regiment. He chose his Regiment.

Morris was short, stocky and normally extremely strong, being described by many as a 'pocket Hercules'. He certainly had the most war experience in the Brigade. As noted earlier, he had fought in India with the 16th Lancers in several skirmishes and at the battles of Aliwal and Sobraon. He was also a thinking soldier, being one of the tiny minority of officers to have studied at the Senior Department of the Royal Military College at Sandhurst. (The intake for 1854 was six.) Now, mounted on his charger 'Old Trumpeter', impatient to be off, he chatted quietly with his long-standing friend, Nolan. He carried in his pocket a letter for Nolan's mother, while Nolan carried a similar note that Morris had hurriedly written for his wife. They had promised each other to deliver them if one survived and the other did not.

The squadron to his right was commanded by Captain Robert White and that to his left by Captain John Winter. On the extreme right of the Regiment was the adjutant, Cornet John Chadwick. At this moment Chadwick and others nearby were astonished to witness the late arrival of the regimental butcher. This was Private John Veigh, a colourful, hard-drinking character who already had 14 years' service (but no promotion), trotting over dressed in his butcher's bloodstained overalls with the sleeves rolled up above his elbows. He had heard of the 'Heavies' earlier attack and had hurried across in time to buckle on the sword and pouch belt of a dead dragoon and ride off on his horse. As he tried to jostle his way into the line Chadwick sent him to join his squadron (White's). He took up a position next to John Lee, close to Wightman.[17]

Also in the ranks were the future VC winners Sergeants John Berryman, John Farrell and Charles Wooden, the latter a none too popular NCO of German extraction who spoke with a guttural accent. Other well known regimental personalities included Corporals Thomas Morley, James Nunnerley and John Penn. To Trumpeter John Brown, who had joined eight years previously as a bandboy, the future probably looked bleak and short. He could not know that he would rise to RSM, have a commission purchased for him by a general, retire an honorary lieutenant-colonel, and live for another 51 years.

On the left of the line were the 11th Hussars, Cardigan's former Regiment, the much publicized 'Cherrybums' – so called from the skin-tight, cherry-red overalls that they wore.[18] Indeed, they made the only splash of real colour in an otherwise rather drab, dirty brigade. Like the rest they wore blue jackets but with their heavily gold-braided pelisses over them. This was an attempt to combat the cold. In common with other hussar regiments, all ranks wore the brown fur busby with its crimson cloth bag hanging down the right side.

Lieutenant-Colonel John Douglas was out in front commanding, with his favourite trumpeter of eight years service, William Perkins, just behind him. Douglas had, until recently, been the long-suffering second-in-command to Cardigan. But there was no second-in-command present now as Major Edmund Peel was sick. The right-hand squadron was under the 19-year-old Lieutenant Harington Trevelyan who, despite his youth, was the most senior lieutenant available, there being only one captain on parade. That was Captain

15

Edwin Cook who was the left-hand squadron commander. The officer destined to win a VC in the charge was Lieutenant Alexander Dunn, commanding 'F' Troop.[19] Loy Smith, his TSM, who was positioned almost directly behind him at the rear, has described Dunn as, 'a fine young fellow, standing six feet three, mounted on a powerful horse and wielding a terrific sword, many inches longer than the regulation.'[20]

Uniquely, an assistant surgeon rode into battle with the Brigade. This was Henry Wilkin, a man whose vocation was surely soldiering rather than medicine as he retired from the Medical Department early in 1855, obtaining a combatant commission as a cornet. He went on to be recommended for the VC (he received a Mention in Despatches instead) during the Indian Mutiny, and was wounded at Lucknow. It is possible he had volunteered to take the place of the adjutant who had been acting as brigade major on the staff. This was the unpopular Cornet John Yates who had been promoted into the 11th Hussars from the position of quartermaster-sergeant in the 17th Lancers a month or so earlier. This caused considerable resentment. TSM Smith had been particularly scathing on this appointment in his diary:

'Unfortunately for us (the 11th) Colonel Douglas allowed Colonel Lawrenson of the 17th Lancers to persuade him that his quartermaster[-sergeant] would make us an excellent adjutant – although at the time our two senior sergeant-majors were both eligible . . . I have heard on good authority that Colonel Douglas deeply regretted this act. If he did not I know the whole regiment did, for a worse rider, a worse drill, a greater humbug never before held the rank of adjutant [it was an appointment not a rank] in the British Army. The 17th might well be glad to get rid of him; they certainly got the laugh of us.'[21]

It seems that a typical old soldier's trick had been pulled on the 11th Hussars.

Smith's forebears were a fine example of a service family of those times. In explaining why he enlisted he wrote:

'I belonged to a family that had seen service both by sea and land – a love of adventure was therefore natural to me. My grandfather served under the Duke of York in Flanders, was present at the Siege of Dunkirk and N. . .? – at the latter place he was wounded in the head by a bursting shell. My father served in the Peninsula; one uncle wore the Gold Medal for Leipsic and the Waterloo Medal, having served in the Rocket Troop; another lost his right arm in boarding a French man-of-war.'[22]

Smith had a reputation with some as a martinet. Private William Pennington thought so. Years later writing about Private Henry Hope, he said:

'[Hope was] an ignorant passionate Welshman of great simplicity of character, very powerful physique, but subject to epileptic fits and with whom a kind word would have accomplished miracles.'[23]

He went on to say that he thought that Hope had been cruelly persecuted by Smith who had him flogged on two occasions. Hope had been a prisoner awaiting court-martial on 25 October, but had escaped from the guard tent in time to grab a spare horse and ride with the Heavy Brigade – for which initiative Cardigan pardoned him. He thus became a unique individual, having fought with both Brigades.

Smith and the other sergeant-majors of the 11th were the only men carrying gun spikes. They were to be driven into the vent of captured cannons with a small mallet to 'spike' them – a crucial task to prevent the guns being used again immediately. Smith had been given his by an ADC at the Alma who had galloped up and handed them to the sergeant-majors with the words, "These are all we have. You may have an opportunity of using them presently."[24]

A short distance to the rear stood the second line – the supports. On the right were the 8th Hussars, dressed like the 11th except that they wore blue overalls with a single yellow stripe and no pelisses over their jackets – they had somehow got lost en route. They were under the command of Lieutenant-Colonel Frederick Shewell, a religious man, an ardent upholder of Victorian morality and with a firm belief in the minor manifestations of discipline. Major Rodolph de Salis had started the day in command, as Shewell, in his late forties and nicknamed 'the Old Woman', had been confined to his tent sadly troubled with gout and sickness . . . besides being old for such exposure.' Nevertheless, a few minutes earlier Shewell had come cantering up to take command, prompting one trooper to remark, "Well, I'm damned if it isn't the Colonel; what do you say to the 'Old Woman' now?"

Shortly afterwards, when the Regiment had mounted, just moments before the advance started, there occurred an incident that is best described by a witness, Private John Doyle:

'I saw as he [Shewell] passed in front of us, that all at once his face expressed the greatest surprise and astonishment, and even anger, and, walking on he broke out with – "What's this? What's this? – one, two, four, six, seven men smoking! – Why the thing is inconceivable! Sergeant! Sergeant Pickworth!" . . .

The truth is we were warming our noses each with a short black pipe, and thinking no harm of the matter . . .

"Sergeant, advance and take these men's names."

. . . leaving the sergeant to find us out, though he couldn't discover any, the Colonel passed on, and halted again . . . It might not be quite according to regulation to be smoking sword in hand, when the charge might be sounded at any moment. . . . He [Shewell] comes up to another now, that [sic] hadn't heard what had been said, and he sings out –

"Sergeant Williams!"

"Yes sir," replies the sergeant.

"Did you not hear what I said about smoking just now?"

"I've not lit my pipe yet sir," answered the sergeant.

"Fall back to the rear, and take off your belts. Farrier, forward and take them, and – why here's another! – To the rear, fall back. I'll have this breach of discipline punished!"

. . . I understand that one was punished the next day, but Sergeant Williams, who was mounted but quite unarmed, as he had given up his sword, belt, and carbine went into the charge with us (it came directly after), and was killed.'[25]

That such trivial matters could be foremost in some military minds in the last few moments before the launching of a major attack is difficult to comprehend today; but perhaps it usefully, if unconsciously, diverted thoughts from the horrors to come.

The 8th Hussars had only one and a half squadrons (three troops) present, as one troop under Captain George Chetwode was permanently detached as Lord Raglan's escort. The complete squadron was commanded by the only captain parading with the Regiment, Edward Tomkinson. The adjutant, Lieutenant Edward Seager, was 42 years old and carrying in his sabretache pictures of his wife and children, a lock of child's hair and a prayer book given him by his mother. He had been the adjutant since 1841 when he had been promoted from RSM. 25 October was to be his last day of duty in that capacity as he was promoted captain the following day.[26] He was then succeeded by the RSM, Robert Harding, who missed the charge as he was unable to ride due to a boil on his leg.

Perhaps the most singular sight in the 8th Hussars was the small rough-haired terrier sitting quietly on the ground behind the rear rank. This was 'Jemmy', the unofficial officers' mess pet. He was to follow, yapping at the heels of the horses all the way down the valley and, although wounded slightly in the neck, scamper safely back.

The fifth Regiment, starting the advance on the left of the second line, was the 4th Light Dragoons. It was the second weakest regiment in the Brigade (after the 8th Hussars) as it had a troop detached to the 2nd Division. Colonel Lord George Paget was not only commanding his Regiment but had just been told in unequivocal language that, as the next senior officer to the Brigade commander, he was responsible for the supports. At this stage the second line was far weaker and shorter than the first, having a strength of some 241 men, compared with 417 in the latter. It was Paget's duty to ensure that this line kept the correct distance from the first, and to try to time its arrival at the objective so as to assist in dealing with enemy reserves, or to the rear of the guns.

Paget had been embarrassed by Sherwell's tirade against smokers as he, Sherwell's senior, was himself still pulling on his freshly lit cigar. As he later wrote:

'The question then arose in my mind, "Am I to set this bad example? (in the Colonel's opinion at least) or should I throw away a good cigar?" . . .

18

Well, the cigar carried the day, and it lasted me til we got to the guns . . .
It was often the subject of a joke between us afterwards.'[27]

As already noted, Major John Halkett, a son of a former governor of the
Bahamas, was present as second-in-command. The right-hand squadron was
under Captain (brevet Major) Alexander Low, a fifteen stone heavyweight of
considerable strength and mounted on a suitably hefty horse, while Captain
John Brown commanded the left. The acting adjutant, their being no ex-ranker
available, was an old Etonian Cornet Fiennes Martyn. The RSM was Henry
Jennings, who was to hold that appointment for a mere six weeks. He had been
promoted RSM on 20 October and was commissioned cornet on 5 November.
The Regiment seems to have got into something of a tangle with their senior
sergeant-major. Jennings' predecessor, John Reilly, had been doing the job
although he had held a cornet's commission in the 8th Hussars since 3
September, 1854. He was about to charge with the Light Dragoons as a sort
of second RSM.

In front of the Brigade was a small group of six horsemen. Cardigan was a
conspicuous, solitary figure well out in front, but facing the Brigade, watching
it settle down and complete its dressing. Facing him, about a horse's length
away, was his trumpeter. On either side of Britten and about level with him
were the members of the Brigade staff – like Britten facing forwards. The senior
was the brigade-major, Lieutenant-Colonel George Mayow, 4th Dragoon
Guards, who had left his sick bed to be present. Then there was Cardigan's
permanent ADC, Lieutenant Henry Maxse of the 21st Foot (eventually to
become the Royal Scots Fusiliers), the only infantry officer to ride with the
Brigade. The two regimental ADCs were Captain George Lockwood, 8th
Hussars, and another old Etonian, Cornet George Wombwell, 17th Lancers.

At about five past eleven Cardigan, resplendent in the full dress uniform of
the 11th Hussars, including the busby, (he much preferred this to the cocked
hat he was entitled to wear as a major-general), was satisfied his Brigade was
ready. All eyes were on him. This was a moment to savour. He certainly looked
the part of a light cavalry commander and he knew his drill-book. There was
a long way to go so the advance must be steady, not too fast, and under tight
control. After a final quick glance left and right Cardigan gave the order,
"Draw swords". Sunlight winked from scores of polished blades now held
upright at the 'carry'. The second line was too far away to react to Cardigan's
order. Paget would give it when he saw the first line move off.

Finally, Cardigan, 'in his strong, hoarse voice gave the momentous word of
command, "The Brigade will advance! First squadron of 17th Lancers
direct!"' This meant the whole line must keep dressing with, and conform to
the movements of, the right hand squadron of the 17th Lancers throughout
the entire advance. It was a key instruction. As his brigade commander
shouted the order Trumpeter Britten put his bugle to his lips – it was obvious
what was coming. Cardigan turned his head slightly to look directly at Britten.
Quietly he said, "Sound the advance." Then, as Britten sounded 'Walk'

19

Cardigan turned his thoroughbred chestnut 'Ronald' to face the distant enemy. Reaching over to his left he drew his own sword. While doing so, he later told *The Times* correspondent Russell, he muttered, half to himself, "Here goes the last of the Brudenells!" (his family name). After a few metres the 'Trot' rang out. The Light Brigade rolled forward into history books.

CHAPTER TWO
The four horsemen of calamity

'The rumour in camp is that someone has been blundering, and that the Light Cavalry charge was all a mistake; the truth will come out some day.'
Sergeant Thomas Gowing, letter to his parents from Sevastopol, 1854.

The match that lit the Crimean conflagration was struck in the Church of the Nativity in Bethlehem, according to tradition built over the site of the stable where Christ was born. In June, 1853, Roman Catholic monks had placed a silver star over Christ's manger. Thus a long-smouldering quarrel between the monks, backed by France, and the Orthodox Church, backed by Russia, finally, and violently, caught light. The Roman Catholics had got possession of the front-door key to the church and seized the chance to install their star. They were resisted physically; a riot developed resulting in the death of several Orthodox monks. Czar Nicholas I protested loudly and passionately, claiming the Turkish police (Bethlehem was then in Muslim Turkish territory, see Map 1) had connived at murder. When no redress was forthcoming within a matter of days, other than the return of the keys, when his demands for recognition of a Russian 'protectorate' over Turkey's 14 million Christians was rebuffed, Nicholas mobilized an army. He sent Prince Mentschikoff, an overbearing bully whose hatred of the Sultan and all things Turkish had been implacable since a Turkish cannon shell had castrated him in the war of 1828, marching into the Turkish principalities of Moldavia and Wallachia (now Romania), heading for the Danube. The real cause of the conflict that developed, however, had nothing to do with religious bickering but everything to do with the geographical and strategic aspirations of powerful nations.

The Crimean War was the direct result of a threat to the balance of power in Europe. The 300-year-old Turkish Empire, which had dominated the Near East from the Persian Gulf to the Adriatic, was in terminal decline. The Czar, who openly referred to Turkey as the 'sick man of Europe', had visions of carving up the Turkish corpse. Specifically he wanted to control Constantinople. This would unlock the narrow Bosphorus channel and release the Russian Black Sea Fleet into the Mediterranean. One hundred and forty years ago this Fleet was based at Sevastopol on the southern coast of the Crimea, its

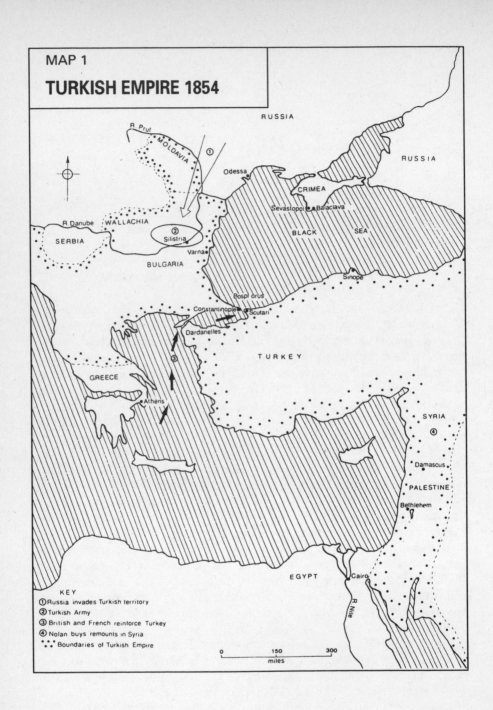

MAP 1

TURKISH EMPIRE 1854

RUSSIA

R. Prut

MOLDAVIA ①

Odessa

RUSSIA

CRIMEA

Sevastopol ● Balaclava

R Danube   WALLACHIA

SERBIA

② Silistria

Varna ●

BULGARIA

BLACK   SEA

Sinope ●

Bosporus

Constantinople ● Scutari

Dardanelles

③

TURKEY

GREECE

Athens ●

SYRIA

④

Damascus ●

PALESTINE

Bethlehem ●

EGYPT   Cairo ●

R Nile

KEY
① Russia invades Turkish territory
② Turkish Army
③ British and French reinforce Turkey
④ Nolan buys remounts in Syria
⋯⋯ Boundaries of Turkish Empire

0   150   300

miles

freedom of action easily restricted by whoever occupied the Bosphorus and the Dardanelles.

Neither Britain nor France were prepared to permit a Russian advance. The 'balance of power' strategy had been the basis of British foreign policy for over a hundred years (and would remain so for another century). When Mentschikoff's troops entered Turkish Moldavia in July a British fleet and a French squadron were already at the Dardanelles. A Turkish ultimatum to the Russians to withdraw was inevitably rejected, and on 23 October, 1853, Turkey and Russia were at war. It was to be exactly a year (bar two days) later that the Light Brigade was launched on its celebrated charge.

On land, at the Danube, the Turkish strength and resistance came as something of a shock to the Czar's Army. They could make no headway. It was to be at sea that the Russians achieved a notable, if one-sided, success. On 30 November they caught a Turkish naval squadron in the harbour at Sinope. The ferocious bombardment sank the ships with the loss of 4,000 Turkish lives, the Russians being virtually unscathed. This perfectly legitimate attack was regarded with considerable horror by the public in Britain and France. After this a war between Russia and both these countries, although long in coming, was inevitable. It was not until 27 March, 1854, after many fruitless attempts at negotiations, that Britain and France found themselves in military alliance with each other against Russia. But it had been more than two weeks earlier that Sergeant-Major Loy Smith had noted in his diary that the 11th Hussars, then in Portobello Barracks, Dublin, had been alerted for 'foreign service'.

For the first time in 200 years British and French soldiers were to stand together shooting at a common foe rather than at each other. Elderly officers on both sides of the Channel had spent their youth fighting in the Napoleonic Wars; many remembered Waterloo. It has been said that during the Crimean War the British commander-in-chief Lord Raglan, caused considerable embarrassment by inadvertently referring to the enemy as 'the French'. Perhaps equally strange, to anyone who gave it thought, was that Christians had allied themselves with Muslims to fight fellow Christians. The public, for sure, was not interested in such subtleties; they were swept up in a patriotic fervour that saw crowds surging through the streets of London wildly enthusiastic for action. The so-called 'massacre' at Sinope must be revenged; the Russians must be taught a lesson.

One of the most crucial, some might say the most crucial, decision that the British government had to take was the selection of senior commanders for the expeditionary force. This meant appointing not only the commander-in-chief and his top staff officers (quartermaster-general, adjutant-general and commissary-general) but the field commanders – in this case four, later five, infantry and one cavalry divisional commanders, plus a chief engineer and an artillery general. As with military campaigns before and since, given reasonable numbers and morale, success would have to be built on foundations of sufficient supplies (logistics), sound command decisions and good staff work, the last being the essential oil that lubricates the military machine.

Unfortunately, these were three areas in which the British Army of the 1850s was seriously deficient. This book is not directly concerned with the first, supply, which was the responsibility of the Commissariat, a civilian organization and a department of the Treasury, although its appalling shortcomings were to have enormously detrimental effects on every aspect of operations. The saga of how the Light Brigade was lost is a tangled tale of commanders' decisions, orders, counter-orders and shoddy staff work. It is also the story of the clash of personalities, of ambitious, opinionated men, two of whom detested each other.

Four men, four horsemen, were the key players in the succession of events that led to that valley north of Balaclava. One issued the order, one delivered it, one received it and one executed it. The result was the virtual destruction of the Brigade as a fighting force. One of the horsemen was killed within moments of the start of the advance and so never got to tell his story. The man who gave the order died in the field eight months later, while the two survivors defended their conduct with much bitterness and acrimony in the newspapers, courts and Houses of Parliament for years afterwards. Needless to say each sought to absolve himself of blame for his conduct that day. To lose a brigade of cavalry to no purpose smacked of negligence or incompetence, probably both. Reputations and professional advancement were at stake. In the century and a half since, authors, with the benefit of hindsight, have failed to agree as to which horseman deserved primary responsibility; each, except for the one who led the charge, has been named as the chief culprit. It is hoped that this book will enable the reader to put himself in the saddles of these horsemen at the moment when they had to make difficult decisions or take immediate action. If this can be achieved then the reasons why the Light Brigade was lost will become apparent. Only then should judgements be made.

Now is the moment to look at each of those horsemen at the start of the war, to understand something of their background, character, military experience, and relationships one to another. With this knowledge it will be easier to understand their thinking and motives as we follow them, and the Light Brigade, through the circumstances that shaped the decisions taken on 25 October, 1854.

*　　*　　*

The horseman who gave the order.

It was early evening on 18 June, 1815, when a young lieutenant-colonel of the 1st Foot Guards walked through the door of a small cottage that had been taken over as a field hospital. The new arrival was Lord Fitzroy Somerset, the chief ADC to the Duke of Wellington; the cottage was just behind the centre of the Allied line at Waterloo. Lord Fitzroy's right arm was shattered; it had been hit by a bullet fired by a French sniper on the roof of La Haye Sainte farm, lost at the height of the battle. The surgeon cut off the sleeve of his jacket, then told him to lie on a table sticky with congealing blood. His assistants knew what was going to happen as the surgeon picked up a saw, and they prepared to hold the wounded officer down. There was no attempt to deaden the hideous torment as the arm was slowly severed between the shoulder and the elbow. Lord Fitzroy did not even murmur. As the surgeon tossed the limb onto the heap on the floor there was a cry of, 'Hey, bring back my arm. There's a ring my wife gave me on the finger!' Forty years on, after Lord Fitzroy had become Lord Raglan, he was appointed commander-in-chief of the British Army in the war against Russia. Nobody had any doubts as to his physical courage or his stoical, almost superhuman, ability to bear pain.

Lord Fitzroy, who was born in 1788, was the youngest son of the Duke of Beaufort. As a boy of fifteen at Westminster School his father purchased a cornet's commission for him in the 4th Light Dragoons. He first saw service under Sir Arthur Wellesley (the future Duke of Wellington) in the 1807 expedition to Copenhagen. His father's influence had secured him a staff post as a very junior ADC. It was the start of the most extraordinary military career, in which he was to rise from cornet to general without ever commanding so much as a company in peace or war. This did not mean that he avoided action or danger in those dramatic years of the Napoleonic Wars – far from it – but that he was always the staff officer never the commander, never having responsibility for operational decisions, always implementing and representing others' wishes.

In 1808 Wellesley sailed for Portugal to take up command in the Peninsula. Again one of his ADCs was the nineteen-year-old dragoon lieutenant, Lord Fitzroy. Despite the 20-year age difference and the enormous gulf of rank, by the time they landed Wellesley had befriended his youthful assistant. It was the start of a friendship, an inseparable working association, almost a father-son relationship that was to endure unimpaired and unbroken for the next 44 years until the death of the Duke of Wellington. For seven years Lord Fitzroy was to have the unrivalled opportunity of watching at first hand, from within the British headquarters, as the man who was to become one of the most illustrious of British generals defeated first Napoleon's marshals in Spain and Portugal, and then Napoleon himself in Belgium; with the nagging proviso on the absence of actual command, surely a military education in generalship par excellence.

Within just over two years Wellesley appointed him to take over as his military secretary on the enforced retirement of Colonel Bathurst. This was a much coveted position. That it was to be held by a man of only 23 was an undreamed-of precedent. With it came promotion – to lieutenant-colonel in the 1st Foot Guards (subsequently the Grenadier Guards). Lord Fitzroy had achieved a rank that normally required long years of purchase, exchange and patronage to achieve. Once an officer reached lieutenant-colonel he was virtually assured of general's rank, if he lived long enough, as all promotions then depended on one thing – seniority. There was not even a requirement to do any further soldiering. The critical factor was the date of promotion to lieutenant-colonel; once there an officer was on an escalator of progressive advancement, even if he never did another day's duty, and remained indefinitely on half-pay.[1]

As military secretary Lord Fitzroy was in effect the senior ADC but with additional responsibilities involving the copying of orders, correspondence, dealing with personnel matters, appointments, purchases of commissions and records. He proved a natural in the job. His writing was clear and he had the trust of the Duke who used him for regular liaison contacts with subordinate commanders. For example, for the attack on Badajoz he amended the Duke's written operation order in his own hand. But his talents and enthusiasm also involved him in tasks of a more active nature.

After the first assault on the three breaches at Badajoz had failed Lord Fitzroy succeeded in getting into the city just before dawn and was instrumental, along with another British officer and a trembling Portugese drummer boy who beat the 'parley', in bringing about the surrender of San Cristobal, a detached fort to which the governor had retreated. By the end of the war he had more battle clasps on his medals than most infantry Regiments had battle honours on their Colours. His Peninsular Gold Cross had Fuentes d'Onoro, Badajoz, Salamanca, Vitoria, Pyrenees, Nivelle, Nive, Orthez and Toulouse; his Peninsular Silver War Medal Rolica, Vimiero, Talavera, Busaco and Ciudad Rodrigo.

With Napoleon exiled to Elba Lord Fitzroy took time off to marry his master's niece, the Honourable Emily Wellesley-Pole, before moving to Paris with the Duke as his secretary when he took up the post of ambassador. This introduction to diplomatic life was shortlived. Napoleon escaped from Elba and the famous 100 days campaign began. Lord Fitzroy rode at Wellington's side at Waterloo until the Frenchman's bullet shattered his elbow.

His wound healed slowly and he spent much time learning to write left-handed and to ride single-handed. He succeeded admirably with both, but was not destined to see another battlefield for almost forty years. Instead he became Wellington's shadow in every appointment the Iron Duke held until his death. When recovered, he rejoined the embassy in Paris as Secretary of Embassy. In 1827 Wellington was made the Commander-in-Chief of the Army on the death of the Duke of York, Lord Fitzroy once again going with him. As Wellington rose so, inevitably, did his secretary. At the age of 39 he was a major-general, ADC to the Prince Regent, MP for Truro and Freeman of the City of Gloucester.

Long years in dusty offices stretched ahead, with the government for ever calling for, and extracting, economies in the Army. Like Wellington, Lord Fitzroy resisted change, being dismayed by the visible weakening of Britain's military power. Both socially and professionally he came to know the notorious Lieutenant-Colonel Lord Brudenell (from 1837 the Earl of Cardigan) and, as Military Secretary, became involved when Brudenell was removed from command of the 15th Hussars in 1834.

For two years, as Brudenell fought for reinstatement with all the public fury and indignation he could muster, Lord Fitzroy was often the reluctant intermediary between him and the prime minister or commander-in-chief. At one time during this period he reported, 'Lord Brudenell favoured me with another of his disagreeable visits yesterday. I confined myself to telling him that you [the C-in-C] would not recommend his appointment to the 11th Light Dragoons.' After a later interview with Brudenell, who vigorously refused to accept 'no' as an answer, Lord Fitzroy commented, with typical restraint, 'The interview was not a very long one nor can I say that it was an agreeable one.'[2] In the end Brudenell, with Court intervention, had his way and secured command of the 11th Light Dragoons – shortly to be renamed the 11th Hussars. Almost twenty years on Lord Fitzroy, then Lord Raglan, was to find the Earl of Cardigan a wilful and unmanageable subordinate.

Lord Fitzroy's private life was happy, except when he lost his eldest son to wounds received in the Sikh War of 1845. His interests outside his duties were hunting, shooting (both of which he accomplished successfully with one hand), good food and socializing. Although an MP for six years he never once spoke in the House. He cared little for the changing world outside the Army. Music, painting, science, even books seemed to bore him. He only ever recorded reading one, *The Count of Monte Cristo*, about which he wrote, 'So far as I have got in it I find it is tiresome – very poisonous.'[3] As he approached old age dignity, charm and gentleness were his most noticeable traits. William Russell, who as a war correspondent at his headquarters in the Crimea observed him closely, described him thus:

> 'There was a personal charm about Lord Raglan which fascinated those around him. The handsome face, the sweet smile and kindly glance, the courteous, gracious, gentle manncr – even the empty coat sleeve that recalled his service in the field under his great master – attracted attention and conciliated favour. . . . his winning ways captivated strangers at once . . . Lord Raglan was the object of the most affectionate admiration and regard.'[4]

A thoroughly likeable elderly gentleman with exceptional talents as a military administrator and diplomat (including the diplomat's desire to compromise, to avoid giving offence). But did he have the robustness, the drive and, if necessary, the bite to command an army in the field, or to control fractious, self-willed subordinates? Even the dangerously shortsighted Crimean War historian Kinglake, who was always a Raglan admirer, had his doubts:

> 'But for more than thirty years of his life Lord Raglan had been administering the current business of military offices in peace-time, and this is a kind of experience which, if it can be very long protracted, is far from being a good preparative for the command of an army in the field.'[5]

In 1852 the 87-year-old Duke of Wellington died. Lord Fitzroy expected to get the C-in-C's post but it went to his junior, Lord Hardinge, who had powerful royal patrons. Instead he was made Master-General of the Ordnance, responsible for equipment, fortifications and barracks, with indeterminate control over the Royal Artillery and Royal Engineers, including their pay and discipline: something of a potpourri, and an undoubted disappointment. Possibly to compensate for ignoring the promotion by seniority rule, Queen Victoria insisted on the waiving of the substantial fees due on acceptance of a peerage (Lord Fitzroy had doubts he could afford it). Thus in October, 1852, he became the first Lord Raglan.

When the government came to selecting a commander for the expeditionary force there was not a lot of choice. Those senior officers with experience of a major war were all over sixty, some over seventy, who had been junior regimental officers when they had last seen a battlefield. Whoever was appointed, it would be equivalent to giving command of the BEF to France in 1939 to a soldier whose last active service, indeed any service with troops, had been as a junior ADC in the 2nd Boer War. There were likely to be problems. Although Lord Raglan was 66 at least he had watched the great Duke at the height of his success; something of his skill must have rubbed off. With such a tutor for all those years surely he had absorbed more of generalship than his equally elderly contemporaries. In addition there was his fluency in French and his acknowledged skill as a diplomat – qualities that would certainly be necessary in dealing with the new ally. It is hardly surprising that Raglan got the job.

*   *   *

The horseman who delivered the order.

Before the invention of the radio or telephone, how did a general transmit his orders on the battlefield, or indeed throughout a campaign? The answer was by horsemen. Orders and messages were carried in the pockets or pouches of mounted officers, normally ADCs. The fate of many a conflict depended on the horsemanship, courage and luck of a young man galloping across a battlefield to find a subordinate commander. An ADC's job in time of peace may have been one of a social secretary, delivering nothing more exciting than an invitation to dinner, but in war he carried out one of the most dangerous and critical duties in the Army. During the battles of Quatre Bras and Waterloo four ADCs to British generals were killed and fifteen wounded. Without proficient ADCs a general could not command, because he could not communicate.

The man destined to deliver the order that launched the Light Brigade and who was to die minutes later, almost certainly the first casualty of the charge, was Captain Louis Edward Nolan. Nolan was an ADC. He was 36 years old and wearing the distinctive uniform of the 15th Hussars – instantly recognizable, as he was the only representative of that Regiment in the Crimea. The uniform was worth over a year's pay. The 15th Hussars dressed in a blue, heavily gold-braided pelisse, dark blue jacket, blue overalls with a broad gold stripe and, most distinctive of all, a red shako (although Nolan always wore the equally distinguishing red and gold officers' forage cap). Nobody could fail to notice Nolan. He was ADC to Brigadier-General Richard Airey, Raglan's quartermaster-general, who was effectively the chief-of-staff of the Army. He was not, however, related to Airey. This made him something of a rarity, as generals choose their own ADCs, often deliberately picking members of their own family, Raglan having no less than five relatives, four of whom were nephews or great-nephews, on his staff. These young men, picked for their family connections and impeccable breeding, often had no worthwhile military experience and consequently knew nothing of their vital duties in the field. Lord Wolseley, who had served as a young officer in the Crimea and considered the staff at all levels inept, was particularly scathing:

> 'They [the generals] were served to a large extent by incompetent staff officers, as useless as themselves. . . . Almost all our officers at that time were uneducated as soldiers, and many of those placed upon the staff of the Army at the beginning of the war were absolutely unfit for the positions they had secured through family and political interest. . . . They were not men whom I would have entrusted with a subaltern's picket in the field. Had they been private soldiers, I don't think any colonel would have made them corporals.'[6]

Nolan may, however, have been one of the exceptions to the rule. His grandfather had been a trooper in the 13th Light Dragoons, his father a captain in the 70th Foot (later the East Surrey Regiment), while he himself had been a cadet in the 10th Imperial (Hungarian) Hussars at fourteen. Nolan and his elder brother were sent to the Engineer Corps School near Vienna to learn the language and other suitable subjects such as equitation, fortification, engineering, mathematics, drawing, fencing and swimming. Excellent reports on their progress were sent back to Milan where their father was the British vice-consul. By the time he was twenty Nolan was an accomplished cavalry officer in the Imperial Army who studied military history, read in several foreign languages and had served with his Regiment in Hungary and the Polish provinces. His confidential report in 1838 commended his 'great zeal and application'. They were characteristics with which the name Nolan was to be invariably associated in the first months of the Crimean campaign. Some were to say that it was this zeal, perhaps misguided zeal, that cost him his life and the Army the Light Brigade.

1838 was coronation year in Britain. Nolan took leave from his Regiment to visit London and, through the influence of the Austrian Ambassador, secured an invitation to one of the Palace levées. It is possible he had a seat in the Abbey for the actual coronation. He certainly watched with riveted admiration as the full pageantry of the British Army paraded at a grand military review in Hyde Park ten days later. The splendour of it all seduced him. While Nolan was still on leave his father was writing to the Military Secretary (Lord Fitzroy Somerset) seeking to obtain a commission for his son in the

British Army. It was the first of several letters from Nolan senior to Lord Fitzroy, at first gushingly grateful for a vacancy in the 4th Regiment of Foot and then ingratiatingly pleading for a cavalry commission. His appeal succeeded. On 23 April, 1839, Nolan, who had by now long overstayed his leave, was gazetted cornet in the 15th Light Dragoons (later to be renamed the 15th Hussars). For seven months, until the 10th Imperial Hussars struck him off their rolls, he held a commission in two armies.

Nolan was on the strength of the 15th Hussars for fifteen years. During that time he served with his Regiment in India, at the Cavalry Depot, Maidstone, and on the staff. He became an acknowledged expert on horses and a self-styled expert on cavalry. He had a book published on each subject expressing his radical and often controversial opinions.

Without horses no army of the 1850s could function, let alone fight. Horses obviously carried the cavalry, which was the eyes and ears of any army, but they also pulled the field and horse artillery's guns; they dragged the carts laden with the basic needs of all soldiers such as ammunition, rations and tents. They pulled the hospital wagons with the wounded; they hauled the heavy siege guns into position, as well as carrying general officers and their staff. Horses were as numerous as men and just as important. The procurement of sufficient numbers of the right size and type, their training and feeding, constituted a major logistic headache on any campaign. Nolan was a horseman par excellence. He became a superb trainer of horses and a fanatical believer in the supremacy of the mounted arm. He came to believe that cavalry could accomplish virtually anything.

His initial posting to his Regiment at Bangalore in India lasted only a matter of weeks as illness necessitated a return to England on two years' sick leave. Hardly a promising start. While on leave he purchased his promotion to lieutenant. Then, in March, 1842, he was sent to the cavalry depot at Maidstone to take the riding master's course in equitation. A riding master had special responsibility for training and equipping remount (new) horses in a regiment. As such it was a key position. Later, remount training was to be the subject of Nolan's first book. Years afterwards his depot commandant at the time was to write of his enthusiasm:

'There are many who soldier to live. Captain Nolan was a man who lived only to soldier. . . . He was a *maitre d'armes* of a very good school [referring to Nolan's Austrian service]; and as there was nobody else of any grade in the place who could fence, I enjoyed the privilege . . . of an occasional bout with Captain Nolan.'[7]

With regard to Nolan's unshakeable confidence that cavalry was invincible the commandant went on to describe an exceptional glimpse into the future:

'Often I have heard him express his conviction that cavalry could accomplish anything. . . . I remember . . . that in putting a case hypothetically of cavalry charging artillery in a plain, Captain Nolan drew with a piece of chalk on the wall of the Quartermaster's store in Maidstone barracks a rough sketch which as nearly as possible represented the relative positions of the Russian artillery and British light cavalry at the battle of Balaclava; the only thing he was not quite right about was the result. He assumed in such a case the certain capture of the guns.'[8]

Nolan rode under instruction on the course with everyone else, including NCOs and privates, one of whom wrote:

'Mr Nolan was a thorough gentleman in every respect. He had one peculiarity, however, a complete absence of anything in the shape of pride, and in all his intercourse with the NCOs and men of the classes he was as unpretending, and I may say as familiar as any of us. Some one has said that "familiarity breeds contempt", but in this case it bred a very deep and lasting feeling of esteem. . . . He was a splendid horseman, supposed to be the best in the Army.'[9]

It was not until mid-1843 that Nolan at last rejoined his Regiment in India. He had finally arrived in what was known as 'the home of light cavalry'. Unfortunately the 15th Hussars were still at Bangalore in the peaceful south. The 1840s saw a succession of campaigns and battles across central and northern India and into Afghanistan, but Nolan's Regiment missed them all. In 1843 Napier conquered the Scinde, at the conclusion of which he sent his famous signal – *'Peccavi'*. At the headquarters in Delhi there was some consternation until a classics scholar could be found who accurately, and with some amusement, translated the solitary Latin word as 'I have sinned'.

In the First Sikh War of 1845–46 invaluable experience was gained of cavalry operations at battles such as Mudki, Ferozeshah and Sobraon, while contemporaries of Nolan, like Lieutenant William Morris, were charging with the 16th Lancers at Aliwal, a great cavalry triumph. The Second Sikh War two years later saw plenty more action in the Punjab. But it was at Chillianwala that there was an ominous foretaste of what could happen to a brigade of cavalry under an incompetent commander, and with misunderstood orders. The brigade, which included the 9th Lancers, was formed up to attack when the order 'Threes right', intended to adjust the frontage, was misheard or misconstrued as 'Threes about', the order to retire. The entire brigade disappeared to the rear pursued by Sikh horsemen who captured several British guns which had become exposed by the cavalry's rout.

Many officers gained enormous practical experience of active service in India but none of them received worthwhile commands in the Crimea. There was, at the Horse Guards and among most senior officers in Britain (including Raglan), an inexcusable snobbery that downgraded 'Indian' officers as second class; they were regarded with suspicion, and their Indian service of no military consequence. It was just not proper soldiering. Nolan, however, was distraught at his Regiment's inaction. His body remained at Bangalore but his mind and spirit were with the light cavalry as it charged and counter-charged across northern India. He made every effort to meet officers who had taken part, testing their experience with his own views. It was during this time that he became a firm friend of Morris, little thinking that a few years on, with Morris commanding the 17th Lancers, he would take his place beside him at the head of that long, open valley near Balaclava.

Remounts were the lifeblood of cavalry. Nolan immersed himself in the minutiae of every aspect of their training and equipping. He debated the merits and demerits of employing geldings instead of stallions for military service. A troop of 67 geldings from the 15th Hussars took part in a test of suitability with a troop of 67 stallions from another regiment. A distance of 439 miles was covered. At the end there was no noticeable deterioration in either troop. Nolan was also intensely interested in the effectiveness of the sword in a mêlée. He had read of how the Nizam's Irregular Horse were apparently able to lop off their enemy's heads, arms, and legs below the knee with remarkable facility. How was this achieved? Was some special training or special weapon used? The reply surprised him. 'We never teach them any way, Sir: a sharp sword will cut in anyone's hand.' The swords were merely weapons discarded by British dragoons, honed to a razor's edge and kept in wooden scabbards.

In October, 1852, Nolan was back at Maidstone, this time as a captain and the commander of the regimental depot troop. During this period he has been described as:

'A mature officer of thirty-four, confident in the exercise of his powers; a soldier who had known success and become accustomed to high society. Professionally, a certain intolerance had appeared; where matters affecting his beloved principles were at stake he could be arrogant and easily roused.'[10]

When the Duke of Wellington died two million people watched the endless funeral procession. Every regiment in the Army was represented in the slow, sombre march to his final resting place under the dome of St Paul's. Nolan rode at the head of the 15th Hussars detachment.

This was the time of the publication of his first book, *The Training of Cavalry Remount Horses: A New System*. It was highly technical and examined every aspect of current teaching, highlighting ways it could be improved. Nolan was an outspoken advocate of combining firmness with gentleness in training. He was absolutely opposed to harsh bits. In his opinion a horse frightened of the bit would never be suitable for cavalry as it could not be properly controlled in battle. Every charge against cavalry should end in a mêlée involving hand-to-hand fighting between riders. To win meant having total control of the horse and, Nolan insisted, the ability to pirouette. To swing your mount round on its haunches gave the rider a considerable advantage in gaining his opponent's weaker side – normally the left. Nolan's second book was entitled *Cavalry* and was primarily concerned with organization, formations and tactics.

He wrote with total conviction but not much modesty:

'After long consideration of the whole subject, I honestly believe that the main principles I propose are right. Without this conviction I would not publish at all, but with it I feel it to be a dereliction not to offer to my brother officers, and the service in general, the results of my practice and meditation.'

Nolan felt the minimum height for a cavalry recruit of 5 feet 10 inches to be ridiculous, preferring shorter men with agility – and less weight. Nor did he like the traditional 'military' seat. This he termed 'Tongs across a Wall' when the rider sat stiffly, legs straight, with heels in line with the shoulders. When the horse trotted the rider had too little control with his legs and 'bumped' in the saddle. There was no gentle rising up and down at the trot in the Army. It would look ridiculous on parade. Perhaps of greater significance, in view of what happened to the Light Brigade, was his realization that once a charge started, once a line was launched, it was virtually impossible to manoeuvre it. In other words it must be facing the right direction to start with. In his view cavalry preparing to charge should deploy the first line only, the supports being in column behind the flanks, and the reserve much further to the rear in a similar formation behind the centre. If the first line was defeated and fell back a second line coming up behind would collide with the first. His last chapter was devoted to how cavalry should, with the aid of horse artillery, always be able to break an infantry square. It was radical stuff. Much of it made good sense, some of it was controversial but it all confirmed Nolan's reputation as a cavalry fanatic.

His books invited criticism by many who felt him to be conceited, rash and intolerant, and far too junior to be lecturing his seniors on cavalry warfare. Nevertheless his reputation and his writing secured him a place in the public eye, and a critical staff appointment at the outset of the Crimean War. No less a person than the Secretary at War, the Duke of Newcastle, wrote on 11 March, 1854, to Raglan:

'I find that a steamer leaves Marseilles for Constantinople upon the arrival of a Messenger who will leave London on Monday at 6 p.m. Will it be possible for Captain Nolan to go by that Boat to purchase horses for the Army? . . . Pray send for Captain Nolan, if you think it possible for him to go so soon.'[11]

It was possible, although Nolan had no time to buy the frock coat and cocked hat that were his entitlement as a staff officer. He was briefed personally by

33

the Secretary of State, the Commissary-General (William Filder) arranged for his funds, and Nolan departed in a rush with responsibility of buying remounts for the entire Army.

*　　*　　*

The horseman who received the order.

The fifty-four-year-old commander of the Cavalry Division received the order – and lived to regret it. The man born George Charles Bingham in 1800, who became the 3rd Earl of Lucan in 1839, was eventually to be sacked for accepting the order, failing to query the order, failing to read it in conjunction with the previous order, but, above all, for misintepreting it and then acting on his misinterpretation. According to him it was an ambiguous order, a confusing, puzzling if not downright stupid order, but nevertheless one that had to be obeyed at once.

Some people, including Kinglake, felt that Lucan had got the command of the Cavalry Division for his ruthless energy as one of the dreaded 'consolidating landlords' in Ireland. It can hardly have been for recent military experience as he had been on half pay, performing no military duties whatsoever, for almost seventeen years. During all this time he had, however, been fighting. He had been fighting a pitiless battle to expel the majority of his penniless peasants from his vast estates around Castlebar in County Mayo. He sought to eject them in order to consolidate the patchwork of tens of thousands of tiny holdings into economic farms. In doing so he spent years fighting his land agent, his neighbours and his tenants. His tactics were browbeating, the demolition of houses, the eviction of the occupants and lawsuits. His battlegrounds ranged from the wild, desolate moors and mud huts of Mayo to the benches of the House of Lords. During these years he became accustomed to being loathed and execrated. He became combative, vindictive, impatient and antagonistic. With everyone turned against him, with his motives misunderstood (he genuinely felt he was working for the long-term good of the people) his character hardened; he had no time for sentiment and no time for public opinion.

Lucan had gone to reside in Ireland in 1837 and was there throughout the harrowing years of the potato crop failures and consequent famines when the poor were forced to flee in their droves, while thousands died of starvation or disease. The situation in Ireland during this period has been described thus:

'The country was a country of holdings so small as to be mere patches. The people inhabited huts . . . built on the bare earth . . . without chimney or window and destitute of furniture, where animals and human beings slept together on the mud floor. . . . The Irish people were religious, their family affections strong, their women proverbially chaste. Early marriages became invariable . . . but religion and ignorance combined to make birth control unthinkable . . . Thus the population [had] spread like

34

an epidemic. . . . The Irish came to live on the "Lumpur" or "Horse Potato", the largest, coarsest and most prolific variety known. . . . They ate this potato boiled and they ate nothing else . . . greens were unknown, bread was unknown, ovens were unknown. The butcher, the baker, and the grocer did not exist.'[12]

Without the potato there was nothing. When the crops were blighted in successive years the mass exodus of destitute refugees gave the United States her Irish American community and saw many English regiments awash with Irish recruits.

Almost unique among the large landowners Lucan lived at Castlebar through most of this period. Exceptionally, he poured money into his estates, prepared to forego rent and invest his capital in machinery, barns and drainage – anything to get the countryside productive. The price to be paid was depopulation. Lucan was a callous practitioner. Nevertheless, detested though he was, Lucan had a few supporters. *The Times* of 16 November, 1849, published a letter from one of its correspondents who had recently visited Mayo:

'Lord Lucan is eminently a practical man; that which he determines to do he sets about at once, suffering no expense of pocket or popularity to interrupt him. . . . It is true . . . he has earned himself the character of "a great exterminator". . . . I saw also, what is not often seen in Ireland, the so called exterminator giving his every effort, at any cost, to lay the foundation of a system of cultivation to give a future generation . . . comfortable dwellings, with fair wages for fair work. . . . Now, Sir, if a Landlord is to be found resident . . . cultivating large tracts of land in the best possible manner . . . he does appear to me to deserve no little credit.'[13]

This was always Lucan's defence against his multitudinous detractors.

Despite all the recent years of military inactivity Lucan had formerly been an enthusiastic soldier. Like so many of his aristocratic contemporaries he had started young. Educated at Westminster, he was a cornet in the 6th Foot (subsequently the Royal Warwickshire Regiment) at sixteen, and within ten years was the Commanding Officer of the 17th Lancers. His meteoric rise was entirely due to wealth. Like his future brother-in-law, the Earl of Cardigan, Bingham, as he then was, was able to manipulate the purchase system to the full and thus secure quick promotion over the heads of others both more senior and more experienced.

Since Cromwell's rule by the Major-Generals, parliament was terrified of officers who were professional military adventurers dependent on soldiering for a living, and with no stake in the land. The answer was the purchase system, introduced into the new standing Army in 1683. To be an officer required the payment of a substantial sum. Such men had property, had land, and thus everything to lose from revolution. The pay of an officer was merely an honorarium. Inconceivable though it may seem today, for 200 years officers' pay remained virtually the same, whereas that of NCOs and soldiers rose. Under purchase it was accepted practice to buy promotion one day and go on half pay the next – Bingham did it twice – while waiting for the chance to buy into a regiment more to one's liking.

By the time he bought the lieutenant-colonelcy of the 17th Lancers, over the head of one of the most able and experienced cavalry majors who was a Peninsular and Waterloo veteran, Bingham had been on the rolls of five regiments. This purchase cost him £25,000 – £20,000 above the official price. It soon became clear that the new commanding officer was a martinet and a perfectionist. As he was to do in Ireland twenty years later, Bingham poured his own money into the Regiment to smarten it up, to give it the best uniforms and the best horses. The 17th Lancers became an elegant, stylish Regiment which soon acquired the nickname, 'Bingham's Dandies'. But all was not well. Bingham was obsessional, seemingly unable to distinguish between the trivial and the important. Parades, inspections and drills followed one another in a remorseless, unending cycle. The slightest fault, the slightest trifling deviation from regulations aroused the wrath of the lieutenant-colonel. Officers were bawled out, soldiers were flogged; all suffered.

Harsh, bad-tempered, arrogant, vindictive and unpopular, at twenty-six Bingham was all these. He was also, however, a worker. He rose before dawn and toiled incessantly, demanding the same from all under him. His efforts were rewarded. Outwardly the Regiment looked good. The Duke of Wellington thought well of Bingham and his 'Dandies'. Uniquely, he studied military history as well as tactics and drill, but he had no experience of war.

For a few months at the end of 1828 and early 1829 the 17th Lancers had some relief from the relentless pressure. Bingham secured an appointment on the staff of the Russian Prince Woronzoff in the Balkans soon after the outbreak of the Russo–Turkish War. This war was fought over the same ground south of the Danube that saw the start of the war in 1854. Bingham became familiar with places like Varna and Silistria, and he came to know something of the Russian military, against whom he would one day have to fight. He acquitted himself remarkably well. He was indifferent to hardship and danger, was always at the forefront and, as usual, worked unflaggingly. The Russians were most impressed. When he returned home he did so with the Order of St Anne 2nd Class.

Of particular significance in the story of the Light Brigade's fate was the hostility between Lucan and Cardigan. Both were proud, jealous, ambitious and overbearing, but Lucan was the more intelligent, and it had infuriated Cardigan that at twenty-nine he was still a captain while Lucan, three years younger, was a lieutenant-colonel. The relationship did not improve when Lucan married the youngest of the seven Brudenell sisters shortly after his return from the Balkans. The brothers-in-law were incompatible, irritating and antagonizing each other whenever their paths crossed. Lucan despised Cardigan as 'a featherbed soldier', while Cardigan considered his sister to have been treated abominably, 'sacrificed to Lord Lucan's farming mania, kept short of money and deprived of suitable enjoyments.' Things were not improved when Cardigan and his wife parted in 1842. The enmity between the two noble earls intensified.

With war imminent Lucan, unusually diffident, but perhaps feeling his long years out of uniform put him at a disadvantage, applied to Lord Hardinge, the commander-in-chief, for a post as an infantry brigadier. Cardigan wrote to his old family friend, Raglan, requesting an appointment with the cavalry. In February Lucan was overjoyed to get command of the Cavalry Division – his cavalry background and previous service in the likely area of operations, albeit 26 years ago, no doubt counting for something, in addition to his ferocious energy more recently displayed in Ireland. On April Fool's Day, 1854, Cardigan was finally given the Light Brigade, serving directly under Lucan. It was no joke. It was a decision that was to cause considerable grief to the high command in the months ahead.

* * *

The horseman who executed the order.

The oldest, at 57, and by far the most outrageously notorious man in the entire Army, was the brigade commander, Major-General the Earl of Cardigan. Cardigan, whose youthfully slender figure still cut a dash in his hussar uniform, had had 30 years of turbulent, indeed tempestuous, service. During none of those long years had he ever heard a shot fired in anger – unless one were to count four shots from duelling pistols on Wimbledon Common. His battles had been fought on the parade ground, in the stables, around officers' mess dinner tables, in the columns of *The Times*, in the Old Bailey, the House of Commons and the House of Lords. His notoriety, the contempt in which the public and the press had held him, were due exclusively to his own scandalous and offensive behaviour.

Entering the Army late (at 27) Lord Brudenell, as he then was, had but one ambition – to command a regiment. To realize such an ambition quickly, and Brudenell was never known for his patience, did not require brains or ability, merely money and influence. Nearly two hundred years before, a Thomas Brudenell had been created the Earl of Cardigan by King Charles II in the banqueting hall of Whitehall Palace. From that moment on the Brudenells had rank, had influence at Court and acquired, so they thought, a divine right to command. In 1832, at a cost of almost £40,000 (several million pounds in today's money) the young Lord Brudenell purchased the lieutenant-colonelcy of the 15th Hussars – the regiment Nolan was to join seven years later.

His tyrannical rule came quickly into the public domain when, the following year, he had a Captain Wathen court-martialled on several ludicrous charges in connection with the issue of stable jackets. The régime of persecution and victimization that he had instituted from the outset of his command came to light. Wathen was acquitted. Of Brudenell *The Times* said, 'Such a man ought

37

never to have been placed at the head of a regiment.' Within days he was dismissed. The court-martial findings were published in a General Order which concluded, 'His Majesty has been pleased to order that Lieutenant-Colonel Lord Brudenell shall be removed from the command of the 15th Hussars.' Then, the ultimate disgrace – the Order was to be read out at the head of every regiment in the Army. Such ignominy would have destroyed most men, but within less than two years Brudenell was back, as commanding officer of another cavalry regiment.

Heartrending entreaties, almost literally on bended knee, by the aged, sickly and disconsolate Earl of Cardigan to King William, plus another £40,000, eventually won his son a second chance. Lord Hill, the Commander-in-Chief who had thus far adamantly resisted any possibility of reinstatement, was summoned to the Palace and persuaded to capitulate. He was later to say, more in hope than expectation, 'I hope that the author of this distress is now sensible that he cannot be permitted to follow the dictates of his ungovernable temper. I trust this lesson has not been thrown away.'[14] There was a public outcry, but a debate in Parliament, which actually voted on the matter, resulted in a resounding victory for Lord Brudenell.

It did not take long for the 11th Light Dragoons (later Hussars) to discover that their new commanding officer had learnt no lesson. His new Regiment was in India, but Brudenell was in no hurry to join it. He and his wife (his marriage was to be another disaster) left London in June, 1836, at the start of a gentle journey via France, Italy, Malta and Egypt. They arrived in Bombay in January, 1837, but it was not until October that the 11th Light Dragoons saw their new commander. Brudenell had dallied and dawdled, making the most of visiting important people, shooting tiger at hill stations and ensuring that 'Society' in India knew he had arrived. It was during this time that he received the news of his father's death and of his becoming the 7th Earl of Cardigan.

After taking command it took him less than four weeks to wreak havoc in the Regiment. In that time he had eight courts-martial and over a hundred men disciplined. Flogging was back on the punishment list. Private Loy Smith, as he then was, well remembered those days. An old soldier who was awaiting his discharge had been confined for drunkenness on picket – an offence that usually merited about eight days in the cells. Cardigan ordered him court-martialled. Smith records in his diary,

'I could never quite forgive Lord Brudenell. . . . When the court-martial was read, to our amazement he was awarded corporal punishment. No one supposed for a moment that Lord Brudenell would be hard-hearted enough to carry it out, particularly when the old man turned round to him and, in an imploring tone said, "My Lord, I hope you won't flog me. I am an old man and just going home to my friends, and I should be sorry for such a disgrace to come on me now." "Tie him up," said Lord Brudenell.'[15]

There was a huge regimental sigh of relief when it was learnt that Cardigan had no intention of accompanying the Regiment when it sailed for England a few weeks after his arrival. Naturally, he took the leisurely, and infinitely more comfortable, route back.

By the time he rejoined at Canterbury Cardigan had been the commanding officer for two years, during which time he had been with the Regiment a month. He resolved to make up for lost time. In six months he held another fifty-four courts-martial. Nevertheless, as the months passed the extravagant expenditure from his own pocket on glittering new uniforms (including the tight cherry coloured trousers), the incessant daily drilling, the insistence on all manoeuvres being carried out at a gallop, coupled with draconian discipline gave the Regiment a reputation for military competence. The 11th Light Dragoons were selected to provide an escort for Prince Albert from Dover to London for his marriage to Queen Victoria. For this honour it was rewarded with the title of The 11th, Prince Albert's Own Hussars.

In the years that followed Cardigan's scandalous and controversial behaviour was such that he could not attend a theatre without such a commotion of booing and hissing that no performance could start until he left. In 1840 the 'black bottle' affair became public knowledge and a popular catchphrase. At a Regimental dinner night an officer had allowed a bottle of wine to be put on the table for a guest instead of champagne as Cardigan had ordered. The resultant row embroiled not only the whole Regiment but the inspector-general of cavalry who had been the guest of honour. In September of the same year Cardigan fought a duel with a former officer of the Regiment, and with his second shot wounded his opponent in the side. This was a criminal offence for which the penalty could be death, or at least transportation for life.

Needless to say neither of these things happened to Cardigan. But before he was arraigned at the Old Bailey Cardigan's pernicious persecution of Captain Reynolds, the 'black bottle' officer, came to a head with Reynolds being court-martialled and reprimanded, and then being driven to the brink of resignation by the continued hounding he received from his commanding officer who regarded him as a useless and troublesome 'Indian' officer. Finally, the military secretary, Lord Fitzroy Somerset (Raglan), became involved, using his persuasive talents to induce Reynolds not to sell out. In October, 1840, Cardigan appeared briefly on the duelling charge at the Old Bailey. Because he was an earl Cardigan elected to appear before the House of Lords in the following February. *The Times* cried out for justice. 'Let his head be cropped, let him be put on an oatmeal diet, let him labour on the treadmill.' It was not to be. The noble lords rallied round their recalcitrant fellow peer, all bar one rising to proclaim, 'Not guilty, upon my honour'. On a totally transparent technicality Cardigan was unanimously acquitted. That night a riot took place at the theatre in Drury Lane where he had had the audacity to appear. In March, 1841, the House of Commons debated Cardigan's behaviour and continued command of the 11th Hussars, but the Duke of Wellington, ever mindful of upholding military discipline, let it be known that the matter should be dropped. It was.

That Easter Sunday Cardigan was at it again. The Regiment had been marched to divine service in the riding school, afterwards being taken back to barracks for a short inspection before returning again to the riding school to witness a flogging. To have the same body of men who a short while before had been praying and singing hymns in the same building, watch somebody have the flesh flayed from his back was, to say the least, insensitive. In fairness, it was not a sentence Cardigan himself had imposed. The prisoner, Private William Rogers, had been tried by a District Court-Martial and Cardigan had been ordered by the Horse Guards to carry out the sentence. He could, however, have waited a day or so. There was renewed and prolonged public uproar. A special cabinet meeting was convened. Cardigan should go, for the second time. The Prime Minister wrote to the Queen, preparing her for a visit from the Commander-in-Chief. Once again Wellington intervened. Such was his influence that the clamour for Cardigan's head withered away.

For sixteen years Cardigan commanded the 11th Hussars. Towards the end of this time his behaviour, if not moderated, was at least less public. At times there could even be amusing moments at his expense. In 1848 Queen Victoria and Prince Albert reviewed the Regiment on Hounslow Heath where the ground was wet and slippery. Several horses fell, which infuriated Cardigan and brought forth dire threats as to what he would do if it happened again. Moments later his horse slithered and fell, depositing Cardigan in the dirt. Loy Smith wrote, 'As he lay on the ground, all smothered in mud, one of the men in the ranks exclaimed in a loud tone, "Behold the Lord is down," which caused a tittering among the men.'[16]

These were the four horsemen who led the Light Brigade to calamity.

CHAPTER THREE
To War

'On weighing anchor I assembled the men on deck . . .
[I] gave the signal for three cheers for the Queen, for a
glorious war and happy return, which was warmly
responded to.'

*Lieutenant-Colonel Lord George Paget, 4th Light
Dragoons, on 19 July, 1854, on board the
steamship* Simla *off Plymouth.*

Raglan had two major problems with the cavalry from embarkation to
Balaclava and beyond, and he was never to solve either of them satisfactorily.
In fact they both got worse as the war progressed. The first was the numerical
weakness of the Cavalry Division; the second the total incompatibility of, and
open antagonism between, the Divisional and Light Brigade commanders.
These difficulties impinged on one another and became exacerbated as
Raglan's attempts to solve them proved ineffective. His solution to the first
was conservation; to the second separation, reinforced by conciliation. Raglan
kept his cavalry 'in a bandbox'. He refused to risk them in battle or even in
pursuit; until Balaclava there had been no cavalry combat. At that battle, even
when he finally launched the Cavalry Division at the enemy, it ended in the
destruction of the Light Brigade.

With his fractious commanders Raglan endeavoured to maintain the peace
by keeping them apart, initially by some 150 miles. At the start he virtually
divided the command of the division. When all else failed he got his staff to
write polite and private letters, always anxiously trying to cool the boiling
water, but he was never the man to remove either of the sources of heat, never
the man to meet his troublesome commanders face to face. All he succeeded
in doing was to alienate the commander of his cavalry.

Raglan's efforts to conserve the cavalry are understandable. The division
had two (four regiment) brigades of around 1250 men each when they left
England in April and May; barely half that at Balaclava despite having, by
then, been reinforced by two fresh regiments. The Crimea, and indeed Turkey
and Bulgaria, was cavalry country. The Russians were known to have huge
mounted forces, including a high proportion of light irregular cavalry, or
Cossacks. Incredibly, the French initially took no cavalry at all due to lack of

41

shipping. Even after the landings on the Crimean peninsula in September it was the British cavalry, specifically the 900-strong Light Brigade, that was responsible for protecting the front, left flank and rear of a combined Allied force of 60,000 men as they marched towards Sevastopol.

Two regiments of the Light Brigade (8th Hussars and 17th Lancers) were the first to sail, followed three weeks later by the 11th Hussars and then the 13th Light Dragoons. Not until afterwards did the Heavies leave.[1] The Light Brigade became the formation responsible for strategic as well as tactical reconnaissance, for advance guards, flank guards and rear guards; for skir-mishing and scouting; for escorting and foraging; for picketing and patrolling and, finally, for fighting. Certainly the Heavy Brigade took on some of these duties, but it was not complete until the arrival of the Scots Greys after the Battle of the Alma on 20 September. Even so, Raglan, remembering how well light cavalrymen had served Wellington in the Peninsula, tended to deploy his on the wider-ranging tasks, although the only differences between the two were their titles and the colour of their uniforms.

Despite the numerical shortfall, despite the acknowledged priority of having fit cavalry on the ground quickly, the British planners made some strange deci-sions when transporting the division to Turkey. First, they intended to march them to Marseilles. To the relief of most, this idea was dropped. Instead they would sail all the way. The majority of the cavalry, however, were not allo-cated the fast new steamers which took a mere two weeks to cover the 3,360 miles to the Bosphorus, but the much smaller and slower sailing vessels which took anything from three to over eight.

Sail had other serious disadvantages. The size of the ships was such that five or six were needed to transport one regiment of about 300 men and horses. If the weather was foul delays could mean units arriving piecemeal with long gaps between troops. A steamship like the *Himalaya* could take an entire regiment whereas the tiny transport *Asia* (721 tons) could barely manage 'F' Troop of the 11th Hussars – three officers, 43 NCOs and men with 46 horses.

Many soldiers, and most horses, make poor sailors. The sooner both reach land the fitter they will be. It was not only the voyage that held its horrors; for the animals embarking and disembarking could be terrifying as well. Private Mitchell of the 13th Light Dragoons has described watching the horses swung onboard:

'Each horse was led up to the ship's side (which lay close alongside the quay); a sling was placed beneath the horse's belly, and fastened to the tackling on the main-yard. The word was given to "hoist away", when about a hundred convicts manned a large rope, and running away with it, the poor trooper was soon high in the air, quite helpless.'[2]

TSM Smith states that it took four hours to get the 46 horses of 'F' Troop down below decks 'as some of the horses resisted violently'. Soldiers had to spend countless hours with their mounts below decks, sometimes struggling to stay upright, often vomiting themselves as they held their horses heads trying to soothe them, restrain them, or coax them to eat. Barrels of vinegar had been shipped to refresh the animals, the men 'sponging their heads well over with it, particularly their nostrils'.[3] In rough weather the noise, stench and sights in the stables could be hellish.

Even the 4th Light Dragoons, who did not join the Brigade until early August and travelled on the steamship *Simla*, had to shoot horses, although for different reasons. Captain Portal wrote home that:

'[Two horses] on the main deck got perfectly mad from the heat, and at last became so dangerous to all the horses near them that they had to be destroyed. . . . Those poor beasts that stand below close to the engines are in perfect steam all day and night too. We ought not to have any horses there at all.'[4]

It says much for the devotion of the soldiers that the division only lost about 130 horses on the voyage out, although the great majority of the remainder arrived in a distressing condition, in no state for arduous campaigning.

As the Allied armies began to assemble in Turkey there was much emphasis on elaborate ceremonial parades, military etiquette and diplomatic dinners. On 27 April the Turkish Seraskier Pasha (C-in-C) was invited to review the troops. Captain Calthorpe, a nephew and ADC on Raglan's staff, described his arrival:

'The Seraskier Pasha came attended by a numerous but very ragged staff: such dirty, untidy looking fellows, and so badly mounted. The Pasha himself is a very fat man, with a bad expression of countenance, and sits like a sack on his horse.'[5]

When Prince Napoleon arrived off Scutari, to take command of the French 3rd Division he received a 101-gun salute (the number specified for royalty) from all the warships in the harbour. After the arrival of the British royal representative, HRH the Duke of Cambridge, who had been given command of the 1st Division which contained the Guards Brigade, a sumptuous banquet was given in his honour. Calthorpe again:

'The guard of honour . . . was furnished by the 93rd Highlanders, with their band, and they astonished many of the diplomatists (sic) with the noise of their bagpipes and the appearance of their kilts.'[6]

There were, however, more serious matters needing attention. Near the top of the pile was a pressing need for more horses, not only remounts for the cavalry and artillery but, literally, thousands more for commissariat duties. One of the first officers to see Raglan when he arrived at the end of April was Nolan. He knew Nolan well by reputation and had a high opinion of his skill and judgement. Nolan's mission to buy remounts in Bulgaria, however, had been a flop. In six weeks he had obtained 40 at most. In his view the local horses were too small for remounts, and, although there were plenty of baggage animals, he had not been authorized to purchase these. Disappointed, Raglan sent him off to Syria to look for more. Despite this lack of success, Raglan's despatch to London dated 2 May commended Nolan for 'the zeal and intelligence for which he is remarkable'.[7]

Three days later Calthorpe met Nolan and described him in a letter home as, 'an officer who, most justly, is very highly thought of by the authorities.'[8] Raglan, always the orthodox soldier, also rejected the suggestion of raising irregular cavalry locally, if only for reconnaissance and foraging. It had been made to him in London; now he used Nolan's opinion to justify his own view. He wanted proper soldiers mounted on proper horses. According to Nolan the irregular cavalry in Turkish pay were 'mounted upon ponies too small and too weak to resist an attack of Russian cavalry, or to be of material use in the field in any way.'[9] The Commander-in-Chief and the Captain were in agreement – but the lack of cavalry, the lack of horses, with all the detrimental implications for operations, remained. The military logic was simple. Cavalry (and horses) are vital to the prosecution of any campaign – there is a chronic shortage of cavalry – I cannot easily replace them – deduction, I must preserve them. The Commander-in-Chief's and the Captain's views were to diverge on this issue.

The strategic situation in mid-May was awkward (see Map 2). The Turkish army, based at Shumla, was confronting the Russians along the Danube while the British and French sat around Scutari, opposite Constantinople. The Turks urged the Allies to move by sea to Varna; the Allies demurred and dawdled as they were awaiting the arrival of the British Cavalry Division, the bulk of which was still at sea, with some regiments still in their barracks at home. While a move to Varna would be possible piecemeal they could not venture further afield without cavalry. On top of this Filder, the commissary-general, claimed he would need thousands of extra baggage animals to supply the army on the move. The Russians then precipitated matters by crossing the Danube and investing the Turkish fortress at Silistria, so it was agreed a start must be made. The only two cavalry regiments available were the 8th Hussars and 17th Lancers, so they and the Light Division from the British Army would sail for Varna. A French infantry division would do likewise. At the end of May the move began, with steamers towing the sailing transports. It would be another three weeks before Raglan arrived at Varna.

From the moment the appointments of Lucan and Cardigan became known speculation on the inevitable clash became rife within the Army and among the public. William Howard Russell, *The Times* correspondent, wrote:

'When the Government made the monstrous choice of Lord Cardigan as Brigadier of the Light Cavalry Brigade of the Cavalry Division, well

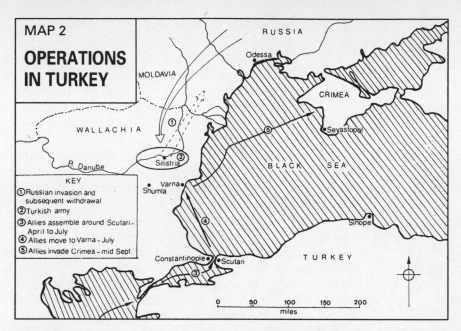

MAP 2

OPERATIONS IN TURKEY

RUSSIA

MOLDAVIA

WALLACHIA

R. Danube

Silistria

Odessa

CRIMEA

Sevastopol

Varna
Shumla

BLACK SEA

Sinope

KEY
① Russian invasion and subsequent withdrawal
② Turkish army
③ Allies assemble around Scutari – April to July
④ Allies move to Varna – July
⑤ Allies invade Crimea – mid Sept.

Constantinople • Scutari

TURKEY

0   50   100   150   200
miles

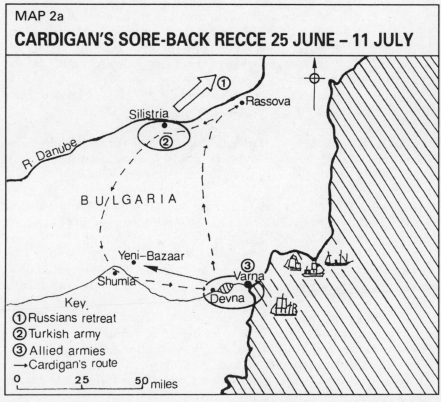

MAP 2a

CARDIGAN'S SORE-BACK RECCE 25 JUNE – 11 JULY

① Silistria

• Rassova

②

R. Danube

B U L G A R I A

Yeni–Bazaar

③

Varna

Shumla

Devna

Key
① Russians retreat
② Turkish army
③ Allied armies
→ Cardigan's route

0   25   50 miles

45

knowing the private relations between the two men, they became responsible for disaster.'[10]

Lucan had been at Scutari for nearly a month when Cardigan arrived on 24 May. Within the week Cardigan threw down the gauntlet. As his two regiments left for Varna he applied to join them, not unreasonably considering they constituted the bulk of his brigade. But, instead of applying through Lucan, he did so direct to Raglan, who immediately compounded the error by agreeing without consulting Lucan. Although it made military sense, with the majority of the cavalry still to arrive, to leave Lucan at Scutari, it also made practical sense in Raglan's mind to keep these two apart. It certainly marked the start of a subtle, if imprudent, 'separation' policy which was maintained even when the justification for it had gone.

Raglan's action reinforced Cardigan in his belief that he had an independent command, that he was responsible to the Commander-in-Chief first and Lucan second. He had gained this notion of independence in London, supposedly from Raglan himself, on appointment. According to Cardigan, although nominally under Lucan as the Divisional Commander, the Light Brigade was to operate separately under his orders and would not be subject to petty interference from his brother-in-law. This was not Lucan's view. He was the major-general commanding the cavalry division, Cardigan a brigadier-general (both received brevet promotion later) commanding one of his brigades. It would be intolerable if his subordinate should ignore his instructions or bypass him in any way. The military hierarchy did not, and could not, work like that.

Lucan got to hear of Cardigan's departure the day before he was to leave through a note from the Deputy-Quartermaster-General (DQMG):

'Lord Raglan wishes Lord Cardigan to proceed to Varna with [on the same ship as] the 17th Lancers tomorrow; will you give him instructions accordingly?'[11]

Cardigan, however, did not feel the arrangements suitable (the ship would be far too crowded) so he wrote back to the DQMG, again ignoring Lucan. He asked to be informed as to

'what vessel I myself and the others of my staff and 22 horses, including our baggage horses, our servants, and baggage . . . are to be embarked in.'[12]

He was immediately allocated the steamer *Jupiter*, 'as a proper vessel to convey a General Officer and his staff.'[12] Cardigan was a firm advocate of the old Army adage that 'any fool can be uncomfortable'.

Lucan was livid, although he restrained himself from involving Raglan, writing unofficially instead to Cardigan:

46

'It is obvious that the service cannot be carried out as it should be, and as in my Division I hope it will be, if a subordinate officer is allowed to pass over his immediate and responsible superior. . . . I write privately, as, though I consider the errors deserving and requiring notice, I wish this, like all other communications between us to be of the most friendly nature.'[13]

Cardigan declined to reply.

Stuck at Scutari he might be, but that did not stop Lucan from driving his staff to distraction with an endless stream of instructions on all conceivable military minutiae. Rising daily before dawn he drove himself and all around him relentlessly, nothing too trivial to escape his attention. This massive dissemination of paper was to continue unchecked from the end of May, 1854, to February, 1855. An intriguing example was dated 2 June, 1854:

'Long hair on the head is most objectionable . . . Moustachios and whiskers are to be allowed to grow, but no officer or private will be allowed to wear a beard. Below the mouth there is to be no hair whatever, and the whisker is not to be worn more forward on the chin than the corner of the mouth.'[14]

Cardigan, once established near Varna, ignored them all but followed his superior's example and bombarded his brigade with equally useless directives from a commandeered house, to which he had retreated from the heat to nurse a bout of bronchitis.

On 9 June the 13th Light Dragoons and 5th Dragoon Guards arrived in Turkey only to sail on to Varna a few days later to swell Cardigan's command. Lucan seethed. He was supposedly a divisional commander but had no troops to command, while Cardigan's independence grew by the hour. On 11 June he wrote an indignant letter to Raglan:

'The Commander of the Forces takes the field . . . and with the larger part of my division, and I am to be left behind. . . . for all I can see without duties. . . . I must require . . . submission from my subordinates; from Lord Cardigan I can scarcely hope to receive it if his lordship is allowed to continue in the opinion he is well known to entertain, that the position of a brigadier is one of independence towards his divisional superior.'[15]

He simultaneously wrote to Cardigan demanding that all reports and returns be sent through him, not direct to the Commander-in-Chief.

His first letter worked; the second did not. Lord de Ros, the DQMG, induced Lucan to withdraw his letter on the understanding that he could join his division and that Cardigan's position as a subordinate had been made crystal clear. Somewhat mollified, but still sensitive and suspicious, Lucan arrived at Varna on 15 June. His reception was disappointing and humiliating.

Raglan was away and nobody else at Varna expected him. The Light Brigade and the Horse Artillery were detached ten miles away at Devna, under Cardigan, and it was also the day that that burdensome brigadier responded to Lucan's reprimand of 11 June. There was no way Cardigan was going to surrender his independence.

'I consider that being sent forward in advance of the Army, and not being very far distant at the present moment from the enemy, that my command may be considered as a separate and detached command, and that I am not bound to anybody except the general officer in command of the forces in the country in which the brigade under my command is serving. . . . It is my intention to take an early opportunity of submitting an appeal to General Lord Raglan upon the subject for his decision.'[16]

This was open insubordination. Lucan forwarded it to Raglan with a vigorous complaint. It was testing time for Raglan as a commander-in-chief. So far there had been no action, but if there was the cavalry would have a key role to play. There was only one cavalry division for two armies, and it was public knowledge that the divisional and Light Brigade commanders were utterly incompatible. One of them should go before damage was done or lives lost. The question was whether Raglan had the moral courage to sack one of them. He did not. Still unwilling to confront them himself, this time Raglan told his adjutant-general, Major-General Estcourt, to write privately and reassuringly to Lucan. Estcourt was described by Calthorpe as, 'one of the most gentlemanlike persons I have ever met. I think him perhaps too lenient for an Adjutant-General whose duties require a stern and rather severe man.'[17] He wrote:

'The misapprehensions which Lord Cardigan has entertained of the nature of his command have already been rectified by private communication from me, written by Lord Raglan's desire . . . you may call for what returns you think necessary. . . . I have not returned you an official letter because, the misapprehension being corrected, it is better to consider the question as never having arisen in formal shape.'[18]

An incensed Lucan refused to let the matter drop. He replied to Estcourt that Cardigan was getting away with gross insubordination and was flouting regulations by going direct to Raglan instead of through him. It was to no avail. Raglan could not bring himself to abandon a lifetime of diplomatic manners; he lacked the ruthless streak essential in senior military commanders. Estcourt again patted Lucan on the head like an upset child, informing him that Cardigan would behave from now on. Unconvinced, another desperate plea came from Lucan to be allowed to join his division at Devna. 'I beg that I should not be any longer totally separated from the division to which I have been appointed.'[19] It was rejected by Raglan on the grounds that, as some detachments of cavalry had still to arrive, Lucan was 'to inspect each troop carefully as it disembarks'.[20] Another degrading snub.

At Silistria, meanwhile, the Turkish defenders had done remarkably well, with the assistance of a handful of British advisers with Indian experience, in seemingly forcing the Russians to abandon the siege. The enemy were withdrawing over the Danube. Raglan now needed strategic information on the Russians' intentions, locations and particularly whether the whole country south of the river was clear. If it was, the war might be over. This was a task for light cavalry. Accordingly, on 25 June, a jubilant Cardigan was despatched by Raglan, with a squadron each from the 8th Hussars and 13th Light Dragoons plus some Turkish cavalry, on an independent mission that was to go down in the history books as the 'sore-back reconnaissance'. Raglan saw no need to tell Lucan.

For seventeen days Cardigan, accompanied by one of Raglan's staff (Lord Burghersh), pushed his command to the limit. He covered some 300 miles in a gigantic loop (see Map 2a, p.45), first heading north until he hit the Danube, then moving east as far as Rassova, then backtracking along the river to Silistria. They arrived to find that the Russians had abandoned the siege three days before. Cardigan's officers examined the siegeworks and defences with interest. The Turkish efforts impressed them; as Captain Tremayne (13th LD) remarked, 'The Turks must have fought like demons.'

Cardigan returned to Devna via Shumla, arriving back on 11 July. He had been away so long that a subaltern's patrol had been sent, unsuccessfully, to find him. Day by day the condition of the horses had deteriorated. Cardigan had ignored the intense heat, the heavy loads, the difficult sandy soil, the fact that the horses were unfit, and insisted on a cracking pace from sunrise to sunset. Food and forage were short, the latter particularly so after a false alarm of Cossacks on the first night had caused everyone to lighten their loads by discarding their hay-nets. As the march progressed horses started to collapse and die; many were too feeble to carry their riders and all but a handful developed raw, suppurating sores on their backs. Mrs Duberley, the wife of the 8th Hussars' paymaster, has described the patrol's return:

'a piteous sight it was – men on foot driving and goading the wretched, wretched horses, three or four of which could hardly stir. There seems to have been much unnecessary suffering, a cruel parade of death.'[21,22]

Major Forrest, who had suffered for years under Cardigan in the 11th Hussars but had wisely exchanged into the 4th Dragoon Guards, wrote to his brother saying, 'a precious business it [Cardigan's patrol] appears to have been and will cost about 50 horses to say nothing of about 150 very bad backs.'[23] In the event nearly a hundred horses were lost. Nolan, who had just returned from Syria with barely enough remounts to replace those Cardigan had lost, was unrestrained and vitriolic in his condemnation of the obvious mismanagement of a critical resource. Captain Jenyns of the 13th Light Dragoons, however, who was on the patrol, spoke favourably to a friend about Cardigan, saying, 'We got tremendous praise from Lord Cardigan, who is a capital fellow to be under at this work.'[24]

On 19 July cholera struck the French camp and within three days had spread to the British lines. An epidemic had hit southern Europe that summer so the disease probably arrived at Varna on a boat from Marseilles. It was to kill far more men than the Russians ever would. Deaths became so numerous and depressing that funeral music was forbidden and burials took place at night. William Cattell, serving as a surgeon with the Heavy Brigade, has written of how frightened men were of catching the disease, and of how, occasionally, there could be a hint of humour in the situation. First the fear of infection:

'Duckworth was seriously ill but bore up with wonderful resignation; his features became so terribly changed that F., the vet, who went to sit with him, became nervous. I met F. in a state of intense excitement rushing out of D's tent – "Oh! I've got it", pressing his hand against his stomach; then – "What is it like?" He was sent to bed, diarrhoea set in and a week after he was buried in the ditch at Varna (Hospital Cemetery).'

Now the humour:

'On one occasion the funeral service was being read by the adjutant (Godman); when he was about to begin the hospital sergeant, Fisher, ran up and said, "Wait a minute, another is almost ready." On being asked, "Is he dead?" the reply came, "Not quite." In a few minutes he was brought out in his blanket and laid to rest.'[25]

It was an awful scourge from which there was no escape. In an attempt to do so the Light Brigade was sent by Raglan another 30 miles inland to Yeni-Bazaar, a remote and shadeless spot detested by the troops. Neither Raglan nor Cardigan bothered to tell Lucan that one of his brigades had been moved. Another discourtesy, another slight to be borne and remembered.

In order to keep their men occupied, and latterly to keep their minds off the horrors around them, both Lucan (who had grabbed the Heavy Brigade and the 4th Light Dragoons) and Cardigan had embarked on a rigorous, daily routine of inspections, drill and field days. In this Lucan was at a disadvantage. It was seventeen years since he had handled a formation of cavalry on parade and in that time the drills and the words of command had changed. Major Forrest, who suffered on these field days, wrote:

'Lord Lucan is no doubt a clever sharp fellow but he has been so long on the shelf that he has no idea of moving cavalry, does not even know the words of command and is very self-willed about it, thinks himself right. . . . If he is shewn by the drill book he is wrong, he says, "Oh, I should like to know who wrote that book, some farrier I suppose".'[26]

He made a fool of himself at a review for the Turkish Commander-in-Chief, getting the 5th Dragoon Guards into an embarrassing tangle. Being the Divisional Commander, he expected the officers and troops to learn the old drill, the old words of command. It was nonsensical. Lucan summoned all the commanding officers to his tent to reiterate his views and insist on compliance with his methods. As the senior lieutenant-colonel, Paget was prevailed upon to speak up and explain that to persist would cause total confusion, and that to relearn the drills was completely unrealistic. Knowing Lucan's explosive nature, he had approached the task with some trepidation. Years later Paget wrote of this meeting:

'It would be impossible to over-colour the tone of consideration and forbearance with which he met this remonstrance – made, I need not say, in the most respectful terms . . . he evinced qualities which . . . set my mind at rest as regards the character of the man under whom I was to serve before the enemy.'[27]

Unusually, Lucan had backed down.

Meanwhile, Cardigan was, according to the adjutant of the 8th Hussars, Lieutenant Seager, 'annoying everyone; he does all he can to knock up both horses and men before the work begins in earnest.' Drills and inspections followed drills and inspections. A favourite of Cardigan's was to advance a squadron and, on a given signal, to get every man to scatter in all directions, whereupon the 'Rally' would be sounded with the squadron having to reform. Perhaps he had a premonition of Balaclava in mind. Captain Shakespear of the RHA was particularly damning of Cardigan at this juncture, calling him:

'the most impracticable and most inefficient cavalry officer in the service. . . . He may do all very well when turned out by his valet in the Phoenix Park, but there his knowledge ceases. We are all greatly disgusted with him. . . . We have field days every other day now. Cardigan kills the horses with pace. . . . We wish we were with Scarlett and his Heavy Cavalry.'[28]

Morale had sunk dangerously low. Drunkenness was the most prevalent offence and flogging the most common punishment. From April to December, 1854, of the 1,245 punishments awarded to soldiers in the field 761 involved flogging. Captain Maude, Shakespear's commanding officer with I Troop of the RHA attached to the Light Brigade, christened the Army in Bulgaria 'the Army of no occupation'. By this he meant meaningful military occupation against an enemy, not drill followed by drinking or, for the officers, wild dog hunting. This was carried out on ponies, the riders armed with lances, and was resorted to as much for amusement as to rid the place of pests. Captain Portal (4th LD) summed up the mood of most when he wrote:

'The men and officers are getting daily more dispirited and disgusted with their fate. They do nothing but bury their comrades; they have no excitement to relieve the horrid monotony of camp life. . . . There is nothing in the world to do but listen to growls and grumbling, from the highest to the lowest, of the dreadful mortality that is decimating our once magnificent Army.'[29]

The news that the Crimea was to be invaded and Sevastopol captured was greeted, not with enthusiasm as would have been the case two months earlier, but by apathy.

* * *

The Russians had left Bulgaria, abandoned the siege of Silistria and thereby acknowledged Turkey's independence without the British or French firing a shot. Both these governments were reluctant to bring their armies home without a prize, without something to show for the suffering and death inflicted by disease. The seizure of Sevastopol, the Russian port, fortress and naval base from which she controlled the Black Sea, would redeem an otherwise messy and seemingly unnecessary expedition. Lord Palmerston, in London, put it neatly, if inaccurately: 'The eye tooth of the bear must be drawn.' Theoretically Raglan, as the field commander, had a choice. He had handed the despatch to Sir George Brown, Commander of the Light Division, for an opinion. He advised Raglan to go as,

'it is clear to me from the tenor of the Duke of Newcastle's letter that they have made up their minds to it at home, and that, if you decline to accept the responsibility, they will send someone else out to command the Army.'[30]

The problems were legion. Time was short if Sevastopol was to be captured before winter set in; it involved transporting 60,000 men, plus animals and supplies, 300 miles across the Black Sea; nobody could agree on a suitable landing place; knowledge of the Crimea was scanty and maps non-existent; the strength of Sevastopol's defences was unknown and estimates of the Russian Army that might oppose them ranged from 45,000 to 140,000. To these obstacles must be added the dispirited and diseased condition of both armies, particularly the French.[31]

When Captain Wetherall, of the DQMG's staff, brought the news to the Light Brigade it was greeted with 'the gloomy silence of men with sickness and death uppermost in their minds'.[32] The brigade commander, however, was immensely cheered as he was given to understand that his *de facto* independence would continue. There was no room on the transports for the Heavy Brigade (it would be brought over later); light cavalry had priority and Cardigan commanded them. The Cavalry Division would still be divided, with

Cardigan scouting and skirmishing in front of the entire army while Lucan kept the Heavies in Varna.

Not even his promotion to lieutenant-general on 18 August could placate Lucan when he was told of the plan. It was unbelievable, unthinkable. Cardigan himself would be in the pocket of the Commander-in-Chief, acting as his adviser and commanding the only cavalry force in the field while he sat on his hands 300 miles away across the sea. He picked up his pen:

'I find myself, as on former occasions, left without instructions . . . except, as I read them, not to accompany the Light Brigade or to inter-fere with their embarkation. . . . During the four months I have been under your lordship's command I have been separated, as much as it was possible to do so, from my division.'[33]

He made it plain that he would sail with the Light Brigade – it was his right.

While awaiting a reply he alleviated his frustration by blasting the Brigade, which had assembled at Varna prior to embarking, at a formal inspection. He could find nothing right; in his opinion, 'It would appear as if the object were that every soldier on service should look as unsoldierlike, slovenly and dirty as possible'.[34] Shortly afterwards Raglan used his considerable diplomatic skills to try to persuade Lucan, at a personal interview, that his presence at Varna was important – the Heavy Brigade needed inspecting. Lucan knew his man and played his ace, quoting a Wellingtonian Peninsula War regulation that a divisional commander might accompany any part of his division he wished. Raglan relented. Cardigan was outraged, writing in his diary: 'From this day my position in the cavalry was totally changed, all pleasure ceased.'

Embarkation took much more time than anticipated. The sea was often unfriendly so the loading was chaotic, and accomplished with a sullen slow-ness that further enfeebled a large proportion of the troops. Units started embarking on 24 August and by the time they disembarked some had spent up to 17 days on board rolling ships in miserable, insanitary conditions where sea-sickness, dysentery and cholera took their toll. Men and horses were jammed into every conceivable space. The transport *Simoon* was not untypical. It was allocated to 1,300 troops of which 200 had to be taken off, it being physically impossible to cram more in. Healthy men came on board, contracted cholera and had to be taken off again. Dead men, weighted down at the feet, were dumped overboard. Scores of these popped up again with bloated faces decomposing in the sun and grimacing horribly.

Shortly before embarkation de Ros had succumbed to sunbathing and been invalided home. Brigadier-General Richard Airey, another officer with no active service but a great friend of Raglan, was now the DQMG. Airey, who had a forceful, decisive character, brought Nolan with him to general head-quarters as his ADC. Despite this change at the top the staff still had little conception of their duties. There was no load planning, so that stores likely to be needed immediately after landing, such as tents, were put on first and buried under items not urgently required. At the last moment, when some of the regimental wives for whom no alternative arrangements had been made appeared screaming on the quayside, it was realized that they were being abandoned with no food, no money, no shelter and no transport. They had to be taken. Over 5,000 officers' chargers and baggage animals, the Army's means of movement that had taken months to acquire, were not so fortunate. They were assembled at a depot where many eventually died of neglect and starvation. When the fleets finally weighed anchor on 7 September to rendezvous at sea off the mouth of the Danube the Allied commanders-in-chief were still uncertain of, and arguing about, where to land.

On the 9th Raglan, on the *Caradoc*, accompanied by the French second-in-command, General Canrobert, and escorted by three warships, steamed off towards the Crimea for a last reconnaissance of the coast (see Maps 2 and 3). A decision must be made. The next day they approached to within two miles of Sevastopol. Calthorpe wrote:

'We remained for upwards of half an hour gazing at the scene . . . excited by the thought that there lay the prize for which we were to fight. . . . The fortifications looked of immense strength, and appeared to bristle with guns.'[35]

They then steamed south-east to Cape Cheronese where low cliffs, beaches and several natural harbours were favoured by some generals. They were rejected as too near Sevastopol. North again, to the mouth of the River Belbec which seemed promising but was in fact also too close, almost under the guns north of the harbour. Next the Katcha River was examined. The soldiers liked it but the sailors did not, considering it far too small for the size of the fleets. Then the Alma, 18 miles from Sevastopol, but with high cliffs and, more importantly, large military camps seen close by. North from the Alma the coast was low-lying, with sandy beaches and entirely suitable for a landing. The decision was taken to land at the northern end of the ominously named Kalamita Bay. It was near a salt lake, close to an unpronounceable place the British were to call Old Fort, and some seven miles north of a stream called the Bulganek. Possible problems identified were lack of water, the distance (30 miles) from Sevastopol and whether the landings might be opposed.

Although Lucan and Cardigan sailed on different ships, that did not stop the feuding. Paget was to describe them as being 'like a pair of scissors, which go snip and snip and snip, without ever doing each other any harm, but God help the poor devil that gets between them!'[36] Cardigan took exception to being told he could not order a court-martial and was to submit embarkation returns on landing. Lucan had his secretary attempt to put him in his place:

'The Lieutenant-General instructs me to add that whilst he knows his own authority he equally respects yours; and that your position as a major-general commanding a brigade in the Cavalry Division will not,

MAP 3

ALLIED MARCH TO BALACLAVA

Eupatoria

Kalamita Bay

① Old Fort

Bulganek R

Simpheropol

KEY
① Allied landings
② Russian force clashes with LB south of Bulganek
③ The Battle of the Alma
④ British led flank march. L – Lucan loses the way
⑤ Russian field army leaves Sevastopol 25th Sept.
⑥ Sevastopol beseiged
⑦ Russian lines of communication
→ Allied route

②

Alma R

③

⑦

Katcha R

Belbec R

④ L McK Fm
⑤

Sevastopol

⑥

Balaclava

Woronzoff Road

0   10   20
miles

55

so far as depends on him, differ from that held by the other brigadiers, [then a twisting of the knife] of whom there are so many in the six divisions of this army.'[37]

Luckily the enemy did not interfere with the landings on 14–16 September. Had Russian artillery shelled the beaches on the first night, when most of the infantry but no cavalry were ashore, a successful disembarkation would have been seriously in doubt. The massive fleet with its 3,000 guns, which was lit up by thousands of masthead lights, would have been hampered by fear of hitting Allied troops. Even when everyone was ashore all was not well. As *The Times* correspondent Russell put it:

'The French, though they had tents, had no cavalry; the Turks had neither cavalry nor food; the British had cavalry, but they had neither tents nor transport, nor ambulances nor litters.'[38]

Later, near the beach, Russell met some senior medical officers, one of whom exclaimed:

' "Do make a note of this! By —! They have landed this army without any kind of hospital transport, litters or carts, or anything! Everything was ready at Varna! Now with all this cholera and diarrhoea about, there are no means of taking the sick down to the boats." '[39]

The men had disembarked without packs (medical opinion was that they were too debilitated to carry them), with a few essential items bundled up in a blanket, carrying three days' rations of cold salt pork, hard tack biscuits and a canteen of water. The landings started on 14 September but it was not until the 19th that both armies began to march south. The French, who had virtually no cavalry, hugged the coast while the British moved on their left with the Light Brigade's 900 horsemen spread thinly to the front, left flank and rear. Left behind on the beach to sort out the mess were the 63rd Regiment, two companies of the 46th and a troop of the 4th Light Dragoons. Stores had been landed but no wagons waited; tents that had eventually been laboriously dragged up the beach had now to be hauled down again; hundreds of sick had to be carried back to the boats while scores of dead awaited burial.

The march started late, but from a distance the long colourful columns with flags flying and bands playing looked and sounded impressive. It was not a long march, ten miles or so, but it was blisteringly hot for men weakened by dysentery and crippled by cholera. Their only food was salt pork and hard, bone-dry biscuits. Their canteens were empty. This dreadful combination wrought havoc among the infantry. Within half an hour men began discarding equipment, throwing off shakos, straggling, staggering and finally collapsing. They could not be moved. There was no water to revive them. Paget, bringing up the rear with the 4th Light Dragoons, wrote: 'The stragglers were lying thick

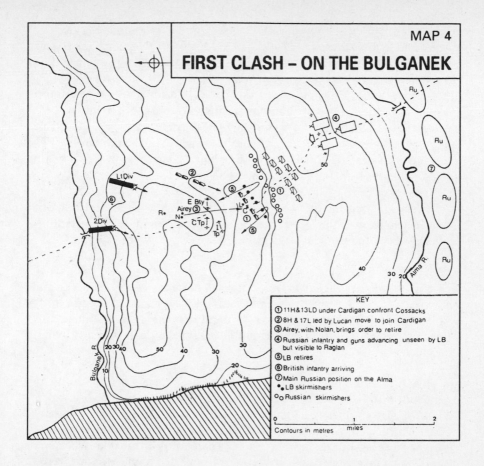

MAP 4

FIRST CLASH – ON THE BULGANEK

KEY
① 11H & 13LD under Cardigan confront Cossacks
② 8H & 17L led by Lucan move to join Cardigan
③ Airey, with Nolan, brings order to retire
④ Russian infantry and guns advancing unseen by LB but visible to Raglan
⑤ LB retires
⑥ British infantry arriving
⑦ Main Russian position on the Alma
•• LB skirmishers
°° Russian skirmishers

Contours in metres

on the ground, and it is no exaggeration to say that the last two miles resembled a battlefield!'[40] It was two o'clock when the leading units struggled to breast yet another undulation to see, miraculously, water. They had reached the Bulganek River. Although only knee-deep, the sparkling water offered release from torment. Only the Highland Brigade had sufficient discipline left to resist joining the stampede. Regiment after regiment broke ranks and rushed down the slope to hurl themselves into the water alongside the horses of the advance guard.

Advancing ahead of the Army was Cardigan with the 11th Hussars and 13th Light Dragoons; on the left flank was Lucan, riding with his old regiment, the 17th Lancers, and the 8th Hussars; at the rear, Paget. Lucan had been leaning on his Brigade Commander so Cardigan was feeling intensely irritated. The affair at the Bulganek, which was about to begin, was the first time the enemy would be met face to face. It was also the first of three incidents involving the cavalry prior to Balaclava that were to widen the gulf between the two cavalry commanders, reinforce Raglan in his view that cavalry must be conserved, and goad Nolan and others into open, outspoken and insubordinate criticism of Lucan, and to a lesser extent of Cardigan. By early October, some three weeks before the Light Brigade charged, Lucan had become widely known as Lord 'Look on'.

Cardigan, on Raglan's direct orders, crossed the Bulganek to reconnoitre what lay beyond the Cossack scouts on the high ground south of the river. 'On gaining the crest of the hill, we came in sight of the main body of the Cossacks, spread out in skirmishing order in a beautiful valley about a mile across.'[41] The 11th and 13th descended the slope in line with skirmishers in front, halting some 200 metres from the Cossack line. TSM Smith recalled:

'It was now that the first shot of the campaign was fired. A Cossack directly in front of the 11th raised his carbine and fired. It was instantly taken up by the whole [Cossack] line. Our trumpets now sounded the "Fire".'[42]

For twenty minutes there was a continuous popping of carbines with little visible effect on either side. During this time Lucan cantered up and, to the exasperation of Cardigan, began to demand explanations and make alterations. As the senior officer on the spot he assumed overall command. Smith continued:

'This [the skirmishing] continued for some little time, when the crest of the hill in front suddenly became lit up with the glitter of swords and lances. Lord Cardigan now gave the order, "Draw Swords – Skirmishers In – Trot". The ground in front of us was uphill. As we moved off we began to throw off our hay nets. . . . We had not proceeded far when an aide-de-camp galloped up with an order from Lord Raglan that we were to retire.'[43]

Both Lucan and Cardigan had been in agreement that a charge was in order but disagreed over details. They were arguing as to how best to proceed when Airey arrived, accompanied, of course, by his ADC, Nolan. Raglan had seen something that was invisible to his cavalry generals in the valley – a mass of infantry and guns over the brow of the low hills to the south. It was in fact a Russian reconnaissance in force of 6,000 men pushed forward from their main position on the Alma, although Raglan was not to know this. He did know that if he did not act urgently there was a danger, almost a certainty, that his precious cavalry would ride into a hugely superior force and could be destroyed. On top of this, with his army exhausted and straggling badly, the last thing he wanted was a general action.

Raglan sent messengers galloping away to bring up two infantry divisions plus the cavalry on the flank to cover the withdrawal of Cardigan's two regiments. The 6-pounders of 'I' Troop RHA with the cavalry screen were outgunned, so Raglan also ordered up 'C' Troop RHA and 'E' Field Battery.

Airey was to 'suggest' (Raglan could never bring himself to be other than diplomatic) to Cardigan that he should now retire. Neither Lucan, to whom Airey spoke as the senior officer at the scene, nor Cardigan were interested in a 'suggestion' to pull back, so Airey, knowing that time was critical, converted Raglan's message into a direct and explicit order to retire at once. That, he stressed, was Raglan's wish. With great reluctance, Lucan complied, passing the instructions to Cardigan to implement, which he did, giving a Hyde Park field day demonstration of the drill involved in retiring squadrons alternately. Nevertheless, it brought forth derisive jeers from the Cossacks who were still only a short distance off. As the 11th and 13th withdrew artillery on both sides came into action and the cavalry suffered the first battle casualties of the war. The cannon balls that came bouncing over were often clearly visible to those at the receiving end. Calthorpe thought the whole affair:

> 'The prettiest thing I ever saw, so exactly as one had done a dozen times at Chobham. . . . If one had not seen the cannon balls coming along at the rate of a thousand miles an hour, and bounding like cricket balls one really would have thought it only a little cavalry review.'[44]

Two light cavalrymen claim the doubtful distinction of being the first casualties of the war, both having a foot blown off. One was Sergeant Joseph Priestley of the 13th and the other Private Williamson of the 11th. The latter, with his foot and stirrup-iron dangling from shreds of skin and tendon and blood gushing from his leg, rode up to his troop commander (Cornet Palmer) and calmly said, 'Sir, I am hit. May I fall out?'[45]

Although undoubtedly a wise decision by Raglan, the cavalry's withdrawal was regarded by many within the Brigade as humiliating. Lucan was, quite wrongly, blamed for turning his back on a glorious opportunity. He was furious with Raglan for interfering in how he handled his command, while Cardigan fretted at Lucan's usurping his authority by his presence with the Light Brigade. Cardigan, however, was consoled by Raglan's despatch to London which contained the sentence: 'Major-General the Earl of Cardigan exhibited the utmost spirit and coolness, and kept his brigade under perfect command.'[46] Nolan, meanwhile, had enjoyed himself by riding forward into the skirmish line and revelling in being under fire for the first time. He commented on what poor shots the Russians were.

Raglan had conserved his cavalry, pushing them back into 'the bandbox' just before they were about to leap out, while Lucan had begun to earn a reputation for hanging back.[47]

The next event that helped to harden the hostile attitudes of the four horsemen towards each other was the Battle of the Alma on 20 September, the day following the Bulganek skirmish. It was a major engagement in which the cavalry took no part. Not much generalship was displayed by the Allies, who merely launched their infantry in a frontal attack on the Russian defensive positions on the hills overlooking the river. Raglan managed to command (watch) much of the action from a hilltop behind the enemy lines – an achievement unique in military history. After a prolonged and bloody struggle in which the Light Division was forced back, the Russians eventually turned and fled south towards Sevastopol.

Pursuit is second only to a charge as the prime function of cavalry on a battlefield. Vigorous pursuit can turn a retreat into a rout, a disorganized force into a rabble. There is much excitement for, and little danger to, the mounted man who rides down escaping fugitives. He has a great moral as well as physical advantage over his opponents, no matter how numerous they be. It is all a matter of timing. Unleash the cavalry too early and enemy reserves may still be steady, too late and the enemy will have made a clean break or rallied behind a rearguard. Good judgement and speed of action are essential to see and seize the right moment. To hold back fresh cavalry that have been sitting as spectators all day when they see the enemy turn their backs is not likely to win great respect within the mounted arm. This is what Raglan did at the Alma, although the blame went to Lucan.

When Raglan sent Airey forward to stop the cavalry pursuing he was looking at the situation through the eyes of a commanding general, not just of a cavalry commander. Lucan, with the Light Brigade, received no orders throughout the battle, although later, on his own initiative, he crossed the river and moved up onto the heights on the left flank from where he could see the enemy streaming away in flight. To Lucan, Cardigan and all those present it seemed a glorious opportunity for the cavalry to show its worth. All within the Brigade were mad with eagerness. Then Airey arrived with the first order for the cavalry – escort the field artillery to the left and right but on no account attack. Almost unable to believe their own ears, Lucan and Cardigan used this instruction as an excuse to dash forward and take some prisoners. Airey recalled them to their duties with a second behest. This too was only partially obeyed. For a third time an order, couched in peremptory terms, insisted the cavalry pull back. Lucan, in disgust, released the incredulous prisoners already taken. Raglan allowed the Russians to flee unmolested.[48]

Afterwards Lucan protested vehemently that he should, as cavalry commander, be allowed some freedom to act on his own responsibility. It was ignored. The cavalry felt themselves chastened and discredited. Later, Major Forrest wrote of this time: 'We all agree that two greater muffs than Lucan and Cardigan could not be; we call Lucan the cautious ass and Cardigan the dangerous ass.'[49] Nolan, visiting Russell the day after the battle, voiced the views of many within the cavalry when he raged:

'There were one thousand British Cavalry looking on at a beaten army retreating – guns, standards, colours and all . . . within a ten minutes' gallop of them – [it's] enough to drive one mad! It is too disgraceful, too infamous. They [the generals] ought to be —!'[50]

Now on his favourite topic, Nolan ranted on:

'The cavalry were sent down into a ravine all surrounded by a wood. A battalion of infantry could have disposed of every man-jack in twenty minutes. Lord Raglan says we ought to be kept in a bandbox. Did any one ever hear of cavalry in a bandbox doing anything?'[51]

That same day Cardigan sheathed his sword and took up his pen again. Out poured all the old complaints. He had been given to understand his command was an independent one; Lucan continued to meddle in the running of the Brigade; Lucan wrote orders to him as though he were a junior subaltern; Lucan ignored him; Lucan insisted on seeing all orders he, Cardigan, received direct from Raglan; Lucan had taken away his command; and why had he ridden with the Light Brigade in the recent battle? Would Raglan now make it clear to Lucan that he was in no way to intrude in the affairs of the Light Brigade? Strangely, Cardigan followed the rules and sent this letter via his superior and Lucan could not forbear to add a covering note in which he refuted Cardigan's claims.

Raglan did not find time to reply for a week, by which time the Army was at Balaclava and beyond. His letter was a combination of a hefty rap on the knuckles for Cardigan and a polite, despairing plea for his two commanders to cooperate. He wrote:

'I consider him [Cardigan] wrong in every one of the instances cited. A General of Division may interfere little or much with the duties of a General of Brigade. . . . His judgements may be right or wrong; but the General of Brigade should bear this in mind, that the Lieutenant-General is the Senior Officer; and that all his orders and suggestions claim obedience and attention.'

That was the rap; now came the plea:

'They [Lucan and Cardigan] must permit me, as the Head of the Forces, and I may say the friend of both, earnestly to recommend them frankly to associate with each other, and to come to such an understanding as that there should be no impression of the assumption of authority on the one side, and no apprehension of undue interference on the other.'[52]

It was a waste of time and ink.

On 25 September the famous flank march was undertaken. During it there occurred the third incident that further estranged the four horsemen of calamity – the cavalry got lost and Raglan, who ended up as almost the leading scout for the Army, came within an ace of capture. On the day before, the 24th, there occurred an instance of stupidity that shocked those of the cavalry who saw it. It involved a totally unsound, irrational, but perhaps typically pigheaded enforcement of discipline by Lucan. Even TSM Smith, himself a tough disciplinarian, was appalled by what he saw. As the 11th Hussars were fording the Belbec stream the desperately thirsty horses were frantic to drink. However, as Smith wrote in his diary:

> 'Opposite the part we were fording sat Lord Lucan, storming and threatening that he would flog any man who attempted to water his horse, so that the men who passed over directly opposite him had great difficulty in forcing their horses through the water, as they plunged their heads into it eager to drink, not having been watered since we left the Alma. What could have been Lord Lucan's reason for this I never could make out, for a greater piece of cruelty I never witnessed. . . . Lord Cardigan sat some little distance from him, evidently indignant for he rendered no assistance in enforcing the order.'[53]

After crossing the Belbec the decision was taken to move the Allied armies round the east of Sevastopol preparatory to attacking the city from the south (see Map 3). It was during this flank march that the leading elements of the British Army, which at the time happened to be the Commander-in-Chief and his 8th Hussars escort, clashed with the tail end of the Russian Army moving east. As Russell was, somewhat cynically, to observe:

> 'The French General was dying [of a heart condition]. The English General . . . in order to take Sevastopol, was marching round it! The Russian General, anxious to save Sevastopol, was marching away from it! Neither of them had the least notion of what the other was doing.'[54]

To get round Sevastopol the armies had to negotiate an area of rough, hilly country covered with stunted woods and thick scrub crossed by an ill-defined tangle of tracks and trails. The only road worthy of the name led generally north-east from the city to Simpheropol. This was the road the Russians used on 24 September when they left Sevastopol. It passed a place called Mackenzie's Farm some seven miles north-east of the city. By coincidence this spot was the objective of the Cavalry Division on the 25th. The Allied plan envisaged Lucan's command, led by the Light Brigade, following a track to Mackenzie's Farm, keeping hidden and reporting on the situation on the road. The guns would follow on the track but the infantry would, for the most part, be required to march across country using compasses as necessary.

Maps were poor, so Lucan was given two direction-finders from headquarters, Captains Wetherall and Hardinge. During the march they took the wrong fork at a track junction and led the cavalry into denser and denser scrub. Meanwhile the artillery and Raglan had taken the correct route and arrived at the Farm just as the Russian rearguard and some baggage were passing. There were a few anxious moments while the escort deployed and messengers dashed off to find the cavalry who were blundering about in the brushwood some 500 metres to the south. Raglan was extremely angry. As Lucan galloped up he actually raised his voice and shouted, so that all nearby heard, 'Lord Lucan, you are late!' Lucan seethed with resentment, but said nothing. Staff officers had been sent, not at his request, to take responsibility for navigation; they had got lost and yet he got the blame. Cardigan took great delight in the situation. When he met Raglan at Mackenzie's Farm and began to get some of the blame he 'simply reminded his Lordship that he did not command the cavalry'.[55]

On the 26th the British took Balaclava without a fight and established it as their base, while the French moved further west and did likewise at Kamiesch. Both Armies encamped and prepared to besiege and capture Sevastopol. During the four weeks remaining before the dramas of 25 October Raglan and Lucan were barely on speaking terms, the endless bickering between Lucan and Cardigan continued with undiminished enthusiasm, while Nolan's contempt for the two cavalry commanders became daily more forthright and indisciplined. For a captain to speak thus of his seniors was unthinkable, yet Nolan did so with impunity, probably because it was such common gossip anyway. Captain Portal of the 4th Light Dragoons had much the same opinion as Nolan. He wrote:

'We are commanded by one of the greatest old women in the British Army, called the Earl of Cardigan. He has as much brains as my boot. He is only equalled in want of intellect by his relation the Earl of Lucan. . . . Without mincing matters two such fools could not be picked out of the British Army to take command. But they are Earls!'[56]

According to camp tittle-tattle the cavalry had done nothing since landing; they always failed to seize opportunities; they never charged; they never pursued; they frequently retired; in other words they were grossly mismanaged. Although it was Raglan who restrained Lucan, who insisted (perhaps rightly) that the cavalry be husbanded, that no risks be taken, it was Lucan who had to live with the scornful, insulting 'behind-the-back' remarks.

By early October Lucan had at last got his whole division assembled. The Heavy Brigade had arrived after a dreadful storm at sea had wrought havoc with the horses. The journey had cost the brigade 226 horses, the Royals having to receive 75 from the Light Brigade in order to mount its second squadron. While Lucan lived in the field suffering the poor rations and numbing cold at night Cardigan received permission to command his brigade first from a ship and then, in mid-October when his yacht arrived with his French chef, from on board. He pleaded ill-health. Nevertheless the situation was ludicrous, a divisional commander living in the field on active service while one of his brigade commanders obtains special permission from the Commander-in-Chief to live in luxury. When Lucan paraded the division half an hour before dawn every day, when he received the daily returns, inspected the men and horses, rode round the pickets Cardigan was still snugly tucked up in bed with Raglan's blessing. There was nothing Lucan could do. The 'Noble Yachtsman', as he was then called, seldom arrived in camp before 10.00 am. During his prolonged absences Paget commanded the Light Brigade.

63

Throughout this period, while the siege got under way, the Army was jittery, ultra-sensitive to its exposed base at Balaclava and to the vulnerable road from the harbour to the camps. Cossack patrols were active, British patrols jumpy. With the exception of Cardigan all endured the cold, the damp and the debilitating lack of sleep. As Paget put it in one of his letters:

'We are now regularly turned out about midnight. . . . Every fool at the outposts, who fancies he hears something, has only to make a row, and there we all are, generals and all. . . . Well, I suppose 500 false alarms are better than one surprise.'[57]

Cornet Wombwell of the 17th Lancers, who was to ride as Cardigan's orderly officer, put it more bluntly:

'The way they keep turning us out is ridiculous. If a heavy dragoon or any other thick-headed individual sees a Cossack he comes galloping into camp and instantly magnifies the Cossack to 500. So of course out we all go and by the time we all get there not a soul is to be seen.'[58]

On one occasion, in broad daylight, a large body of Russian cavalry appeared across the Chernaya River and manoeuvred in a provocative manner in full view of Lucan who had his division deployed to confront them. Despite protestations from some of his officers, Lucan was not drawn into attacking. Raglan's instructions would not permit it. Eventually a few shells from Maude's battery of the RHA drove off the enemy, which had included infantry and artillery. To Nolan, who had been present with the 17th Lancers, it was yet another wasted opportunity. According to Cecil Woodham-Smith an angry scene took place between Nolan and Lucan and that, according to Nolan, he told Lucan to his face that he had neglected his duty. Afterwards Cardigan, who had been on his yacht at the time, flew into a rage with the officers of the 11th Hussars, whom he considered should have charged anyway, shouting at them that they were 'a damned set of old women' – for which he later apologized. Paget, however, wrote of the incident:

'but from all I have gleaned from those whose opinions are worth anything . . . it appears that a more unjust accusation [against Lucan] never was made.'[59]

Be that as it may, this incident, on 7 October, resulted in some wag giving Lucan the nickname of Lord 'Look-on'. It was unjust, but fitted perfectly and stuck.

A few days later Raglan once again tried to separate his ungovernable cavalry commanders. Paget commented:

'There are Lucan and Cardigan again hard at it, and it is found desirable to separate them. Cardigan must needs be ordered up here [the Sapoune Heights] to command the 4th and 11th, both of which are usefully placed with their divisional generals, and all this must needs be upset to part these two spoilt children.'[60]

The four horsemen, in whose hands the fate of the Light Brigade was to be placed, could hardly have been more estranged or hostile to each other.

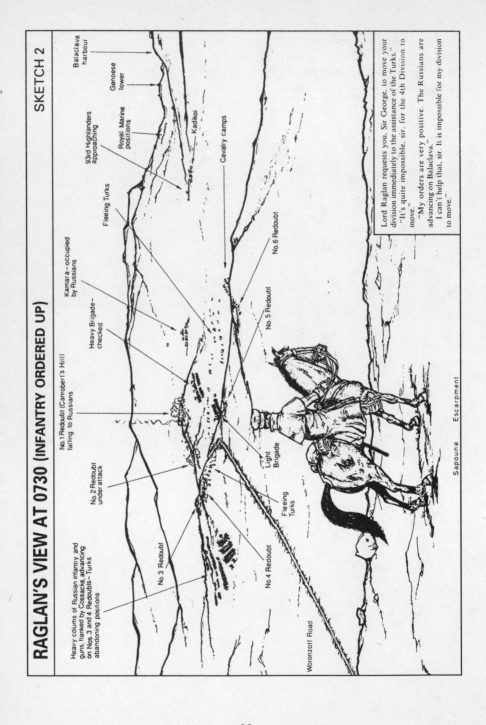

RAGLAN'S VIEW AT 0730 (INFANTRY ORDERED UP)

SKETCH 2

Heavy colums of Russian infantry and guns, flanked by Cossacks, advancing on Nos. 3 and 4 Redoubts—Turks abandoning positions

No. 1 Redoubt (Canrobert's Hill) falling to Russians

Kamara – occupied by Russians

Balaclava harbour

Genoese tower

93rd Highlanders approaching

Royal Marine positions

Kadikoi

Fleeing Turks

Heavy Brigade – checked

No. 2 Redoubt under attack

Cavalry camps

No. 3 Redoubt

No. 6 Redoubt

No. 5 Redoubt

No. 4 Redoubt

Light Brigade

Fleeing Turks

Woronzoff Road

Sapoune Escarpment

Lord Raglan requests you, Sir George, to move your division immediately to the assistance of the Turks."

"It's quite impossible, sir, for the 4th Division to move."

"My orders are very positive. The Russians are advancing on Balaclava."

I can't help that, sir. It is impossible for my division to move."

66

II

THE ORDERS

CHAPTER FOUR
Orders for the Infantry

'Well sir, if you will not sit down in my tent, you may
as well go back to Lord Raglan and tell him I cannot
move my division.'
*Lieutenant-General Sir George Cathcart, commanding
the 4th Division, to Raglan's ADC early on the
morning of 25 October, 1854.*

Captain Ewart was one of several ADCs who were pounding madly back from
the edge of the Sapoune Heights shortly after 7.30 am that morning. They carried urgent messages for the 1st and 4th Divisions and the French
Commander-in-Chief, General Canrobert. After frequent false alarms it
appeared that the Russians were mounting a major attack on Balaclava. Both
infantry divisions were needed at once to prevent the likely loss of the British
base. As he was leaving, Airey, on Raglan's instructions, had called Ewart back
to say that he must tell Cathcart on no account to use the Woronzoff Road.

Riding hard, Ewart took some ten minutes to cover the mile and a half to
Cathcart's tent. He saluted the elderly general who had fought at Waterloo:

'Lord Raglan requests you, Sir George, to move your division immediately to the assistance of the Turks.'

'It's quite impossible, sir,' he snapped back, 'for the 4th Division to
move.'

'My orders are very positive. And the Russians are advancing on
Balaclava.'

'I can't help that, sir. It is impossible for my division to move, as the
greater portion of the men have only just come from the trenches. The
best thing you can do, sir, is to sit down and have some breakfast.'

It was not a situation within the experience of a young captain – a lieutenant-
general refusing to obey orders. Instructions given through an ADC were to

be followed as if given by the general himself, in this case the Commander-in-Chief. It was all clearly laid down in Army regulations; without adherence to it the system of command was meaningless. Although shaken, Ewart still did not give up, politely trying a more persuasive approach.

'No thank you sir. My orders are to request that you will move your division immediately to the assistance of Sir Colin Campbell [responsible for the close protection of Balaclava]. I feel sure every moment is of consequence. Sir Colin has only the 93rd Highlanders with him. I saw the Turks in full flight.'[1]

Cathcart remained unmoved. His final words are those quoted at the start of this chapter.

This was getting ridiculous. If the commander of the 4th Division was determined to stay put there was nothing a mere captain could do other than report back to headquarters. Ewart saluted, ducked out of the tent and mounted his horse. Having ridden a short distance he halted. He had seen the situation developing in the Balaclava plain, knew full well the importance of getting infantry reinforcements, knew the urgency with which Raglan and Airey had despatched the necessary orders. He could not return with a refusal to move. With commendable courage he returned for a second confrontation, this time stating firmly, but still respectfully, that his instructions had been to come for the 4th Division, that the situation at Balaclava was critical and that he would not leave until the division was ready to march. Cathcart listened and, to Ewart's intense relief, gave in.

'Very well, I will consult my staff officers and see if anything can be done.'

A little later bugles shrilled as the infantry paraded. Ewart passed on the message that the Woronzoff road was not to be used. It was after 8.30 am that the regimental columns moved off, late and without much evidence of urgency. To reach Campbell's position they had a five-mile march.

By contrast Captain McDonald, the messenger to the 1st Division, had an easy task. 'There's a row going on down in the Balaclava plain and you fellows are wanted,' was all he said. It was enough for the Duke of Cambridge.

Balaclava was a cavalry battle; first the Russians, then the British, then the British again and finally the French, charged – but it was not intended to be that way by Raglan, who wanted it to be a predominantly infantry affair. The cavalry was meant to delay things until the infantry arrived; the cavalry still had to be preserved; only infantry and guns could see off a determined assault on his Balaclava base. His first orders were, therefore, to the infantry.

In order to put the reader in approximately the same saddle as our first horseman (Raglan), as he stared anxiously east and south-east from the lip of the Sapoune plateau, we need to know what he knew and, as far as possible,

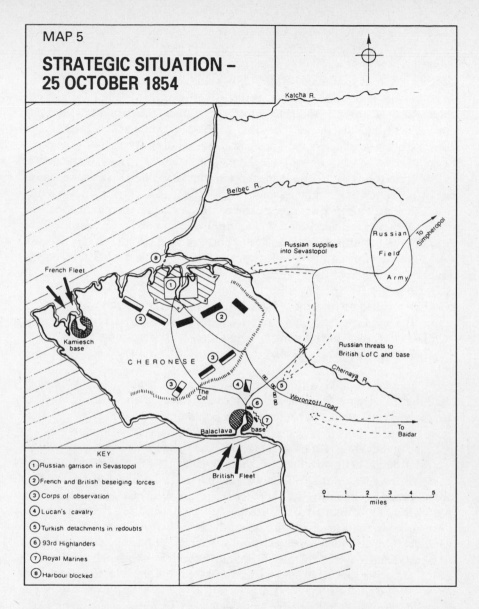

MAP 5

STRATEGIC SITUATION – 25 OCTOBER 1854

Katcha R.

Belbec R.

Russian Field Army

To Simpheropol

Russian supplies into Sevastopol

French Fleet

Kamiesch base

CHERONESE

Russian threats to British L of C and base

Chernaya R.

The Col

Woronzoff road

To Baidar

Balaclava base

British Fleet

KEY

1 Russian garrison in Sevastopol
2 French and British beseiging forces
3 Corps of observation
4 Lucan's cavalry
5 Turkish detachments in redoubts
6 93rd Highlanders
7 Royal Marines
8 Harbour blocked

0 1 2 3 4 5
miles

see what he saw. His was a magnificent viewpoint, one from which he never moved for some five hours. It was from here, over 400 feet above the Balaclava plain, that he issued all his orders. From this spot he, and all his staff, including Nolan and the other ADCs, watched events unfold as they might watch a play from a box in the theatre. The strategic and tactical situations known to Raglan are shown on Maps 5 and 7 respectively, while what he could actually see through his special telescope (mounted on a skeleton rifle stock to facilitate one-armed use), is reconstructed as nearly as possible in Sketch 2.[2]

As the alarm was raised that morning Raglan's strategic situation was not promising. The Allied war aim was to take Sevastopol. Everything else was subordinate to that objective; all battles would be fought either to raise the siege or ensure it continued. Map 5 shows this situation. With regard to the siege there was a glaring defect in the Allied position – the city was not surrounded. It was garrisoned by a mixture of over 20,000 infantry, gunners and sailors (the Russian fleet had been deliberately sunk across the harbour entrance after the seamen and guns had come ashore). It could not therefore be supplied by sea, but instead relied on a land route to the north and north-east for provisions and reinforcements. This line of communication was never cut by the Allies. Sevastopol was not truly invested as there was always a life-line for the defenders, always an escape route. The Allies maintained that the terrain in the area and deficiencies in manpower prevented their sealing off the city, or even of cutting the critical Simpheropol road. Whatever the reasons, it was a major flaw in the Allied strategy, evident to the lowliest of soldiers. Private Pennington of the 11th Hussars was later to write:

'Our so-called investment of Sevastopol was practically no investment, for here was Liprandi's command free to take the field at will, and a road always open by which the enemy could receive reinforcements. This will admit of no dispute.'[3]

Having failed to make a quick assault in early October, which was when the defences were weakest, the Allies faced a protracted and costly struggle of attrition with the winds of winter already beginning to blow.

Raglan had another pressing problem directly related to maintaining the pressure on Sevastopol, a problem not affecting the French to anywhere near the same degree. His own lines of communication were exposed and vulnerable. An army whose supply lines are cut first ceases to fight and, unless they are reopened or alternatives found, then begins to disintegrate for lack of food. Again, Map 5 makes this clear. The Allies relied on their fleets to bring supplies to their bases, the French at Kamiesch, the British at Balaclava. The French had a short, direct and well protected route to the siege line, their exposed flank (the east) being covered by the British, and their rear by French and Turkish troops along the almost cliff-like eastern and southern slopes of the Cheronese. At no point were their lines of communication threatened. Raglan was not so fortunate. British supplies had to be hauled some six miles, either up on to the Woronzoff road or along the southern route which climbed the Sapoune escarpment up onto the Cheronese at a place called The Col. East of the escarpment these routes were wide open to threats from the Russian field army. If the enemy took Balaclava the campaign was finished; even if they only cut the roads below the Sapoune Heights they would have to be reopened or the British would also be facing defeat.

Raglan simply did not have the men to maintain the siege and guard his communications adequately. On 3 October he wrote saying he had a mere

16,000 effective soldiers available, with infantry divisions struggling to muster more than 2,000 men. Cholera and sickness had continued to take a grim toll. With manpower dwindling, winter looming and Sevastopol untaken it is hardly surprising that he concentrated his efforts on the siege, entrusting the defence of his base and supply routes to a widely dispersed handful of Royal Marines, the 93rd Highlanders, two battalions of Turks and the 1,500-strong Cavalry Division.

By contrast the Russians' strategic situation was encouraging. They had refused to allow themselves to be bottled up in Sevastopol. Leaving a strong garrison, they had marched out and established a field army on the Allied flank and rear. This army not only protected their own lines of communication back to Simpheropol from attack, but posed a powerful and disquieting threat to the British base. If the Russians exploited their strategic advantage the British must react. At that stage tactics would take over from strategy.

Throughout the preceding weeks the British cavalry's patrols and pickets can hardly be described as wide-ranging or productive. With numbers so small, duties demanding and sickness so prevalent it is not surprising that few of Lucan's horsemen had got beyond the Chernaya River or the Kamara hills. Tactical information was scanty, but rumour was rife. Never a day went by without the appearance of the ubiquitous Cossacks. Private Farquharson, 4th Light Dragoons, has given us an insight as to what it was like in the days immediately prior to the battle:

'The nights were awfully cold, and the heavy dews would almost drench us, till the blood felt like ice, and what with "outlying" and "inlying" pickets, almost always in the saddle, and never undressed, sickness, want of food – and I've gone three entire days without food – we were very queer indeed. . . . Every day now the Russians loitering or moving in great masses about the Chernaya . . . began to keep us on the alert morning, noon and night. If we came in from picket fagged, cold and hungry, we might hear the trumpet sound "boot and saddle" at any moment.'[4]

The frequency of false alarms, the fruitless stand-tos and repeated bogus reports of enemy advances, had served to sap the strength of the troops and the credulity of the generals. On 14 October, a Sunday, when the 4th Light Dragoons provided pickets it had a mere five officers, eight sergeants and five dragoons on church parade. A week later the all-night vigil in the freezing wind that had finished off Major Willett had also infuriated Cathcart. The 4th Division had been required to send Brigadier Goldie with 1,000 men and artillery down to the Balaclava plain to assist Campbell as a strong attack was supposedly imminent as guns had been fired and bands heard playing. No attack materialized; Campbell abruptly told them they were not wanted and neglected to thank them for their efforts. So back the brigade marched up The Col.

Toiling in the trenches all day followed by futile marching all night was not good for the infantry's morale. In addition, Cathcart did not get on with Raglan. He held what was known as a 'dormant commission' to take over from Raglan should he die, although Cathcart was not the next senior general. Despite this, Raglan never consulted him, never called him to conferences, in fact did his best to avoid him. Cathcart felt that Raglan was messing his men, and himself, about. All this went some way to explaining his extraordinary behaviour with Ewart.[5]

As the terrain, and particularly what could be seen by our various horsemen when they made decisions or took action, is a key factor in unravelling what happened and why, it merits careful consideration. Map 6 is a contour map of the area with only key reference points shown so that the reader can become orientated with the ground before the complex events leading to the charge are described.

East of the Sapoune escarpment, which had been entrenched and was manned by French infantry and gunners from General Bosquet's 2nd Division, was the Balaclava plain. From west to east it stretched for over 6,000 metres to the hills that hid the Chernaya River and the assembly area of General Liprandi's force around the village of Chorgun. North-south it was about 4,000 metres across at its widest part. It extended from the tangled contours of the Fedioukine hills, where the British Army had bivouacked during their flank march, to the heights overlooking Balaclava and the sea. These often precipitous hills were penetrated by the Balaclava gorge, near the head of which was Kadikoi village.

The feature that played the predominant role throughout the battle was the low ridge (seldom more than 100 feet above the plain) that bisected the area into two shallow valleys. It ran roughly NW–SE and carried on its shoulder the Woronzoff road and has been likened to a causeway – hence its name, the Causeway Heights ('Heights' being another misnomer). At no point were the Causeway Heights totally inaccessible to cavalry. Interspersed along its length were a number of knolls, rising perhaps 20 or 30 feet above the rest of the ridge. On these the Turks, under British supervision, had dug some rudimentary fortifications which assumed the grandiose (and undeserved) title of redoubts. They were numbered, as on the map, from 1–6. The first, on the highest knoll, which was not really a part of the ridge, was later to be known as Canrobert's Hill. General Canrobert had succeeded St Arnaud as the French Commander-in-Chief and upon this hill he had first met with Raglan to survey the surrounds of Balaclava.

For ease of reference the valleys to the north and south of the Causeway Heights are termed the North and South Valleys respectively. A horseman in the South Valley could not know what was happening in the North Valley, and vice versa, because he could not see from one to the other. The only way to know what was going on in both valleys simultaneously was either to be on the Sapoune Heights (like Raglan and his staff), or to be on top of the dividing ridge itself.

With the exception of the charges of the Light Brigade and the French

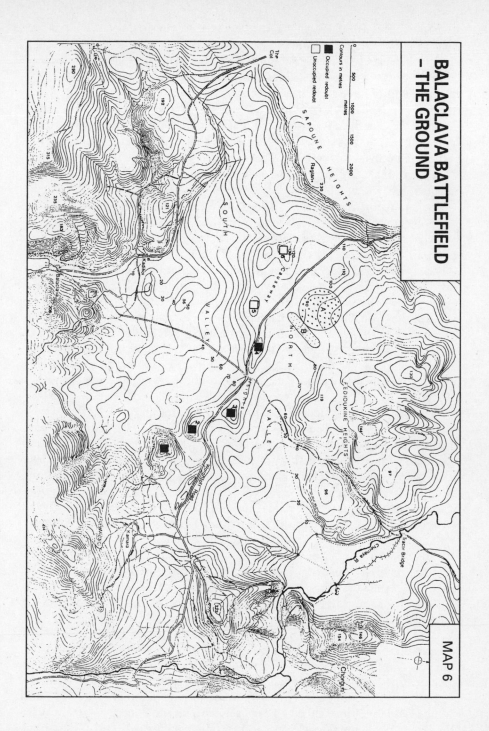

BALACLAVA BATTLEFIELD
– THE GROUND

MAP 6

Chasseurs the Battle of Balaclava was fought in the South Valley, which sloped steadily down from the Causeway Heights towards Kadikoi and the hills around Balaclava. There were undulations, but generally the land fell away from the north to the south. Vineyards occasionally obscured the view of the infantry and hindered the movement of horsemen. The North Valley was different. The general slope of this valley was from west to east, as it descended down to the Chernaya River. From the foot of the western escarpment to the banks of the river was some 7,000 metres (4½ miles) and over this distance the ground fell only about 350 feet. To the casual observer, therefore, the valley floor seemed flat. This presupposed a clear view up and down the valley – but the lie of the land was deceptive.

There were no artificial obstructions, no trees, or even vineyards, but there were slight dips, slight swellings, one or two of sufficient height to prevent even horsemen in certain spots from being able to see or be seen. The most striking example of this is the area near the head of the valley, marked 'A' on the map. Cavalry positioned here could not be seen by anyone on the valley floor to the east. Similarly they could not see what was happening in the rest of the valley. Only from the area marked 'B' was there a really clear line of sight down towards the river. The river itself could not be seen as the valley gave a sudden turn half left (NE) about 2,000 metres from its banks. In other words there was not a straight line of sight down the whole length of the North Valley, rather the line followed a shallow 'V' shape. These features are important and have been ignored by many historians when drawing maps of the Light Brigade's charge.

Finally, the road system was significant. Armies need roads, or at least reasonable tracks, in order to keep themselves supplied. Wagons, carts and heavy guns can seldom move across country, while, if speed of movement is essential, infantry also benefit by marching along roads. Assuming Balaclava to be the objective of the British and the Russians, the former to protect it, the latter to attack it, both could use two approach roads. They had the option of using both simultaneously or only one. Raglan could get into a position to cover his base by using either the Woronzoff road as far as the Causeway Heights crossroads and then turn south, or he could use the road via The Col. The Russians had the choice of coming up the Woronzoff road from the east (Baidar) or of crossing the Tractir Bridge, cutting through the Fedioukine Heights, cresting the Causeway Heights at the crossroads, with Balaclava then in sight. Three out of four of these approaches met at the crossroads on the Causeway Heights.

On the Causeway Heights were constructed the redoubts mentioned above. They were intended as the outer line of defence for the Balaclava base. For this role they had serious defects. Firstly, their garrison was of questionable quality. Not that Turks could not fight (they had seen off the Russians around Silistria), but rather these were untried colonial troops, 'Bono Johnnies' as the British called all Turks, from Tunisia. Some 200–500 of these soldiers were manning each of Redoubts 1–4, numbers 5 and 6 being undefended as they had yet to be completed. These men were rough, illiterate infantrymen who knew nothing about gunnery. As artillery fire would be a crucial element of the defence of these redoubts Raglan had insisted that nine 12-pounder guns, with a range of around 1,200 metres with round shot (more with shrapnel), be shared out among the four redoubts. Three went to No. 1 and two each to the others. 'W' Battery, equipped with four 9-pounders and two 24-pounder howitzers, which was supporting the 93rd Highlanders, was told to detail off five gunners to supervise the firing of these redoubt cannons. Theirs was not an easy or popular task, and although the battery-sergeant-major could hardly have been expected to pick his best men, they all did their duty admirably on 25 October. It was these guns, or rather the loss of them, that finally set in motion events that led to the charge of the Light Brigade.

The next, and more serious, problem was the isolation of the redoubts. Their defence could only be expected to delay, not defeat, a strong attack. Raglan was gambling on their holding out for sufficient time to allow him to get infantry support up on to the Causeway Heights. But there was no immediate reserve available apart from the 93rd, who were far too weak, and the Cavalry Division, which was not a suitable force to confront a powerful infantry assault backed by artillery. Also, the gaps between each redoubt could easily be penetrated by advancing attackers, and they were just out of range of the heavier artillery deployed in the entrenchments along the hills north and east of Balaclava – the so-called second line of defence for the base.[6] Russell quotes the opinion of a Prussian engineer called Waagmann with whom he spoke one day as he was watching the Turks wielding their spades on the redoubt parapets:

' "No! I say no, Sir!" quoth Waagmann. "It is all a mistake! Dere should be a second line of redoubts! Dese should be entrenched, and dere should be a brigade or two of infantry behind them. Vat good de cavalry do? Dese redoubts far away from every one! Dey are too far from Balaclava, and dey are too far from de French up dere [on the Sapoune Heights]. It is a mistake!" '[7]

The siege took priority. Raglan knew the risks he had been forced to take, knew the vulnerability of his base and knew precisely what he had to do if it was ever put in jeopardy – hence the speed of issuing his first order to the infantry. From his vantage point what he could see is set out on Map 7 and in Sketch 2. He and his staff had arrived in time to witness the loss of Redoubt No. 1 and the flight of the surviving Turkish defenders (that is a little before 7.30). Even so Russell felt that Raglan had been somewhat tardy in getting to his observation post:

'I could never understand why they were so long in turning out at Headquarters. The advance of the Russians was detected as soon as it

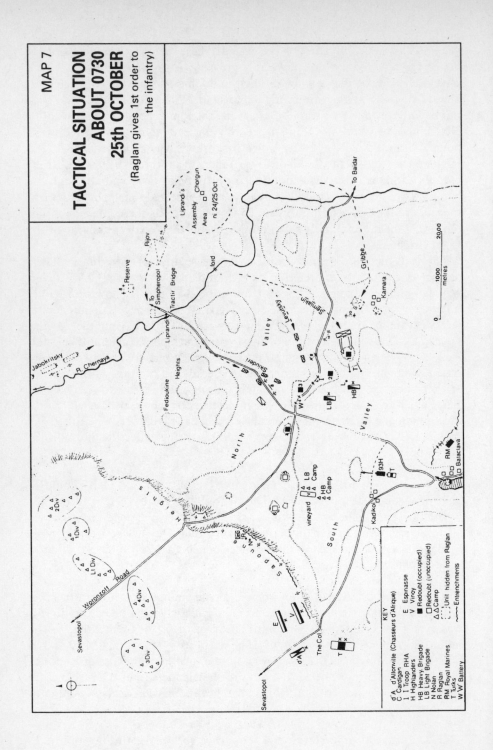

MAP 7

TACTICAL SITUATION
ABOUT 0730
25th OCTOBER
(Raglan gives 1st order to
the infantry)

KEY

d'A. d'Allonville (Chasseurs d'Afrique)	
C Cardigan	E Espinasse
I 1 Troop RHA	V Vinoy
H Highlanders	■ Redoubt (occupied)
HB Heavy Brigade	□ Redoubt (unoccupied)
LB Light Brigade	△△ Camp
R Nolan	⊂⊃ Unit hidden from Raglan
R Raglan	
RM Royal Marines	Entrenchments
T Turks	
W W Battery	

76

was daylight. A quiet canter would have taken one from Headquarters to the edge of the plateau in twenty minutes.'[8]

As Raglan watched he saw 'W' Battery and 'I' Troop RHA in action on the Causeway Heights against large columns of infantry advancing across the North Valley to assault Redoubts 2 and 3. He saw the Cavalry Division close up on the reverse slope, just west of the redoubts, and he saw the 93rd Highlanders hurrying to a position to block the direct route to Balaclava through the gorge. That much was set out below him, like some gigantic chessboard. His opponent was making the opening move; now it was his turn. But what even his elevated position could not reveal was the massing of Rijov's Hussar and Ural Cossack squadrons, all twenty of them, en route to the Tractir Bridge. Nor were the long columns of Jaboritsky's infantry led by Don Cossacks visible, as they moved south-east along the east bank of the Chernaya, also heading for the Tractir Bridge and, ultimately, the Fedioukine Heights. At the moment those hills were bare.

It was crystal clear to Raglan and those around him that his base was in peril. This was no false alarm. If the redoubts fell, and they were falling in front of his eyes, Balaclava was surely the Russians' next stop. All that was deterring them from going the extra mile were Lucan's cavalry and Campbell's Highlanders, plus a few sick and a handful of jittery Turks. There was still, however, a nagging uncertainty in his mind. Was this attack to be the prelude to a sally from Sevastopol as well? There was no way of knowing. What was unquestionable, what was imperative, was the need for infantry and guns on the plain quickly.[9]

ADCs, including Ewart, spurred away. They went to the French Commander-in-Chief to alert him to the situation; to the 1st and 4th Divisions with orders to march at once to Campbell's assistance; to the 3rd Division to put it on standby in case of an attack from within the city. Without specific orders, Captain Brandling, with 'C' Troop RHA, who was with the Light Division, moved out of camp as soon as news of the Russian advance was confirmed. He headed down the Woronzoff road – the shortest and quickest route. Unlike the infantry divisions Brandling had received no instructions not to use this road.

Raglan was to issue five significant orders that day, one to the infantry and four to the cavalry. Had his first order to the infantry been carried out with urgency and speed there is little doubt that the last two, which sealed the fate of the Light Brigade, would never have been given. Raglan's plan to get infantry to block or deflect the danger to his base was thwarted by less than clear thinking and a dilatory, almost disobedient, subordinate. As described above, Cathcart was responsible for the latter, Raglan himself was to blame for the former.

If the Turks could hold out for a while (those in Redoubt No. 1 had been under attack for well over an hour when Raglan saw them flee), if the cavalry and Highlanders could delay things, then the infantry had a chance of arriving in time. In fact the Turks in Redoubt No. 1 had fought well against daunting odds and given Raglan at least an additional hour. The layout of the British infantry divisions' camps is shown on Map 7. By far the fastest and most convenient, indeed obvious, route for either the 1st or 4th Divisions to take was straight down the Woronzoff road. It led directly on to the Causeway Heights, it led to the line of redoubts, it led to the critical crossroads in the centre of the outer line of defences and, perhaps most important of all, it led on to the flank of the Russian attack as Raglan saw it at 7.30 am. The maximum distance to be marched by either division was about two and a half miles. Either, possibly both, could have made their presence felt in under an hour. Yet Raglan specifically forbade the use of the Woronzoff road. Why he did so has never been specifically explained. The simple solution was to send the 1st Division down the Woronzoff road to take the enemy in flank and re-occupy the Causeway Heights while the 4th took the southerly route to help Campbell in the close defence of Balaclava. Perhaps this was his intention and was the reason he insisted on Cathcart not using the Woronzoff road – he wanted it free for the 1st. If so, it never happened that way.

Historians seldom mention this strange start to the battle. A general whose base is under serious threat sends his reinforcements by the longest route to the battlefield. If the Fedioukine Heights had been swarming with troops and guns perhaps Raglan might have felt it too risky, but they were deserted at that early hour. Even if they had been occupied, the Woronzoff road was out of effective military range from even the forward edge of the Fedioukine Heights. It remains a puzzle.[10]

Coming down via The Col almost doubled the distance for the 1st Division and added a mile to that of the 4th. Despite this the 1st overtook the 4th and eventually arrived in the South Valley first. Unfortunately it was Cathcart's troops (he had a complete division, whereas the 1st did not) that Raglan had in mind to lead the way to the defence of Balaclava and the retaking of the redoubts. Their general's slowness, however, resulted in their being at the back of the queue at The Col.

When Raglan issued his order he had no reason to doubt that the 4th Division, which was the nearest, would lead. Assuming the division got on the move by eight o'clock and marched the three and a half miles via The Col a rough time and space calculation must have led Raglan to expect to see some infantry below him between 9.00 and 9.30 am. What he could not know, what he failed to allow for, was that the friction of war would intervene to frustrate his intentions. First there was Cathcart's behaviour. Then the French decided to send infantry and cavalry to the South Valley by the same route. Two infantry brigades, the equivalent of a cavalry regiment (Chasseurs d'Afrique) and some Turks were much closer than the British formations, so they got there first. Something of a log-jam developed at The Col. Those at the back had to wait, except for Brandling and his guns for whom the Chasseurs d'Afrique gallantly gave right of way when they both arrived together. Brandling had reached the Sapoune escarpment on the Woronzoff road at about 8.15, seen no British troops along the way, seen nothing but Russians crossing the North Valley, so decided to turn south and go bounding across country for The Col.

To recap on the four horsemen of calamity at 7.30 am on 25 October. The first was up on the Sapoune Heights, and had just given his first order; the second (Nolan) was close by him, watching and listening with mounting excitement; the third (Lucan) was well forward in the South Valley and under fire; the fourth (Cardigan) was about to start cantering up to take command of his brigade from his yacht in Balaclava harbour.

The day before, Russell had ridden down to Balaclava, and on his way back stopped at the Light Brigade camp for a gossip:

'I was told that "the Ruskies were very strong all over the place", that reports had been sent to headquarters that an attack was imminent, and that Sir Colin Campbell was uneasy about Balaclava. As I was leaving Nolan overtook me. . . . The evening was chilly. He remarked that I ought to have something warmer than my thin frock [coat], and insisted on my taking his cloak – "Mind you send it back to me tomorrow; I shall not want it tonight." Nor did he next night or ever after! I bought the cloak at the sale of his effects, and I have it to this day. All the way back he "let out" at the Cavalry Generals, and did not spare those in high places. "We are in a very bad way I can tell you." '[11]

Like Nolan, Russell lived at headquarters and has left us with some interesting tit-bits of life with the generals and staff:

'Our meals were not luxurious, but they were sufficing (sic); ration beef or mutton, served up as steak or stew, accompanied by the preserved vegetables of the "Maison Chollet" . . . and (if our soldier cook was tolerably sober) a suet pudding, hard biscuit, rum and water, and a pipe of tobacco; and so, after a little talk over the news of the day, to earth!'[12]

On 24 October Russell mentioned his recent conversation with Nolan. 'He is an inveterate croaker,' muttered one of the staff, 'I wish he was away in Jericho with his cavalry. What do we want with cavalry here? We are not going to leave this till Sevastopol is taken.'[13] Nolan's obvious high opinion of himself, and the outspoken way in which he criticised his seniors with impunity, did not always improve his popularity with his peers.

Until 14 October Lucan had sole responsibility for the protection of the Balaclava base and the roads to it. On that day, however, Raglan made a significant change. Campbell, a highly experienced officer with years of active service in India to his credit, and until then in the 1st Division, was pulled out of the siege lines with the 93rd Highlanders and given the same task. Under him he had, in addition, the redoubts, 'W' Battery RA under Captain Barker with four 9-pounders and two 24-pounder howitzers, two battalions of Turks, 1,000 Royal Marines and the heavy guns in the entrenchments along the hills north-east of the town. Lucan kept the cavalry. Campbell expressed a willingness to serve under Lucan but was told his command was independent. The lieutenant-general commanding the cavalry had no authority over the brigadier commanding the infantry, yet both had the same duty – keeping the enemy out of Balaclava. It was a typical Raglan fudge. Fortunately Lucan and Campbell got on remarkably well and when the long-expected attack materialized both consulted and cooperated fully.

On the evening of 24 October the weather was miserable. Wind and rain lashed the flapping canvas of hundreds of tents and soaked the sentries huddled in their cloaks as they struggled to maintain some sort of vigilance in the darkness. It was impossible to keep weapons dry. Men and animals suffered equally from those raw elements so hated by soldiers – the cold and the wet. It was weather that favoured an attacker. Out of the gloom a sodden figure was brought before the Turkish commander in Balaclava. He was a spy with alarming and impressive intelligence. According to him the Russians were massing that night around Chorgun; they were over 20,000 strong; they would attack tomorrow, and they would advance towards Balaclava from the east and north-east. The Turk was taken to Lucan and Campbell for close questioning, with the aid of Lucan's Turkish-speaking interpreter John Blunt, a young man from the British Consular Service. What they heard impressed them. Lucan sent the information immediately to Raglan by the hand of his son, Lord Bingham.

When this young officer arrived Raglan was in conference with Canrobert so the message was sent in to him. Bingham waited, confident that he would have more work to do. He was disappointed. The Commander-in-Chief's response was a terse, 'Very well!' In disbelief that this was Raglan's final word Bingham hung around until the meeting broke up. His reward was to be told, when Raglan later spotted him, 'Let me know if there is anything new.' That was the answer he took back to his father.

To the Commander-in-Chief this was yet another cry of 'Wolf!' Another report that the enemy were threatening to attack. He had reacted to so many of these rumours recently and he remembered clearly Cathcart's unfavourable comments on his men being marched fruitlessly around at night only three days before. Yes, he was worried about his base; yes, he knew the Russians were bound to make a move sometime soon; and yes, he knew he was gambling. Nevertheless, it would take more than a message from a spy to get him to mobilize his exhausted infantry and march them down the hill in the darkness and drenching rain. His decision led to the Battle of Balaclava starting with the odds decidedly in favour of the Russians.

A task force is the best way to describe the mixed bag of formations the Russians had assembled. It had been building up, none too secretly, over the previous ten days and was now encamped noisily, uncomfortably and sleeplessly around the village of Chorgun (see Map 7). The commander of the 12th

Infantry Division, Lieutenant-General Liprandi, was the task force commander. He issued his final orders for a three-pronged infantry assault, supported by a heavy artillery bombardment, on the afternoon of the 24th. The assaulting units were to start moving to their forming-up places at 4.00 am the next day. The attack would start at 6.00. Phase one would be the capture of the Causeway Heights and Redoubts 1–4. Phase two was to push a strong force of cavalry down the South Valley to threaten Balaclava itself. It was an attack on the most sensitive part of the British military anatomy. If successful the siege of Sevastopol would disintegrate. Even just the securing of the Causeway Heights could not be ignored, so pressure on the city would ease. With the British lines of communication cut or, at least seriously threatened, and with winter only weeks away, the Allied position could become untenable. Such was the valid military logic behind the Russian plan.

The three assaulting columns for phase one, each with a company of riflemen attached to act as skirmishers, were provided by Liprandi's own division. On the left would be Major-General Gribbe with the bulk of the 24th Dnieper Regiment (3 battalions), Colonel Jeropkine's lancers of his Uhlan Regiment, a *sotnia* (squadron) from the 53rd Don Cossacks plus ten guns (6 light guns and 4 guns of position). He would move south along the river to the Baidar Valley area, turn west over the bridge to take the village of Kamara, there to set up his artillery to reinforce the bombardment of No. 1 Redoubt. As he had the longest distance to cover Gribbe would be the first to leave the assembly area.

The central column was to be coordinated by Major-General Semiakin, himself taking personal charge of the left echelon of attack, directed at No. 1 Redoubt. For this task he had four battalions of the Azov Regiment under Colonel Kirdner, one battalion of the 24th Dnieper Regiment, plus ten guns (6 light guns and 4 guns of position). Moving as part of the centre column but diverging from it to form the right echelon of Semiakin's attack was another smaller force under Major-General Levutsky. Its objective was Redoubt No. 2. For this purpose he had three battalions of the 23rd Ukrainian Regiment, supported by eight guns (4 light guns and 4 guns of position). Semiakin had no cavalry. The central force would take the rough track leading south-west from Chorgun to Kadikoi which crossed the Chernaya at a ford.

On the right was another column under Colonel Skiuderi. Despite his comparatively junior rank Skiuderi commanded a strong force, as his mission was to storm Nos. 3 and 4 Redoubts. For this he had all four battalions of the 24th Odessa Regiment and, because his right flank would be 'in the air' and the nearest to the Sapoune Heights from which the British reaction was anticipated, he was given three squadrons of the Don Cossack Regiment. His artillery amounted to the eight guns of No. 7 Light Battery. His route would take him over the Tractir Bridge and then through the Fedioukine Heights and across the North Valley using the Balaclava road as his axis. Unlike the other columns, Skiuderi had no navigation problems in the dark.

Over 14,000 infantry were allocated to the critical phase one assault. They would have the support of 36 artillery pieces (12 guns of position and 24 light guns). Their target was 1,500 Turks with nine guns. Their tactics would be to pound the redoubts with artillery and then assault with the infantry in dense formations that had not changed since Napoleonic times. The Russian artillery was the élite arm with better quality recruits and training; the infantry were hardy, backward peasants subject to endless drill and draconian discipline. Unlike the British who (except for the 4th Division) had the new Minié rifle, the Russian infantry musket was the old smooth-bore, muzzle-loading weapon that had been converted from flint-lock to percussion firing. Although it fired a massive .700 inch spherical bullet capable of tearing off an arm, it had no range sight and was of little use above 200 metres. The firing rate was ponderously slow at around one shot a minute, and reloading could only be done while standing. As a result the Russian infantryman was taught to rely on his bayonet. He would endeavour to advance shoulder to shoulder with measured step, the muskets of all ranks except the first at the shoulder.

The second phase was given to the mounted arm. Liprandi had been given the 6th Hussar Cavalry Brigade under Major-General Rijov. This was an impressive formation in terms of size, outnumbering Lucan's division by some 500 men. Allowing for wastage it was close to 2,000 strong, made up of eight squadrons of the 11th Kiev Hussars, six squadrons of the 12th Ingermanland Hussars and six of the 1st Ural Cossacks, the latter being lance-armed. Rijov also had eight guns from the 12th Light Horse Battery and another eight from No. 3 Battery of the Don Cossacks. The plan envisaged this brigade following Skiuderi's route over the Tractir Bridge, up on to the Causeway Heights on the right flank of the attack, and then advancing down into the South Valley to threaten Balaclava or deal with interference from the British cavalry.

The final part of the plan was semi-defensive. It involved securing the Fedioukine Heights which would in turn protect the Russians' right. For this purpose Major-General Jaboritsky was to march down from the mouth of the Chernaya, cross the Tractir Bridge behind the cavalry and deploy on to the Heights. This was a powerful force of some 6,400 men composed of the 31st Vladimir and 32nd Sousdal Regiments, two companies of riflemen, one of Black Sea Foot Cossacks, two squadrons each of the Ingermanland Hussars and 60th Don Cossacks together with fourteen guns (4 light guns and 10 guns of position).

There was not much wrong with the Russian plan. It employed concentration on a massive scale against the weak point of the British position; a tactical success would bring about huge strategic advantages, which could even win the war; it was very much a combined arms operation with infantry, artillery and cavalry cooperating; it guarded the exposed flank, and the night approach march, followed by a dawn attack, could well achieve surprise. The only weakness was the lack of sufficient reserves. The 23rd Ukrainian Regiment, one rifle company and eight light guns were all that Liprandi had seen fit to hold back in this role. Nevertheless, merely to secure a line stretching from Kamara

1. General Lord Raglan, the British Commander-in-Chief who gave the fatal fourth order (IWM)

2. Captain Louis Edward Nolan, 15th Hussars, the ADC who delivered the order

THE FOUR HORSEMEN OF CALAMITY.

3. Lieutenant-General Lord Lucan, the GOC the Cavalry Division, who received the order (IWM)

4. Major-General The Earl of Cardigan, the Light Brigade Commander who executed the order (IWM)

5. Brigadier-General Richard Airey, the Quartermaster-General at the British HQ who wrote out all Raglan's orders for despatch (IWM)

6. Lieutenant-General Sir George Cathcart, a Peninsular War veteran and GOC the 4th Division. It was his deliberate dilatoriness that caused his infantry to arrive so late on the battlefield. Had he arrived sooner the Light Brigade would probably never have charged. By way of contrast, eleven days later at Inkerman he placed himself in the forefront of the action and was shot dead for his temerity (IWM)

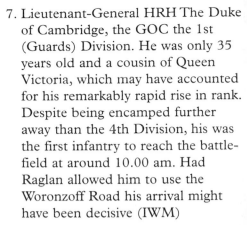

7. Lieutenant-General HRH The Duke of Cambridge, the GOC the 1st (Guards) Division. He was only 35 years old and a cousin of Queen Victoria, which may have accounted for his remarkably rapid rise in rank. Despite being encamped further away than the 4th Division, his was the first infantry to reach the battle-field at around 10.00 am. Had Raglan allowed him to use the Woronzoff Road his arrival might have been decisive (IWM)

8. Brigadier-General James Scarlett commanded the Heavy Brigade at Balaclava (IWM)

9. Major-General Sir Colin Campbell, Officer Commanding the Balaclava defences - including the redoubts, the 93rd Highlanders and the Royal Marines (IWM)

10. Omar Pasha, the Turkish
    Commander-in-Chief (IWM)

11. Lieutenant Henry Maxse,
    Cardigan's ADC, the only infantry
    officer to ride in the charge (IWM)

12. The heavily bearded and beefy
    Major Alexander Low. He com-
    manded a squadron of the 4th
    Light Dragoons in the charge. As
    duty Field Officer he aroused the
    sleepy picket near Kamara village
    just as the Russians were advanc-
    ing to start their attack (IWM)

13. Colonel Lord George Paget. He commanded the 4th Light Dragoons together with the second, supporting, line during the charge while he continued to smoke a cigar. He was subsequently highly critical of Cardigan's behaviour after the battery had been overrun (IWM)

. Lieutenant-Colonel Frederick Shewell who commanded the 8th Hussars in the charge. His was the only Regiment to maintain the trot for most of the advance, with the result that the 8th Hussars dropped back and never really charged anything (IWM)

15. 'Old Jack Penn', a hard-drinking troublemaker in barracks but a boon on the battlefield. He rode as a corporal in the 17th Lancers - typically taking time out to loot souvenirs from a Russian officer he had killed (IWM)

16. Captain George Maude, the officer commanding I Troop RHA at the outset of the battle. He was very severely wounded by shellfire early in the morning (IWM)

17. Brigadier-General d'Allonville who commanded the brigade of Chasseurs d'Afrique. It was the 4th Chasseurs who successfully charged the Russian guns on the Fedioukine Heights, thus facilitating the Light Brigade's withdrawal (IWM)

18. Balaclava harbour, the British base at which all supplies were put ashore, in 1855. The view is looking north towards Kadikoi village and the South Valley where the 93rd Highlanders made their stand. It was amongst this tight tangle of shipping that Cardigan moored his yacht (IWM)

19. Approximately the same view today. One of the lasting impressions on a first visit is the narrowness and smallness of the harbour (Sharpe)

20. A view of Balaclava harbour from the opposite direction of that shown in number 18. It shows the northern extremity of the harbour in 1855, with the distinctive Genoese towers on the skyline (Robertson - Royal Archives Windsor Castle)

21. Raglan's HQ on the Sapoune Plateau. The Commander-in-Chief's quarters were in the farm building, while his junior staff, ADCs, orderlies and escort were housed in the tents and outbuildings (IWM)

across the Causeway Heights to the Fedioukine Heights could make the entire British position untenable without ever reaching Balaclava.

Captain Low, of the 4th Light Dragoons, was shivering as he struggled to pull on his boots. An hour or so of fitful sleep had done nothing to dispel the weariness that permeated his massive frame. It was the early hours of the morning of 25 October and Low was the duty field officer for the Cavalry Division. As such it was one of his more onerous duties to visit the outlying pickets. Although the rain had ceased the wind still cut through his thin jacket and cloak as, accompanied by a single orderly, he started out into the blackness on his long ride east.

The Light and Heavy Brigades shared the duties of providing pickets. On this occasion the picket close to Kamara was from the Heavy Brigade. Pickets were under the command of an officer, usually with a sergeant and about thirty junior NCOs and men. Their tasks revolved around early warning of enemy approach and the passing back of information. *Horse Guards' Regulations for the Instruction, Formations, and Movements of the Cavalry* dated May, 1851, stress that, 'Picquets must never for an instant forget that the honour and safety of the whole Army frequently depends on their alertness and vigilance'.

On arrival in the area in which the picket was to operate, in this case Kamara, a suitable base was selected, screened from the enemy. From there the officer placed his vedettes in an outer chain covering all possible approach routes. A vedette consisted of two horsemen and was in effect a pair of mounted sentries. Each vedette was supposed to be able to see those to its right and left. If enemy were observed the horsemen were to circle their horses at a walk, trot or gallop (canter) according to the number of the approaching force. If it consisted of cavalry then they would circle to the right, if infantry only then to the left, and if mixed then one rider circled right the other left. The whole chain of vedettes would take up the signalling. It was a system that worked satisfactorily by day in open country with good visibility. By night and in close country it was fraught with difficulties.

At around 5.30 am that morning, at the same time as Low was nearing Kamara from the east, a *sotnia* of Don Cossacks, followed by the leading elements of Jeropkine's lancers, was approaching from the west. They were the advance guard of Gribbe's force. In the village, near to which the vedettes had been withdrawn, was the none-too-alert picket. Somehow the Russians had crossed the river at the bridge, only a little over a mile from the village, unobserved. The Cossacks and lancers pushed up the slope to approach Kamara from the high ground to the east. It is uncertain who heard who first. There is reason to suspect that Low heard or saw the enemy before he reached the village and arrived just in time to arouse a dozing picket before the Cossacks reached them. What is clear is that the picket had to scramble for their lives and retreat in an undignified manner towards Redoubt No. 1. Kamara fell to Gribbe. He lost no time in setting up his guns on high ground opposite Redoubt No. 1. It was – almost H-hour – 6.00 am.

By this time the Cavalry Division, which had been stood-to in front of their tents for well over an hour, were hoping to be dismissed to water their horses and get breakfast. Men and animals were to be disappointed. Every morning without fail Lucan had his division paraded an hour before dawn, the only permanent absentee being Cardigan. It was not a popular precaution. Twenty-one-year-old Private William Pennington of the 11th Hussars was later to write with feeling:

> 'I found these early morning musters most depressing, for it is at the hour immediately preceding daybreak that the pulse of Nature would appear to beat most faintly. The coldest hour in all the twenty-four is that before the dawn; and fasting even from the meagre allowance of the coarsest ration, standing drowsily at our horses' heads in the bleak morning air sweeping fitfully across the plain, are not conditions favourable to the most hopeful and heroic frame of mind.'[14]

An anonymous soldier of the 8th Hussars recalled getting ready for the parade and inspection that morning:

> 'We groomed and saddled our horses as well we could, wiped the dews off our swords and scabbards, which were red and rusty despite all our care, got our cloaks, equipments, and so on, in order, with thirty rounds of cartridges in our wallets and thirty more in our pouches, and moved up to the brow of the hill.'[15]

As commanding officers received the parade states of their regiments and squadron commanders checked their men and horses Lucan prepared to go on his usual ride to the front. He took with him the majority of his staff, which included Lord Paulet (Assistant Adjutant General), Major McMahon (Assistant Quartermaster General), Captain Walker, Captain Charteris, Cornet Bingham, Mr John Blunt (interpreter) and, as his field trumpeter, Trumpet-Major Joy from his old Regiment the 17th Lancers. Joy was a favourite of Lucan's who had enlisted into the 17th as a bandboy in 1833 and in 14 years had risen to trumpet-major and had subsequently been selected to take charge of the band playing at Wellington's funeral.[16] It was a dank and misty morning as they set out eastwards, with dawn just beginning to break behind the Kamara hills.

As they passed the Light Brigade lines Paget, who, with Cardigan still in his bed, was yet again the acting brigade commander, decided to join them for the ride. He fell in with Paulet and McMahon who were walking some fifty metres behind Lucan's group. They were heading for Canrobert's Hill (Redoubt No. 1), about 3,000 metres away, which made an excellent viewpoint from which to see what daylight would bring. By the time they got to within 300 metres of the hill it was close to six o'clock and they could make out the black silhouette of the newly erected flagpole against the lightening sky. From it flapped two flags. Paget has recorded the moment:

84

' "Hello," said Lord William [Paulet], "there are two flags flying; what does that mean?"

"Why, that surely is the signal that the enemy is approaching," said Major McMahon.

"Are you quite sure?", we replied. We were not kept long in doubt!

Hardly were the words out of McMahon's mouth, when bang went a cannon from the redoubt in question. . . . Off scampered my two companions to their chief while I turned round and galloped back "best pace" to my brigade, which I at once mounted.'[17]

The first cannon shot of the battle had been fired by Turkish gunners under the direction of two British artillerymen. Either the fleeing picket had warned them or the Russians had been spotted forming up – perhaps both. Within moments of this single shot Gribbe's guns responded with a continuous cannonade, audible on Cardigan's yacht and at the British headquarters on the Cheronese.

Lucan's reaction was instantaneous. Charteris was sent back to alert Raglan. He had five miles to cover, mostly uphill and with a stiff climb at The Col, so could not be expected to reach headquarters until around 6.30 am. In reality no reinforcements could be expected on the plain for two or three hours. Until then Lucan had his cavalry, Campbell his Highlanders – and there were the Turks. Another officer was despatched to warn Campbell and to bring forward the Cavalry Division.

The Light Brigade had no sooner climbed into their saddles than they were ordered to advance. Just before moving off the Commanding Officer of the 11th Hussars, Colonel Douglas, shouted out a few words of advice:

'Eleventh, attention. Now men, in all probability we shall meet the enemy today. When you do, don't cut but give them the point, and they will never face you again.'[18]

The Division advanced at a brisk trot with the Heavy Brigade on the right, and in advance of the Light Brigade, which was echeloned to its left rear. When the Light Brigade halted it was within 400 metres of Redoubt No. 2. By this time the struggle for No. 1 was at its height, with the Turks showing no sign of wavering under the weight of shot and shell descending on them. Lucan remarked to Blunt, 'Those Turks are doing well.' The crashing of the cannonade and the popping of musket fire was continuous. Later TSM Smith would record that he also heard the shouts of the Russian infantry as they clambered up the far side to storm the entrenchments. It would have been Colonel Krudener's Azov Regiment making its final charge.

The cavalry's view was obscured by the hills and, to a lesser extent, by the smoke. Despite this they had to remain watchful – on the lookout for 'overs'. The Russian guns were firing a lot of roundshot – cannonballs – a number of which missed the redoubt and came sailing over the hill into the cavalry division. Because they were reaching extreme range they were slowing and were therefore visible to the sharp-eyed soldier at the receiving end. Although they were still horribly lethal they could be dodged. Paget saw a man in the leading squadron of the 4th Light Dragoons struck by one. '[It] completely whizzed him round, and I can well remember the slosh that sounded as it went through the centre of his belly.' Officers out in front of their squadrons were better able to get out of the way as they had room to move quickly, a facility denied to the men bunched together in the ranks. Paget himself had a close call by moving into, rather than away from, the line of fire:

'I was standing with my side to the redoubt, and clear of the front of the brigade, when all sorts of gesticulations and cries of "Look out, Lord George!" met my ears. Bewildered, of course, I moved my horse two or three paces, which had the effect of bringing me into the line of the round shot, which they saw coming, and which bounded actually between my horse's fore and hind legs, bursting a cloud of dust up into my face.'[19]

Like soldiers before and since, who see their officer in the same sort of trouble as themselves, his orderly thought it highly entertaining:

'The first knowledge I had of the danger I had passed was a laugh from my rollicking orderly, "Ah, ha! it went right between your horse's legs;" responded to by me, "Well, you seem to think it a good joke; I don't see anything to laugh at." '[20]

For the first three hours the Battle of Balaclava was primarily an artillery duel. It was an unequal duel, with the Russians heavily outgunning the British and using their advantage to get their infantry successfully into Redoubts 1–4. It was because he was fearful of losing artillery that Raglan later issued the orders that sent the Light Brigade on its charge. Artillery was the real villain, as far as the Light Brigade was concerned, before, during and after its charge. A basic knowledge of how artillery of the day worked, and what it could realistically achieve, is essential if a proper understanding of what happened to the Light Brigade is to be gained.

The British went to war in the Crimea with cannon that had been used at Waterloo. The Russian artillery was equally dated. The new Minié rifle, with which the bulk of British infantry was armed, could shoot further than the cannons. In general terms, like any missile weapon, the shorter the range the greater the chance of hitting. The effective battlefield range varied with the calibre and type of ammunition used, but a rough rule of thumb (for guns) firing roundshot was that the weight of the shot multiplied by 100 would give the effective range in metres. (With shrapnel longer ranges could be achieved.) Thus a 6-pounder's maximum effective range would be 600 metres, a 9-pounder 900, a 12-pounder 1,200. The guns were flat trajectory, muzzle-loading, line-of-sight weapons firing three types of ammunition: roundshot, which often bounced off the ground like a hard-hit cricket ball; shrapnel shells filled with powder and bullets fused to explode after a pre-set time; and case-shot which was a tin of bullets used at ranges below 200 metres like a giant shotgun cartridge. Of these the most commonly used was roundshot because it was a multi-target weapon, being equally effective against buildings, walls, entrenchments, wagons, enemy guns and personnel. To witness the ease with which it could slice men in half or disembowel horses was a demoralizing experience. The noise was often as frightening as the sight. At the Alma a group of fifteen men were killed by the same roundshot while standing precisely in its path.

It is usually not appreciated how large was the establishment for a 6-gun troop. 'C' Troop RHA embarked for the Crimea with 188 men of all ranks, 192 horse (57 riding and 135 draught), nine ammunition wagons, three stores wagons, a rocket carriage and a forge. Maude's 'I' Troop RHA, which was permanently attached to Lucan, under his orders as his divisional artillery, was of a similar strength. He had four 6-pounder guns and two 12-pounder howitzers. The howitzer had a slightly shorter range than the gun as it used a higher elevation to produce more of a lob – in this respect being more akin to a spin bowler whereas the gun was the flatter trajectory pace bowler. The howitzer never fired roundshot. 'C' Troop, under Brandling, had four 9-pounders and two 24-pounder howitzers. Liprandi deployed heavier guns in bigger batteries. His guns of position batteries had six 12-pounders and six 18-pounder howitzers, light batteries six 6-pounders and six 9-pounder howitzers, while horse artillery batteries, such as the one the Light Brigade would charge, had four 6-pounders with four 9-pounder howitzers.

Rates of fire were of crucial relevance. Sponging, loading, firing and realigning the gun by hand after every recoil had thrown it back several feet was exhausting work for the detachment. Normally two rounds a minute was the expected rate of fire. This could be cut in emergencies, when the lives of the gunners depended on speed. The prospect of impending death strengthens the muscles and sharpens the mind. The usual way of speeding up the drill was to sponge out the barrel only after every fifth or sixth shot. It risked an explosion in the gun if any burning embers of powder remained in the bore after the previous firing. However, it meant that perhaps as many as six or seven rounds could be fired in a minute.

The first element of the Cavalry Division into action was Maude's 'I' Troop RHA. He galloped his guns up on to the Causeway Heights between Redoubts 1 and 2. From here virtually the entire Russian assault could be seen moving across the North Valley. He came under fire from the artillery of both Semiakin's and Levutsky's columns. Apart from being outnumbered, he was also outranged and his ammunition was limited to the rounds carried in the gun limbers. This was due to his wagons having been sent, as usual, to Balaclava under his second-in-command, Captain Shakespear, to collect shells for the siege. Meanwhile Campbell had also sent forward 'W' Battery under Captain Barker with 9-pounder guns to take position near No. 3 Redoubt. On arrival he found that there was insufficient room to deploy, so he moved the battery on to the ridge just west of No. 3 Redoubt and opened fire on Skiuderi's advancing battalions.

That the artillery contest was an unequal one was attested by one of Maude's officers who commented, 'What could 6-pounders do against 18-pounders at 1800 yards [the range at the start of the firing]?' Within twenty minutes disaster struck 'I' Troop. The watching cavalry, all of whose eyes were riveted on the redoubts, saw a sort of 'spread eagle' against the horizon, with, according to Paget, 'the splinters of broken guns, horses' legs, etc., shooting up into the air. . . . Five minutes later the mangled form of poor Maude . . . was carried past us to the rear.'[21] A shell had hit the shoulder of his horse, the explosion almost ripping off Maude's arm. Lieutenant Dashwood now commanded the troop. He was down to less than ten rounds when he reported to Barker to seek permission to withdraw. This was given and two guns, under Lieutenant Dickson from 'W' Battery, were sent to the vacated position.

The British cavalry was impotent. It had committed its six guns; it could make theoretically threatening movements towards the enemy but without infantry it could do nothing effective. It was compelled to sit watching as Maude's guns were withdrawn, then Canrobert's Hill, followed in quick succession by Nos. 2, 3 and finally No. 4 Redoubts, fell. Before quitting their posts the British gunners from 'W' Battery spiked the guns in Nos. 2, 3 and 4 Redoubts, for which three were later awarded foreign decorations.[22]

The linchpin of the line had been Canrobert's Hill. Once that went, the Turks, who until then had fought fiercely, lost interest in everything except flight and poured down the South Valley 'as if the Devil were at their heels' (as indeed he was). Lucan had no alternative but to pull back slowly westwards. He also sent his interpreter (Blunt) dashing down to head off the fleeing Turks to try to get them to rally on the Highlanders. Blunt intercepted an officer and explained the orders as a mob of Turkish soldiers gathered round. Many were exhausted and desperately thirsty. There was a chorus of complaints. Why had they received no support? Why were the guns in the redoubts short of ammunition, and why was some of the wrong calibre? Why were rations so meagre and why was there so little water to drink? Blunt had no answers, but an old sweat with a bandaged head and smoking a two-foot-long pipe did. With the stoicism of many Muslims he said, 'What can we do, Effendi? It is the will of Allah.'[23]

TSM Smith had witnessed the collapse of No. 1 Redoubt.

'As the last of them [the Turks] came over the parapet, I noticed the Russians were hard at their heels. . . . As they gained the plain, a number of Cossacks swept round the foot of the hill, killing and wounding many of them. Some of them being unarmed raised their hands imploringly, but it was only to have them severed from their bodies. . . . Had a dozen or two of us been sent out numbers of these poor fellows might have been saved.'[24]

As 'I' Troop retired towards Kadikoi it met up with Shakespear and the missing ammunition wagons. Shortly afterwards Cardigan cantered up, ignorant of what was happening but full of pompous confidence. 'Where are you going, Captain Shakespear?' he demanded. 'Who gave you leave to retire?' 'We are going for more ammunition, my lord,' Shakespear explained.[25] Lucan, spotting the belated arrival of his Light Brigade commander, rode over and pointedly told Shakespear to continue his withdrawal. Then the Cavalry Division retired slowly westwards by alternate brigades and regiments and halted near their camp. The Light Brigade was to the north of their camp site, on the Causeway Heights between Nos. 5 and 6 Redoubts. On arrival Cardigan gave the order, 'Draw swords'.

All four horsemen were now in place.

CHAPTER FIVE
Two Orders for the Cavalry

'Those damned Heavies! They will have the laugh on us this day.'

Lord Cardigan, to nobody in particular but to anyone within earshot, as he watched the Heavy Brigade attack.

'The dead soldier of the 4th Dragoon Guards had red or fair hair, which was cut as close as possible, and therefore well suited to show any wounds. His helmet had come off in the fight, and he had about fifteen cuts on his head, not one of which had more than parted the skin. His death wound was a thrust below the armpit, which had bled profusely.'

The only man in his regiment to be killed in the Heavy Brigade attack – Lieutenant Fox Strangways RA, quoted by Colonel F.A. Whinyates in his book From Coruna to Sevastopol, *1884.*

By 8.00 am Liprandi's first phase objective, the redoubts on the Causeway Heights, had been secured. He was within a whisker of cutting off the British base from the army in the siege lines. From Canrobert's Hill there was an uninterrupted view westwards for over four miles of the undulating South Valley, right up to The Col. Apart from a small smear of red near Kadikoi there was not a steady infantryman in sight. A stream of panicky Turks was disappearing into the Balaclava gorge heading for the harbour. The laggards were being skewered by some of Jeropkine's jubilant lancers or Don Cossacks. A few, however, had some fight left in them, as Private Mitchell of the 13th Light Dragoons recalled:

'Two Cossacks came over the ridge together. One of them lanced a Turk in the back, who uttered a loud scream and fell. Another Turk being a short distance ahead, they both made for him, but before they could reach him, Johnny, who had his piece loaded and bayonet fixed, turned suddenly and fired at the foremost, knocking him off his horse. The other coming up made a point, but whether it touched the Turk I cannot say;

but in an instant he had bayonetted the Cossack in the body, and he also fell from his horse. Johnny resumed his journey at a walk.'[1]

Russian infantry was in possession of Redoubts 1–3. No. 4 had also been captured and the British guns smashed before it was abandoned. Russian artillery had chased the British guns and cavalry back towards the mostly flattened tents of their encampment. The window of opportunity was wide open. Liprandi had but to make one more push and Balaclava was surely his.

Whether or not Liprandi really ever intended to take Balaclava, or whether his objective was some mythical artillery park near Kadikoi as some Russian accounts indicate, is not of great significance. The crucial thing was what Raglan perceived as the Russian intentions; what, to him, were they now capable of doing. At eight o'clock what Raglan saw unfolding below him was disaster. According to Russell he made no attempt to conceal his agitation:

'Lord Raglan was by no means at ease. There was no trace of the divine calm attributed to him by his admirers as his characteristic in moments of trial. His anxious mien as he turned his glass from point to point, consulting with Generals Airey, Estcourt, and others of his staff, gave me a notion that he was in "trouble". Perhaps he alone, of all the group on the spot, fully understood the gravity of the situation.'[2]

His view at that moment has been reconstructed in Sketch 3, while Map 8 shows the situation on the ground at approximately the same time. His first line of defence to his base had gone; the Turkish garrison was in flight; the Cavalry Division had withdrawn slowly to a position just east of their encampment; the two British batteries in the field had been outgunned and compelled to retire with the cavalry; the only reliable infantry in the right place were some 550 Highlanders. If he looked half-left at the low lump of hills that were the Fedioukine Heights he would have noticed they were empty. All this he could see. He also knew that the French infantry were being alerted and that he had ordered two infantry divisions to march to the South Valley. He knew they would be a long time arriving, but he could not know that the 4th Division (the troops he wanted to spearhead the retaking of the redoubts) would be late starting out, or that their pace would lack urgency. With this information to hand Raglan gave his first order to the cavalry.

Turning to Captain Wetherall, the navigating officer Lucan blamed for the cavalry taking the wrong route on the flank march, Raglan issued the order:

'Cavalry to take ground to the left of second line of redoubts occupied by Turks.'

Riding hard, via The Col, Wetherall took perhaps 20 minutes to reach Lucan. The cavalry divisional commander did not understand or like what was required. He demanded explanations from the ADC. Lucan then insisted that

91

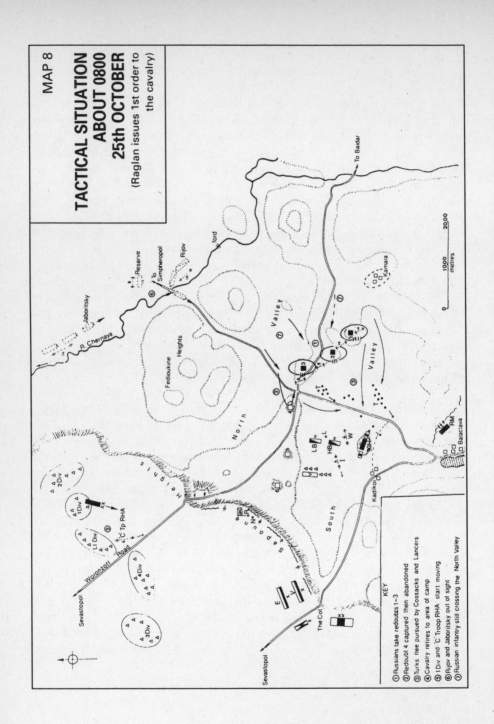

MAP 8

**TACTICAL SITUATION
ABOUT 0800
25th OCTOBER**
(Raglan issues 1st order to
the cavalry)

To Baidar

To Simpheropol

Rijov

Reserve

ford

Jaboritsky

R. Chernaya

Fedioukine Heights

North Valley

Valley

Kamara

North Heights

Causeway Heights

Sevastopol

Woronzoff Road

'C' Tp RHA

2 Div

1 Div

Ll Div

4 Div

3 Div

The Col

Sevastopol

Kadikoi

Balaclava

RM

KEY

① Russians take redoubts 1–3
② Redoubt 4 captured then abandoned
③ Turks flee pursued by Cossacks and Lancers
④ Cavalry retires to area of camp
⑤ 1 Div and 'C' Troop RHA start moving
⑥ Rijov and Jaboritsky out of sight
⑦ Russian infantry still crossing the North Valley

0   1000   2000
metres

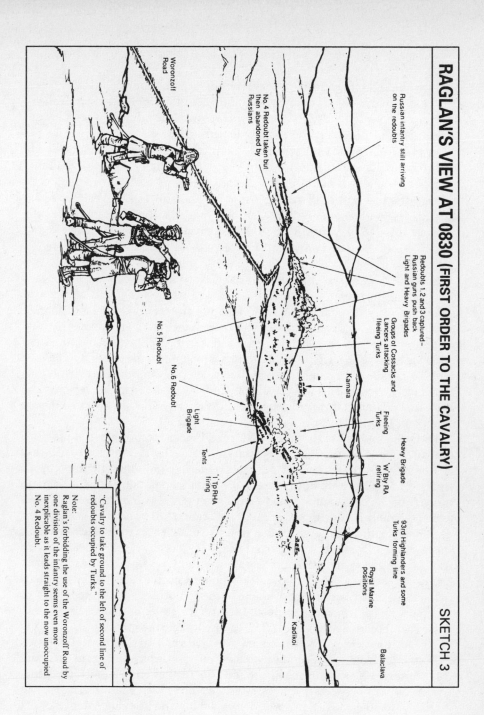

RAGLAN'S VIEW AT 0830 (FIRST ORDER TO THE CAVALRY)

SKETCH 3

Russian infantry still arriving on the redoubts

Redoubts 1, 2 and 3 captured - Russian guns push back Light and Heavy Brigades

Woronzoff Road

No. 4 Redoubt taken but then abandoned by Russians

Groups of Cossacks and Lancers attacking fleeing Turks

No. 5 Redoubt

No. 6 Redoubt

Light Brigade

Tents

'I' Tp RHA firing

Kamara

Fleeing Turks

Heavy Brigade

'W' Bty RA retiring

93rd Highlanders and some Turks forming line

Royal Marine positions

Kadikoi

Balaclava

"Cavalry to take ground to the left of second line of redoubts occupied by Turks."

Note:
Raglan's forbidding the use of the Woronzoff Road by one division of the infantry seems even more inexplicable as it leads straight to the now unoccupied No. 4 Redoubt.

93

Wetherall remain until he had seen the cavalry comply exactly with this first order.

With hindsight it is possible to unravel Raglan's thinking behind this order. He wanted to protect his base with infantry, but they would be some time in arriving. He wanted the cavalry to cooperate with the infantry when they did arrive. He did not want the cavalry embroiled or endangered before the infantry arrived (he was still reluctant to release them from 'the bandbox'). He had lost confidence in Lucan's ability to handle the cavalry to his liking. Lucan's advancing without orders at the Alma, his hazardous attempts to pursue after that battle, his getting lost on the flank march, had culminated in Raglan's deliberately not putting Campbell under Lucan's command for the defence of Balaclava. Now he thought the cavalry were too far forward and exposed to the renewed advance he expected and feared – hence his incomprehensible order.

Only the first four words made sense, 'Cavalry to take ground . . .' meant that the cavalry were to occupy ground – but where, why, and what to do thereafter? Viewed from Raglan's elevated position (see Sketch 3 and Map 8) the line of redoubts stretched away into the distance from approximately west to east, or at least Nos. 3–6 did, with 1 and 2 swinging slightly south-east. So taking ground to the 'left' of the redoubts would mean the cavalry going into the North Valley. Raglan had committed a cardinal sin, revealing his and his staff's inexperience. This was the use of words like 'left' or 'right', which mean completely different things to people in different positions looking in different directions, instead of using points of the compass.

To Lucan, in the South Valley, 'left' might mean west of No. 6 Redoubt if he chose to look north, or in the North Valley if he looked towards the enemy in Redoubts 1–3. There is the possibility that, at the time, the unoccupied Redoubts 5 and 6 were not referred to by numbers at all. If this is true the message becomes even more muddled. Then there was the phrase 'of second line of redoubts occupied by Turks'. What second line? There had never been a second line; there had only been a first line of defences for Balaclava which was the line of redoubts. And what did 'occupied by Turks' mean? When Raglan sent the order the Turks were not occupying any redoubts and they had never occupied Nos. 5 or 6. Little wonder Lucan found it all rather hard to follow. Wetherall led the division up to the gap between No. 6 Redoubt and the Sapoune escarpment and positioned the brigades as shown on Map 9. Lucan had been wise to get the ADC to disentangle the message.

The position selected by Raglan did not have much to recommend it. It was 2,000 metres from the 93rd Highlanders' position which left them far more isolated than before, with the likelihood of a further Russian advance penetrating the gap between the infantry and cavalry. Lieutenant Calthorpe, an ADC and relative of Raglan's who was watching and listening, later recorded in his *Letters from Headquarters* that;

'Lord Raglan ordered the brigade of light cavalry to take post on the

94

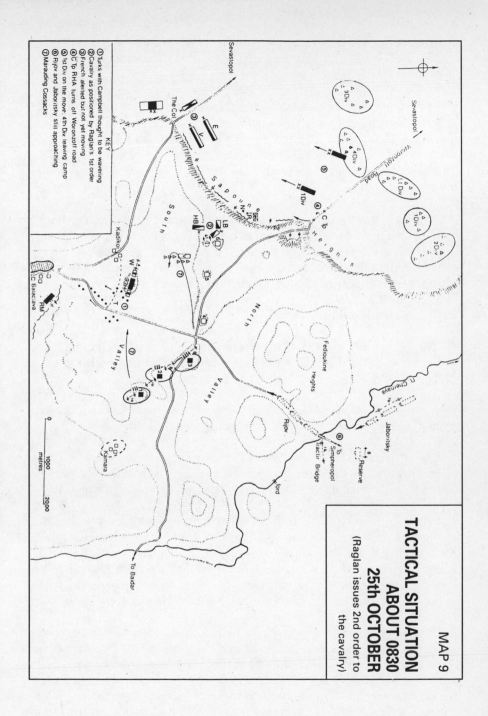

MAP 9

TACTICAL SITUATION
ABOUT 0830
25th OCTOBER
(Raglan issues 2nd order to the cavalry)

ridge, just at the foot of the plateau where we were standing. From this point they could watch and take advantage of any movement on the part of the enemy.'[3]

Thus the Light Brigade supposedly had the task of watching the North Valley to guard against an enemy approach from that direction, but was so placed that it could not see to fulfil that task. While they could see into the top, or western, end of the North Valley the view of the rest was obscured by the knoll on which No. 6 Redoubt was sited. TSM Smith makes this important point in his description of the position:

'Shortly after [the loss of the redoubts and the cavalry's withdrawal] we took up a position on the left rear of our encampment with our backs to Balaclava, facing the opening at the top of the valley through which it was thought the enemy cavalry would come. . . . We were out of sight of the enemy, but expected every moment to see them.'[4]

Paget has amplified this:

'Our next movement was a flank one to our left, to the rising ground that forms the end of the ridge of low hills, on which were situated the line of Turkish redoubts.'

Paget goes on to explain that the Light Brigade was split into two groups one 'facing the end of the ridge of redoubts [east]' and the other

'placed in somewhat of a hollow and facing more to the left front [north], looking towards the lower ground that separated us from the Woronzoff road, as it wound down from the old telegraph station on the plateau.'[5]

It is clear from both these descriptions that visibility was strictly limited, with one half of the Brigade staring at the eastern slope of No. 6 Redoubt hill and the other half at the Woronzoff road where it descended the escarpment. The Light Brigade could neither look towards, or be seen from, the east. These statements pinpoint the position taken up and make it apparent why Rijov's cavalry force was later able to use the North Valley unseen by the Light Brigade.

The Heavy Brigade was deployed a little further south with its freedom of movement to the south-east hampered by a vineyard and the divisional camps.

Until then Lucan had acted independently and competently, cooperating with and consulting Campbell. He had gone forward personally; he had alerted his Commander-in-Chief; he had advanced his division in echelon, with the Light Brigade supporting the Heavy, up into the shadow of Canrobert's Hill; he had brought his artillery into action at the outset; he had demonstrated as boldly as was possible against the enemy; and he had, with reluctance, skilfully retired the cavalry to a position from which he could quickly fall on the flank of a direct advance on Campbell's position. All this had been achieved with minimal casualties. There was little for Raglan to take exception to. That Lucan was exasperated, confused and resentful of Wetherall's appearance and the brief but garbled message is not surprising. He was, he believed, being interfered with again without justification. Once more he was not to be allowed any latitude in his handling of the cavalry. Lucan now found himself in a position selected by somebody who could not visualize the ground accurately from where he stood, and dangerously divided by an enormous gap from the infantry with whom he should cooperate.

The Light Brigade had been concealed in what amounted to an ambush position. They could not see, or be seen by, enemy in the North Valley until that enemy was almost level with their location. Perhaps this was part of Raglan's intentions. At any rate he had a right to assume that his cavalry commanders would implement basic, routine security precautions. In other words, if the main body of a cavalry division or brigade could not see what was happening on the other side of a hill the deployment of scouts would keep commanders informed. From the Causeway Heights, and particularly from the eastern slope of No. 6 Redoubt, an excellent view of most of the North Valley was guaranteed. Seemingly, no such scouts were deployed as, when Rijov's cavalry debouched into the North Valley and moved across and up it, not a single British soldier east of the escarpment was aware of it.

Hardly had the Cavalry division taken up its new position when Raglan changed his mind. He had been peering anxiously at the Turks through his telescope and had seen the redoubts fall one after the other; now he was concerned that the Turks with Campbell would follow suit. Fear and panic are infectious. If a few men retreat others are likely to follow, possibly resulting in an unstoppable rush to the rear. Lucan's division was now a long way (2,000 metres) from Campbell; to the Turks the British cavalry's withdrawal would enhance their feeling of abandonment. It was annoying; Lucan had been right to start with. A compromise was called for. The Light Brigade could stay where it was and he would send most of the Heavies back to Campbell's area (see Map 9). He lowered his telescope and turned to Captain Hardinge, the next orderly officer for duty, to give his second order. It read:

'Eight squadrons of Heavy Dragoons to be detached towards Balaclava to support the Turks who are wavering.'

Hardinge galloped away at about 8.30 am. When Raglan sent this order the North Valley was clear of large formations of Russian cavalry. Raglan was justifiably worried that the enemy on Causeway Heights would press on, that the Turks would rout completely and that Campbell would be swamped. If Rijov's cavalry brigade was, at that moment, moving into the North Valley Raglan's second order would have had something more to focus on than wavering Turks.[6]

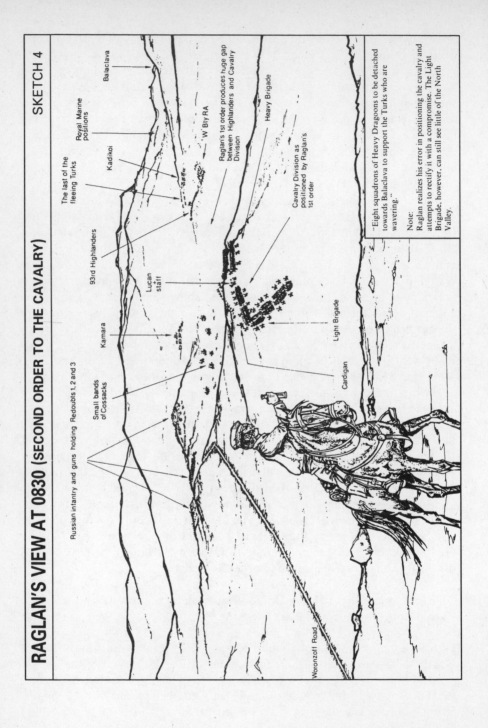

RAGLAN'S VIEW AT 0830 (SECOND ORDER TO THE CAVALRY)

SKETCH 4

Balaclava

Royal Marine positions

The last of the fleeing Turks

Kadikoi

'W' Bty RA

Raglan's 1st order produces huge gap between Highlanders and Cavalry Division

Heavy Brigade

93rd Highlanders

Cavalry Division as positioned by Raglan's 1st order

Lucan + staff

Kamara

Light Brigade

Small bands of Cossacks

Russian infantry and guns holding Redoubts 1, 2 and 3

Cardigan

Woronzoff Road

"Eight squadrons of Heavy Dragoons to be detached towards Balaclava to support the Turks who are wavering..."

Note:
Raglan realizes his error in positioning the cavalry and attempts to rectify it with a compromise. The Light Brigade, however, can still see little of the North Valley.

For Lucan this second order, arriving within a matter of minutes after he had taken up his new positions, was the final blow. He was in command of the cavalry in name only. Raglan issued meaningless orders, placed his brigades out of immediate reach of the troops he was supposed to be supporting, then countermanded them and sent, not all his division, but half his division back. Impossible to fathom was why eight squadrons? Why keep two squadrons of Heavies with the Light Brigade? Once these instructions had been fulfilled his command would be split by over a mile. Seemingly Lucan resolved to do exactly as he was told; it was pointless to argue. If Raglan wanted eight squadrons, not six, not ten, he would have them. Brigadier-General Scarlett was told to move his men back towards the Balaclava area.

* * *

The Tractir Bridge was a substantial stone structure with two large arches. It had strategic and tactical significance for both sides. It carried the road from Balaclava across the Chernaya River to join the main road to Simpheropol near Mackenzie's Farm. The British and French armies had used this bridge on their flank march to Balaclava and now the bulk of the Russians intended to use it for their advance on the same place. The Chernaya was the only obstacle lying along the exposed flank of the British base and their routes forward to the siege lines. The Tractir Bridge was a mere 7,000 metres from Balaclava harbour. There were fords across the Chernaya but, while infantry and cavalry could use them with comparative ease, guns and wagons much prefer, indeed probably require, bridges. Given the option soldiers will always keep their feet dry; given a convenient, unguarded bridge they will use it.

In the damp, chilly darkness before dawn Skiuderi's leading company had approached the bridge with considerable caution. It was the obvious place for a British cavalry picket. If nobody was at the bridge then surely there would be vedettes near where the Fedioukine hills on the other side came down almost to the road itself. Not a shot was fired. The Russians' night march over the river was undetected. The Kamara picket had been surprised, the Tractir picket did not exist. Greatly encouraged, Liprandi arrived at the bridge to watch his troops crossing. It became something of a review as he greeted the passing regiments with the traditional, '*Zdorovo!*' (Good luck to you!) to which the response was the somewhat noisy chorus of '*Zdraviia zhelaem!*' (And the same to you!). After the approach of some squadrons of Rijov's cavalry Liprandi disappeared southwards to offer words of encouragement to the troops shortly to attack the redoubts.

Major-General Rijov was in his mid-sixties. He was proud to have fought in the Napoleonic wars and to have his present command of the 6th Hussar Brigade. His opposite number in the British Army was Cardigan. Both men commanded light cavalry brigades, although Rijov's force was at least three times the strength of Cardigan's. The Russian brigade was composed of three regiments. The strongest was the 11th Kiev Hussars with eight

99

squadrons, then the 12th Ingermanland Hussars with six and the 1st Ural Cossacks, also with six. Twenty squadrons, with allowances for campaign wastage, meant about 2,000 men – not quite the enormous force that some writers have described, but having a comfortable superiority of around 500 over Lucan's two-brigade division. They were armed with the sabre and carbine, with the Cossacks (and Ulhans) having the lance as their primary weapon. Marching with the brigade were the eight 6-pounders of the 12th Light Horse Battery, and the eight guns destined to confront Cardigan's command some four hours later – the No. 3 Battery of the Don Cossacks.

The Russian cavalry force was the largest in the world. A huge mounted army of sixty regiments with a paper strength of over 50,000 men. Unfortunately the regular cavalry (discounting the Cossacks) suffered from the same drawbacks as the infantry. They looked magnificent on parade, they wheeled with perfect symmetry at reviews, where even trotting was frowned on, but tended to come apart if required to undertake prolonged periods of serious soldiering. The sleek, overgroomed and overfed horses quickly deteriorated when overloaded and underfed on campaign. Lucan had fought alongside them against the Turks 25 years earlier. His opinion of the Russian cavalry was not high. He described them as 'being as bad as could be, but the Cossacks could be damnably troublesome to an enemy, especially in a retreat.' The Turks who fled from the redoubts would probably have agreed.

The Cossacks were unique. They were recruited from the settler hosts that had guarded the frontiers of Old Russia, with special terms of service and pay scales. Military duty was a way of life; to be refused acceptance by the Army was to be disgraced. The Cossack's father provided his son with horse, saddlery, uniform, sabre and lance; only his carbine came from the government. He was a frontiersman – cautious, cunning, not amenable to rigid formal discipline, self-reliant and tough, all the attributes of a soldier required for scouting and to harass, hinder and delay an enemy. The Cossack wore what he pleased, seldom groomed his scruffy little pony which was broken in from the wild steppe herds, and shared with his mount an extraordinary endurance. Their method of fighting resembled that of the wolf pack, never charging home on a strong adversary but always hanging on his flanks waiting for an unguarded moment, a sign of weakness or a wound, and then dashing in for an easy kill.

At about 8.45 am Rijov's force started to arrive in the North Valley and change from column into an attack formation more suited to the open ground. Three *sotnias* of the 53rd Don Cossack Regiment joined the brigade at this time. Adopting a rectangular formation with a frontage of some three squadrons (200 metres) the whole force moved at a trot across the North Valley towards No. 4 Redoubt. Rijov's route after crossing the bridge had been simply to follow the road. He moved not so much up the valley but diagonally across it. It made sense to use the road as it led exactly where Rijov wished to go – up on to the Causeway Heights on the flank of the Russian infantry. In more modern military jargon he used the road as his axis. Once in the vicinity of the causeway crossroads he would be perfectly placed either to push down the slope towards Balaclava or to protect the infantry's flank from the British cavalry.[7]

At about the same time 'C' Troop RHA with its 9-pounders was moving fast across the South Valley from The Col. Brandling, its commander, distinctly made out a party of seven or eight enemy horsemen trotting along the skyline between Redoubts 2 and 3. They were obviously a commander and his staff. The likelihood is that this was Liprandi moving forward to assess the situation and decide how best to implement phase 2. What he saw was encouraging. Russian guns from the Causeway Heights were firing spasmodically at long-range targets presented by some Turks and a handful of Highlanders east of Kadikoi, while the British cavalry had pulled right back under the Sapoune escarpment. Viewed from No. 3 Redoubt they had pulled back almost to the point of invisibility, with the Light Brigade completely out of sight. Liprandi sent a galloper to find Rijov; the door to Balaclava's backyard was ajar, if not wide open.

The climax of the Battle of Balaclava was fast approaching. Thus far the Russians had won half a victory. One more heave was needed to complete it, then the British would be facing disaster. It is doubtful if Rijov (although Raglan should have) realized the full significance of the next half an hour on the outcome of the battle, the siege of Sevastopol, or indeed the war. These moments are seldom recognizable by participants at the time. Afterwards, with hindsight, the military historian can often see the critical few minutes in an action when resounding victory or devastating defeat hang perilously in the balance, when a slight push by somebody, somewhere, on either side is all that is required to tip the scales. Such a moment was when Rijov's leading squadrons pulled up the slope towards the causeway crossroads. It was a little before nine o'clock on a bright, clear morning.

An interesting issue that is avoided in many accounts of 25 October, 1854, is why Raglan never warned Lucan that Rijov's force was in the North Valley. He and his staff could see them, but not a single soldier in the South Valley could. The arrival of the Russian hussars on the ridge was about to come as an unpalatable shock to all troops to the south of them. Yet, seemingly, no messages were sent from headquarters, no special orders given. Not one of the surviving players mentions issuing, delivering or receiving any instructions or warning as to the large threat looming in the North Valley. One or two modern authors have got round this problem by slipping an unidentified ADC into their narrative to warn Lucan. Had such a message existed it is inconceivable that neither Raglan, Lucan, Campbell, Cardigan or Calthorpe would ever speak of it in the months and years of controversy and recriminations that followed that day.

There are two possible explanations. The first is that Raglan himself did not see the threat, that he had perhaps moved his position further south along the escarpment to get a better view of events in the South Valley. This possibility can be ruled out on the grounds that there is no evidence that he did move, and that, even if he had, Rijov's force would still have been visible. The height and contours are such that not even the knoll of No. 6 Redoubt blocks the line of sight from the Sapoune Heights.

The second involves timings. If Rijov had, as many accounts suggest, moved slowly up the North Valley from its eastern end and then, as it came level with No. 4 Redoubt, turned left (south) it would have been visible for a considerable time. A warning would undoubtedly have been sent. However, when Raglan issued his second order to assist the 'Turks who are wavering' the North Valley was clear. Hardinge rode off with this message at about 8.30. Allowing fifteen minutes for delivery (the cavalry were at the foot of the escarpment just below Raglan), the Heavy Brigade squadrons would be departing at about the time Rijov was coming into the North Valley from the Fedioukine hills. It could have been another five minutes or so before Raglan or his staff spotted the growing threat. By that time Rijov would have been trotting diagonally across the valley and Scarlett well on his way towards Campbell. A second ADC with a warning or fresh orders would arrive too late. Also Raglan, from his lofty viewpoint, may have forgotten that events that could be seen by him would not necessarily be so obvious (or even visible) to his subordinates 400 feet below. And anyway, surely Lucan would have scouts on the ridge!

Long after the battle Lucan gave an 'explanatory statement' to Kinglake which has frequently been misinterpreted. It reads:

'This advance of the Russian cavalry [Rijov] was no surprise, nor did I ever hear it so described. From the time that they descended into the valley they moved very slow, and should have been seen by General Scarlett when still one mile distant. I saw them before they crowned the heights, and found time to travel over double the extent of ground, and to halt, form, and dress the attacking line before it had traversed more than half the breadth of the valley.'[8]

What Lucan was referring to was Rijov's advance into the *South* Valley, not the North. 'This advance . . . was no surprise' means the advance down the slope into the South Valley, which was certainly no surprise – the Russians were poised on the ridge for just such a move; 'descended into the valley they moved very slow'; they did precisely that when they came down the slope towards Scarlett, whereas in the North Valley they had moved at a far brisker pace; 'should have been seen by Scarlett when still one mile distant'; from the top of the causeway ridge to Scarlett's squadrons was about 1,000 metres initially. As to Lucan's claim to have seen them before they crowned the heights, this may be correct in that he saw them moments before they reached the crest. He certainly had time to gallop to Scarlett yelling warnings and having his bugler sound off before the Heavy Brigade attacked, but how much the organizing of that attack was due to his efforts rather than Scarlett's is open to question – as will be seen.

102

*   *   *

The only British infantry regiment to have the battle honour 'Balaclava' woven in gold thread on its Colours is the Argyll and Sutherland Highlanders – formerly the 93rd Highlanders. Their action that day, when they stood in a two-deep line and saw off an attack by some of the Ingermanland Hussars, won them the epithet of 'The Thin Red Line'. Since then this has come to mean any action in which, in the face of overwhelming odds, a unit has stood alone and victorious in the defence of itself or some critical position. What actually happened that morning, on a small hillock a mile north of Balaclava, appears to fall some way short of the illustrious eminence that history has accorded it.

It is relevant in unravelling the circumstances of that day to appreciate that the attack on Campbell's Highlanders and that by Scarlett's Heavies took place within minutes of one another. They were almost simultaneous. A number of accounts, including one as fascinating as Cecil Woodham-Smith's in *The Reason Why*, have separated these two actions in time, and thus distorted the sequence of events. Campbell himself, in his official report to Raglan on 27 October, makes this point abundantly clear:

> 'One body of them [the Russian cavalry], amounting to about 400 men, turned to their left, separating themselves from those who attacked Lord Lucan's division, and charged the 93rd Highlanders.'

The first riders to reach the causeway crossroads area belonged to the 53rd Don Cossacks and the Ingermanland Hussars. Below them, but at that moment out of sight, were 550 Highlanders, 100 invalids scrambled from their sick beds in the town, Captain Barker's 'W' Battery of six 9-pounders, plus 1,000 'wavering' Turks. They were all that physically stood (or more accurately lay) between the Russians and the British base. Just moments before the appearance of the Russians Campbell's men had suffered a few casualties from long-range, ricocheting round shot from the causeway and he had consequently got his men to lie down on the reverse slope of his hillock. None of the Turks did the same, though some decamped. Still exposed to view, however, were Barker's guns.

Ewart, the officer who had had so much trouble with Cathcart earlier, spoke to a Russian officer at Simpheropol in 1856 who had charged the 93rd. The Russian stated that the object of the charge was to capture some guns that they could see near Kadikoi. These were Barker's battery, which seemed totally isolated and unprotected. On the Russians advancing, the 93rd, who had not previously been seen, stood up and opened fire. Finding the guns to be guarded by infantry the hussars abandoned their attempt. That was the Russian version. Rijov, seeing just six guns between him and Balaclava, immediately sent four squadrons of the Ingermanland Hussars to take them while his main force continued up on to the ridge near No. 4 Redoubt to locate the British cavalry. Rijov had seen what he thought to be a splendid opportunity.

As the Russians came forward Campbell had his line stand up and rode along the front saying, 'Men, remember there is no retreat from here. You must die where you stand.' Advancing towards them were 400 horsemen. The Russians were surprised by the sudden appearance of the infantry and, on their receiving a volley of musketry fire at long range, wheeled half left (SE). No saddles had been emptied by the Highlanders' fire, although it wounded a handful of horses and a few men. For a moment, as the Russians swept forward at speed, it looked as though their intention was to swing round the right flank of Campbell's force and head straight for the Balaclava gorge. Had they done so they had only to face the fire from the Royal Marine guns before reaching the town 2,000 metres away. Campbell over-estimated his adversary. Turning to his ADC he said of the Russian commander, 'Shadwell! That man understands his business.' The right hand grenadier company of the 93rd faced half right and fired another volley into the flank of the enemy. This was enough to force them to wheel away to the east and head for the redoubts, chased by shells from the RM Artillery.

It was all over in five minutes. Six hundred and fifty British infantry, supported by artillery and a few hundred Turks (not all fled), had seen off less than half their number of Russian cavalry without suffering a single casualty. No charge was pressed home on the Highlanders' position, their fire had inflicted minimal loss but the Russian hussars had no stomach for a serious fight. The enemy disappeared, not back the way they had come, but to the north-east, away from the rest of Rijov's force as it prepared to attack the Heavy Brigade. Such was the minor skirmish that was to give the British infantry in general, and the 93rd in particular, a lasting and legendary reputation for dauntless defence against the odds.

Brigadier-General Scarlett was the well-liked commander of the Heavy Brigade. His appearance, however, belied his good nature. He was the perfect model for a 'Colonel Blimp' caricature. Overweight and over fifty (he was 55) his face had that flushed, florid hue associated with an unstinting partiality for port. This was accentuated by the snowy whiteness of his walrus moustache and long fluffy sideburns. He was also notoriously shortsighted. He had formerly commanded the 5th Dragoon Guards and insisted on wearing one of their helmets in action, although he made his ADC (Elliot) wear a staff officer's cocked hat instead of a forage cap. As he had been only fourteen when Waterloo was fought he had missed the Napoleonic Wars. His active service experience since was zero. To compensate for his lack of dexterity with a sword he rode a huge horse of sixteen hands. Special responsibility for his security was given to his orderly-cum-minder Sergeant Shegog, an Irish giant with a powerful sword-arm. Scarlett, unlike his fellow senior commanders, acknowledged his limitations and had unofficially taken two experienced 'Indian' officers, rejected by other generals including Raglan, on to his staff. One was Colonel Beatson, the other Lieutenant Elliot, who had dropped in rank to come to the Crimea. The last member of Scarlett's personal entourage was the brigade trumpeter, Trumpet-Major Monks, from his old Regiment.

On receipt of Raglan's second order from Lucan, Scarlett had directed eight squadrons to march towards Campbell. Like the Light Brigade the 'Heavies' had five regiments each of two squadrons, with a total strength of around 800 sabres. The order of march was Inniskillings, Scots Greys, 5th Dragoon Guards and, at the rear, the 4th Dragoon Guards. The Royals were to stay in the old position. They had about a mile to go, and, as there was no enemy in sight and none had been reported, the march formation was less than tactically sound. Some regiments were in 'column of threes', others in 'open column of troops'. No scouts or 'flankers' were moving on the Causeway Heights to their left. It was a non-tactical move during which the horsemen had to negotiate a number of obstacles such as a vineyard and the tangle of collapsed tents and guy-ropes that were the cavalry camps.

By the time the leading squadron had reached the half-way point the Brigade was not only strung out over 800 metres but had split into two roughly parallel but separate columns. This had resulted from the leading Inniskilling squadron going right round a vineyard while the following one went left. Scarlett, with his command group, was riding on the left of the Brigade when Elliot happened to glance up the rise to his left (NE). There, bobbing up and down on the skyline, was a line of lance heads. As he looked the lances lengthened and horsemen appeared on the ridge. A shouted warning attracted Scarlett's attention. Peering in the direction his ADC was indicating the brigadier could make out little but a blur. It was a few seconds before he was convinced that the Russians had not only arrived but had caught him unawares, in a poor formation, in broken ground and that they had the advantage of the slope. His answer, the only one possible, was to get the Regiments to 'left wheel into line'. At about this time the crash of cannons firing from up on the Sapoune escarpment was heard.

It was this gunfire that had first alerted Mitchell, sitting with his troop with the Light Brigade, that something was happening in the North Valley. As he later wrote, 'Suddenly the French artillery on the plateau opened fire over our heads down the North Valley.'[9] It was at extreme range even for the heavier guns and unlikely to do much damage. Lucan, however, had spotted the enemy a few moments earlier. He galloped over to confront Cardigan who had just given I Troop RHA permission to advance slightly to engage Rijov's flank. There are two variations of what was said. According to Lucan his instructions to Cardigan were:[10]

'I am going to leave you. Well, you'll remember you are placed here by Lord Raglan himself for the defence of this position. My instructions to you are to attack anything and everything that shall come within your reach, but you will be careful of columns or squares of infantry.'

So saying, he turned his back on his brother-in-law, dug in his spurs and, followed closely by his staff, careered madly down the slope to warn Scarlett. Cardigan's recollection of what he was told to do was, inevitably, different:

105

'I had been ordered into a particular position by Lieutenant-General the Earl of Lucan, my superior officer, with orders on no account to leave it, and to defend it against any attack of the Russians; they did not however approach the position.'[11]

Both these versions were given after the event, when the two commanders were each anxious to divert the blame for the Light Brigade's inaction before, during and after the Heavies' attack on to the other.

If Rijov had just rolled down the hill as soon as he saw the thin, ragged line of the Heavy Brigade below him there can be little doubt he would have swept the British away by sheer weight of mass and impetus. Alternatively, had he brought up his guns from the rear and shelled the Heavies this might have aborted their attack. To his lasting discredit he did neither. Instead he hesitated, he advanced slowly, he halted momentarily, he continued at a walk, he stopped and fired carbines and pistols, then he came on again, but never at speed, never at a gallop. With five minutes of fumbling and indecision Rijov managed to snatch defeat, or something pretty close to it, from the jaws of victory. Those five minutes were all Scarlett required to organize his command and launch an attack himself.

Having got the first squadron of Inniskillings and the Greys into line Scarlett realized that the vineyard would prevent the 5th Dragoon Guards coming up abreast of them and joining the line. More space was needed – which meant moving further east. So the leading regiments turned right again and continued on before making a second 'left wheel into line'. It was while this was happening that Lucan and his staff came thundering up from behind. Because the Heavies were again moving east there was doubt that they had seen the threat. The Divisional Commander was yelling at the 5th Dragoon Guards to get into line and TSM Franks of that Regiment heard a staff officer shouting desperately, 'Scarlett! Scarlett! Look to your left'.

When Lucan found Scarlett he agreed that the only option was an immediate charge, later claiming that the decision and order to do so was his. As the senior officer on the spot the responsibility was Lucan's, but the reality was that he merely sanctioned what Scarlett was preparing to do. Lucan, however, quickly lost patience with the delay, ordering his trumpeter (Joy) to sound the 'Charge'. The call rang out clearly and urgently, twice – but nothing happened; it was ignored. The reason for this unprecedented incident was the insistence of the Inniskillings and the Greys in completing their dressing (parade ground alignment). As though awaiting a formal inspection, the officers sat in front of their men, with their backs to the enemy, while the sergeant-majors shouted and checked each squadron line for straightness. The soldiers shuffled their mounts, a little bit left, a little bit forward, a little bit back. When the RSM was satisfied, when the adjutant was happy, the second-in-command gave the command 'Eyes front!' The squadron officers and the Commanding Officer turned their horses to face the front. All eyes were once again focused on the enemy. The regiments were ready. Scarlett turned to Monks. The 'Charge' was

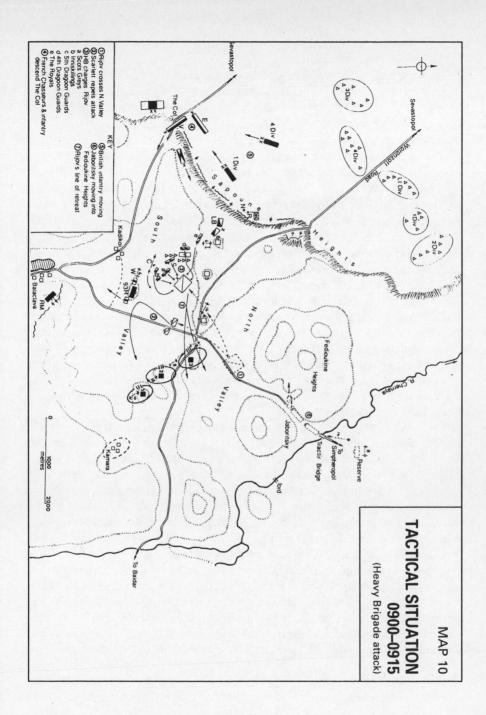

MAP 10

TACTICAL SITUATION
0900–0915
(Heavy Brigade attack)

KEY

① Rijov crosses N Valley
② Scarlett repels attack
③ HB charges Rijov
a Scots Greys
b Inniskillings
c 5th Dragoon Guards
d 4th Dragoon Guards
e The Royals
④ French Chasseurs & infantry
   descend The Col
⑤ British infantry moving
⑥ Jabovitsky moving into
   Fedoukine Heights
⑦ Rijov's line of retreat

sounded for the third time. Three squadrons of the Heavy Brigade moved forward.

It is not within the scope of this book to relate in detail what happened during the actual attack. More relevant is to point out a number of myths that have so far been accepted into history without much scrutiny. The impression is sometimes given of the Heavy Brigade, en masse, engaging vastly superior numbers in a full-blooded charge and routing them. There are a number of inaccuracies in this picture. Map 10 depicts what happened in outline.

Initially only three squadrons advanced, with Scarlett and his personal staff fifty metres ahead of them. The first squadron of the Inniskillings and the two of the Greys hit the Russians roughly in their centre. Then the second Inniskilling squadron came in on the enemy's left (eastern) flank. Next, the 5th Dragoon Guards disentangled themselves and joined the attack in the centre. They were followed by the 4th Dragoon Guards who had an exceedingly frustrating few minutes scrambling through not only a vineyard but also the wreckage of the Light Brigade camp. They moved round to a position opposite Rijov's right rear and went in from there. Finally, the Royals, who had been left out of the equation entirely, advanced on their own initiative and attacked the Russians' right front. It was a piecemeal affair; there was no coordinated brigade attack by the entire force acting together. Rather, regiments, and even squadrons, advanced when they were ready, when they reached an appropriate starting point. In practice this *ad hoc* attack worked well, with the enemy being struck from a variety of directions at different times. But this was more by good luck than good planning.

Then there is the question of numbers. Some authorities put Rijov's force as high as five times that of Scarlett's; most take the view that the Russians outnumbered the British by at least three to one. Both these estimates would seem to be exaggerations. Russian sources clearly put Rijov's strength at 20 squadrons. Allowing for campaign wastage, from which the Russians were no more immune than the British, a generous figure per squadron would be 100 men. So Rijov started with 2,000. He picked up a few Cossacks in the North Valley, but no account has them involved in a charge, or mêlée; they tried to avoid that sort of thing. What is usually forgotten is that four squadrons (400 horsemen) of the Ingermanland Hussars had attacked Campbell a few minutes earlier and had disappeared eastwards, never to be seen in the South Valley again. Therefore, when Rijov moved so painfully slowly down the slope he had some 1,600 men with him. Scarlett initially attacked with 300, but by the time the whole Brigade was committed perhaps 750–800 British cavalrymen were involved – odds of around 2:1 against Scarlett, not quite so devastating as some would make out.

Next, the 'charge' itself. This chapter has consistently referred to the Heavies' exploits as an 'attack' not a 'charge'. This is because no charge took place. The 'Charge' was sounded, but more in hope, and for encouragement, than in expectation. It was difficult for most riders to raise a trot, let alone a gallop. As any rider will know it is virtually impossible to get a horse to gallop up a slope. To do so from a standing start is out of the question. The military manuals describe the sounding of the 'Charge' without the preliminary calls of 'Walk', 'Trot' etc., as a 'direct gallop depart', something only attempted in extraordinary, if not dire, circumstances – and never uphill.

In this case it was not just the gradient that was against them but obstacles as well. Paget, who watched the whole affair from near the Light Brigade position, explained the situation clearly:

> 'Anyone who has ridden, or attempted to ride, over an old vineyard will appreciate the difficulties of moving among its tangled roots and briars, and its swampy holes. . . . The Heavy Brigade had only just time to scramble over the dry ditch that usually encircles the vineyards when they came in contact with their foes. . . . This has been called a charge! How inapt the word! The Russian cavalry came at a smart pace up to the edge of the vineyard, but the pace of the Heavy Brigade never could have exceeded eight miles an hour during their short advance across the vineyard. They had the appearance (to me) of just scrambling over and picking their way through the broken ground. . . . They [the Russians] stop! The Heavies struggle – flounder over the ditch and trot into them!'[12]

And from a participant, Major Forrest of the 4th Dragoon Guards (one of Cardigan's old enemies from 11th Hussar days), who wrote to his brother two days later:

> 'The Greys charged at a trot and our pace was but very little better, but we had very bad ground to advance over, first this vineyard and over two fences, brush and ditch and then thro' the camp of the 17th and we were scarcely formed when we attacked . . . we did not go in at as good a pace as we might have done.'[13]

Rijov has often stood accused of meeting the British attack at the halt. There is no disputing that he threw away the enormous advantage of being above his opponent, but whether his entire force was actually stationary at the moment of contact is not so certain. The Russians are also credited with halting and extending their flanks so as to envelop Scarlett's central squadrons. Observers in Brandling's 'C' Troop RHA cast considerable doubt on both these common contentions:

> 'At this moment the Russian cavalry halted, and, whether from unwieldiness or bad timing of the orders, it so happened that their central officers or leaders in front pulled up before those on the flanks, but only to the extent of from ten to twelve yards. The flanks, however, did not curve round . . . there was no such thing as prolonging the front ranks of the column, though it might have such an appearance from other parts of the field; indeed there was no time for it. . . . After a moment the

Russians halted again, and there were some pistol or carbine shots discharged from their central squadron . . . the Russians moved on, but their pace was so slow that it could hardly be called a trot. The first Squadron of the Greys now quickened their pace, and in a moment or two had passed into the space where the central squadron of the Russians had hung back.'[14]

Then there is the issue of whether it really was the Heavy Brigade attack that scattered the Russian cavalry, forcing their retreat into the North Valley. Could 800 men, trotting uphill into a force twice their size, inflict a resounding defeat on their opponents? The British cavalry were involved in a brief mêlée lasting no more than five minutes; there was no shock action, no impetus behind their attack, which arrived piecemeal. A series of individual fencing matches ensued, there was plenty of cutting, but precious little thrusting. The result, in terms of deaths or even empty saddles, was extremely modest. Some 40–50 Russians were knocked off their horses, or around one man in thirty. Enough to cause the other 1550 to decamp in haste? Perhaps, but doubtful.

The British losses were much lighter. The Heavy Brigade had one officer and nine men killed during the entire day, which included those occurring earlier from artillery fire from the redoubts, and later when they supported the Light Brigade. Forrest felt the whole affair was pretty bloodless:

'I don't know what the loss of the Brigade was but we [the 4th Dragoon Guards] had only one man killed and 5 wounded and I did not see more than two or three English on the ground. We had one horse killed and three wounded, there were about 20 Russians on the ground or perhaps rather more.'[15]

The reason was that a lot of hacking and slashing was taking place with blunt swords – on both sides. As illustrated by the quotation at the beginning of this chapter cuts seldom killed, whereas thrusts invariably did. Elliot, Scarlett's ADC, had good reason to be thankful for the neglect of the Russians to put an edge to their weapons. He was hit no less than fourteen times, several about the head and face, but survived them all to be listed as 'slightly wounded'. It is probable that his cocked hat attracted more than his fair share of the enemy's attention, while the old fellow in the brass hat (Scarlett) was far less conspicuous. The slightly better edges to the British swords seems to have been largely negated by the thick greatcoats worn by their adversaries. Cornet Fisher-Rowe of the 4th Dragoon Guards claimed that their swords 'jumped off like india-rubber'. Witnesses among 'C' Troop RHA who inspected the field immediately afterwards agreed:

'It was stated that they could make no impression on the Russians, owing to the thick overcoat they wore; but it is believed our men's swords were

110

not sufficiently sharp [no doubt Nolan would have felt vindicated had he lived], and as for those of the Russians no attempt whatever had been made to sharpen theirs. The few picked up on the ground were as blunt as could be.'[16]

So what finished off Rijov's brigade if it was not the impact of a vigorous charge that inflicted chaos and numerous casualties? The answer, seldom recorded in modern accounts, is artillery fire. Rijov's force was pushed back by the Heavy Brigade, was surprised at the audacity of what their British opponents had done, but they were not routed by them. Strenuous efforts were made by the Russian officers to rally their squadrons near the top of the ridge. Many of their men had not crossed swords with the enemy; they were disorganized but not defeated. Had they not, at that moment when the British attack had run out of steam, been subjected to sustained shellfire the Russians might well have reformed. Forrest admitted in one of his letters that it was guns not swords that won the day:

'The Russians stood a few minutes and then retired precipitously, *but reformed in good order on the top of a hill* [author's emphasis], but on getting one shot from our Horse Artillery away they went.'[17]

Forrest was over-generous to his gunner comrades. It took more than one shot. Colonel Whinyates' account is more plausible:

'The Russians, however, halted short of the ridge, and their officers could be seen holding up their swords and endeavouring to rally them, and get them into order, which they very soon would have done, but 'C' troop now came into action, and fired forty-nine shot and shell at them, at a range of between 700 and 800 yards, with admirable results, the 24-pounder howitzers making splendid practice.'[18]

Six guns firing eight shells each in two or three minutes into a compact target of horsemen did infinitely more damage than 800 blunt swords in a similar time. Rijov's force turned east and withdrew in some confusion, not back the way they had come but along the top of the South Valley, just under the crest of the Causeway Heights, for some distance before disappearing over into the North Valley to reform eventually at its eastern end. Their day was not yet done as it was there, about two hours later, that they would face the Light Brigade.

The Heavy Brigade and the gunners had certainly achieved something that most impartial observers would have thought impossible ten minutes earlier. The immediate menace to Balaclava had evaporated as quickly as it had appeared. But in truth the Heavies' success had only been possible because of the extraordinary hesitation and dilatoriness on the part of Rijov at the crucial moment. Forrest seems to have agreed:

'For my part I think the Heavies might have done much better. . . . Once in [the mêlée] we did better, but the confusion was worse than I had expected, the men of all regiments were mixed and we were a long time reforming. If we have to do it again I hope we shall do it better.'[19]

Up on the Sapoune Heights everyone was ecstatic. ADCs were sent flying with congratulations. Russell recorded the moment when a desperate situation had been redeemed against all expectations:

'It was a marvellous sight! There arose a great shout from the spectators. Curzon was sent down by Lord Raglan with a condensed eulogy "Well done!" Officers and men, French and English, all greatly elated, clapped their hands with delight.'[20]

Among the staff officers spurring away was Nolan. He was far from happy. Although pleased to see what cavalry could do he was fuming that the Light Brigade had not been involved. He was a light cavalryman, and to see them standing idly by a few hundred metres away watching the Heavies get the glory was too much for him to bear. From where he had sat it seemed that a once-in-a-lifetime opportunity had been thrown away. The Light Brigade had not attacked the Russians' flank or rear and, far worse, had allowed the enemy to retire unmolested. It was disgraceful. Those imbeciles Lucan and Cardigan were to blame! Some time later George Higginson, the adjutant of the Grenadier Guards (part of the 1st Division who were then near The Col), had a brief meeting with Nolan on his way back to headquarters:

'I met Captain Nolan riding up from the lower ground, apparently in search of the adjutant-general [Estcourt]. . . . I do not cast any reflection on his memory by recording the impression he gave me during the short conversation we held together, that under the stress of some great excitement he had lost self-command.'[21]

*   *   *

'My lord, are you not going to charge the flying enemy?'
    'No,' replied Cardigan, 'we have orders to remain here.'
    'But my lord, it is our positive duty to follow up this advantage.'
    'No, we must remain here.'

Captain Morris, the acting Commanding Officer of the 17th Lancers, seethed with frustration. He tried once more. 'Do, my lord, allow me to charge them with the 17th. See, my lord, they are in disorder!'
Cardigan's voice was edged with anger when he repeated for the third time,

'No, no, sir, we must not stir from here.'

Distraught and disgusted, Morris turned his horse away remarking to his officers nearby, 'Gentlemen, you are witness of my request.'

Private Wightman, in the front rank of his squadron, remembered seeing Morris talking 'very earnestly' to the Brigade Commander and heard Cardigan's 'hoarse, sharp' words, 'No, no, sir'. As Morris pulled up in front of the Regiment he was slapping his leg furiously with his sword scabbard and muttering, 'My God, my God, what a chance we are losing!'[22]

Even private soldiers in the Light Brigade were shocked at their inaction. Mitchell later wrote:

'All this time we were expecting an order to pursue, but no order came, and soon the opportunity was lost. We all felt certain that if we had been sent in pursuit we should have cut up many of them.'[23]

Undoubtedly a fleeting chance was missed; those watching from the escarpment thought so, many in the Light Brigade agreed, and Lucan was angry enough afterwards to send his son with a sharp rap over the knuckles for his uncle. Cardigan denied (predictably) ever receiving it, just as he denied that Morris ever urged him to attack. Its contents made it plain that he (Lucan) was disappointed that the Light Brigade had not supported the Heavies and

'he desired that Lord Cardigan would always remember that when he [Lucan] was attacking in front it was his [Cardigan's] duty to support him by a flank attack, and that Lord Cardigan might always depend upon receiving from him similar support.'[24]

These were prophetic words. Ironically, less than two hours later their roles would be reversed, with the Light Brigade undertaking a frontal attack with Lucan controlling the Heavy Brigade in support.

Cardigan was a cardboard soldier. Few looked finer on a review, few were more dedicated to drill, and fewer still were more fanatical over mess rules or the tightness of overalls. Put such a man, with no training and no active service experience, in an important command in battle and it is hardly surprising if he cannot recognize a tactical opportunity when it stares him in the face. In this case the opportunity was forcefully pointed out to him by an officer very junior in rank. Cardigan was arrogant, self-opinionated and despised all 'Indian' officers. For an experienced officer like Morris who, though a mere captain, had distinguished himself in several battles in India to have the gall to tell Cardigan what to do in front of the Brigade was insufferable. Cardigan chose to shelter behind those parts of his instructions that suited his mood – 'placed here by Lord Raglan . . . defence of this position' – and ignore the bit about attacking anything that came within reach. It was typical Cardigan. He was no coward and did not in any way shrink from combat, but he could be deliberately obstructive and he loathed his brother-in-law. So he was quite capable of riding around putting on a show of impatience, muttering about the 'Heavies' having the 'laugh on us' and in the next breath refusing to do anything about it. His motives were almost certainly an unpleasant potpourri of umbrage at being told what to do by a junior 'Indian' officer, sheer bloody-mindedness and the thought that Lucan would get the blame if things went wrong.

CHAPTER SIX
Two More Orders for the Cavalry

'There, my lord, is your enemy; there are your guns!'
Captain Nolan to Lord Lucan, when the latter had the
temerity to query the order just brought by him from
the Commander-in-Chief.

The four horsemen of calamity were about to come together in a welter of muddled messages, confusion, misinterpretation, insubordinate behaviour and personal animosity. The result of this lethal cocktail was the ruination of the Light Brigade. The answer as to how the Light Brigade was lost lies in the thinking and actions of these four men during the two hours from 9.30 to 11.30 that morning. But the way they thought, the way they reacted to each other in the time of crisis was the product of weeks and months of misunderstandings, pent-up frustrations and ill-concealed contempt. Raglan, the kind, gentlemanly general who was forever seeking to pour the proverbial oil on troubled waters, was saddled with subordinates whom he distrusted but dared not dismiss. Orders must be issued, delivered, received and executed. Each stage is of equal substance, and each is dependent on the other. If there is weakness at any stage an operation is put at risk; if there is weakness at every stage calamity is virtually assured.

It was about 9.30 that Raglan issued the third of his now infamous four orders; it was at 11.30 that the bloodied remnants of the Light Brigade staggered back to the western end of the North Valley. Not surprisingly, the series of events that preceded this gallant but ruinous charge (unlike the Heavy Brigade's attack, it was a proper charge) are alive with 'ifs'. If the infantry had not been so slow arriving; if Raglan had not been obsessed with the possibility of losing a few cannons; if he had issued clear orders; if he had not selected Nolan; if Airey had been a competent staff officer; if Nolan had not been so filled with misguided resentment and disrespect for the cavalry divisional commander; if Lucan had but paused to reflect rationally, and if Lucan and Cardigan had not loathed each other so passionately, the Light Brigade would never have done what it did. This lengthy list is far from exhaustive.

At a quarter past nine the elation in the Sapoune Heights was genuine and justified. What a few minutes before had looked like impending catastrophe had been transformed by the flight of Rijov's light horsemen. It was now incumbent on Raglan to make a decision and take the initiative. What he saw

115

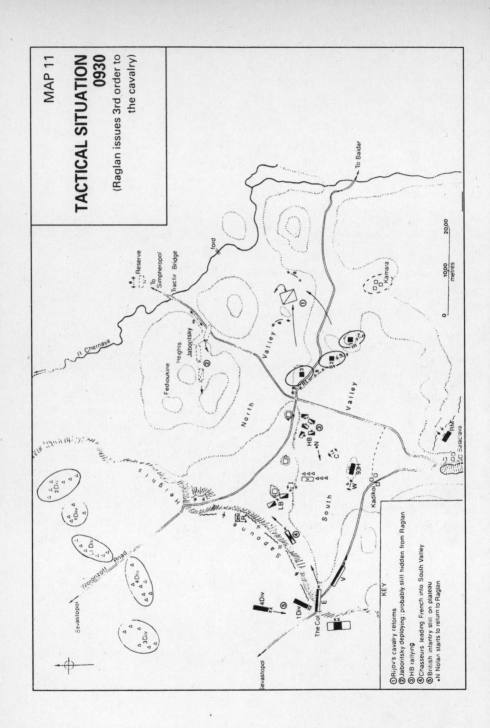

MAP 11

TACTICAL SITUATION
0930
(Raglan issues 3rd order to
the cavalry)

KEY

① Rijov's cavalry reforms
② Jabqritsky deploying : probably still hidden from Raglan
③ HB rallying
④ Chasseurs leading French into South Valley
•N Nolan starts to return to Raglan

To Baidar

R. Chernaya

ford

To
Simpheropol

Tractir Bridge

Reserve

Fedioukine
Heghts

Jabqritsky

North   Valley

Kamara

Valley

RMt

C. Balaclava

HB
N

C
W
93H
Kadikoi

South   Valley

Causeway Heights

LB

Sevastopol

2Div
1Div
1Div
4Div
3Div

Woronzov Road

Sevastopol

4Div
1Div
The Col
E
V
XX

0   1000   2000
metres

116

spread out far below him resembled the situation shown in Sketch 5, and on Map 11. To understand the 'how' and the 'why' of his orders the reader needs to see what Raglan saw, and think as Raglan thought when he made up his mind as to his next move. First, the seeing.

The general impression for the observer on the escarpment was that he was suspended high above a broad plain. Shallow valleys, low hills and ridges existed but they were flattened by height into much more gentle, sometimes imperceptible, undulations. Careful study through a glass was necessary to follow the line of the Causeway Heights, which seemed to blend deceptively at its eastern end with the much higher range of hills that filled the horizon. A glance at the modern photograph (No. 24) taken from near Raglan's position will help in understanding this circumstance.

Instinctively Raglan turned his telescope to search for the enemy. He was looking for movement, for the telltale flash of sunlight on metal, or the great giveaway of all nineteenth century battlefields – smoke. When he looked half-left into the middle distance he was gazing down on the patchwork of stunted scrub that covered the slopes of the Fedioukine Heights. They formed the northern boundary of the North Valley. The centre of these hills was 4,000 metres from where he sat which meant that much of the ground there was hidden from his view. He could not see the Tractir Bridge; he could not see the road that bisected the Heights, so the bulk of Jaboritsky's force was still invisible. Possibly some movement caught his attention as the screen of Don Cossacks with Jaboritsky pushed south-west, but it is doubtful. Another half an hour would pass before the Russian occupation of these Heights became obvious, even longer before guns were deployed.

Looking almost due east there seemed to be no such thing as the North Valley; the Fedioukine Heights and the Causeway Heights appeared to rise so gradually from the plain as to be a part of it, rather than define its limits. Straight ahead, at a distance of over 7,000 metres, the plain (valley) was blocked by a distinctive, rounded hill (which maps show as nearly 700 feet above sea level). In front of this hill Raglan could make out the 'dark mass' of Rijov's cavalry. There was movement; they appeared to be rallying after their defeat; they were certainly not disappearing from sight down towards the Chernaya River. Nearer still, in front of the cavalry, some guns were being brought forward and deployed across the valley facing roughly west.

With a fractional movement of the telescope to the right Raglan could study the redoubts. No. 4 was empty, abandoned shortly after being taken. Nos. 1 to 3 were, he knew, held by infantry and guns. He could just make out the infantry, and the guns nearby could be identified by the puffs of smoke when they fired. The rounded summit of Canrobert's Hill (No. 1 Redoubt) was easily distinguishable, although it was a good 5,000 metres distant. In line with it, and about half the distance away, was No. 4. The South Valley was devoid of enemy.

Next, he quickly identified his own troops in view. The Light Brigade was still tucked up, almost at his feet, under the escarpment near No. 6 Redoubt where he had put it. Its only movement had been a slight adjustment so that the leading ranks could better see the Heavies' action. Scarlett's brigade was reformed in the area close to where they had fought Rijov, on the upper slope of the Causeway Heights just south-east of No. 4 Redoubt. Here they were within range of the Russian artillery around and between Nos. 2 and 3. An occasional shot landed among them and Brandling's guns, which were close by. Campbell and his Highlanders were still in their old position. Of the French reinforcements he could see, as yet, nothing. The curve of the escarpment obscured the view of their descent via The Col on to the plain. Any troops, such as the 4th Chasseurs d'Afrique, that kept to the foot of the escarpment remained invisible to observers on the top.

Now the thinking. Raglan sensed the opportunity. It was clear that the enemy had suffered a serious setback; their morale should be shaky; forceful offensive action now could well result in the recapture of the redoubts. If that could be achieved his base would be completely secure and he would have every right to claim a splendid victory. What was needed was action – but by whom? It would have to be the cavalry initially, as there was still no sign of the infantry. Why they were taking so long he could not imagine. The 4th Division had only got 4,000 metres to march before they should be visible below him, but they were not there after an hour and a half. It was inexplicable and inexcusable.

His original plan had been to get the infantry to the Balaclava area to block the supposed thrust towards his base. Cavalry had always been a scarce commodity and its role was still to delay, but not to get embroiled in a major fight. As his young nephew, Calthorpe, put it perfectly:

> 'Lord Raglan's object was to place the cavalry in a position of safety, and
> at the same time prevent a general action coming on until the arrival of
> the 1st and 4th Divisions.'[1]

Tradition normally requires the cavalry to arrive at the last moment to save the day. Raglan had it the other way round. If Rijov had not been so hesitant at the crucial moment this plan would probably have failed. But it was now an outdated plan, overtaken by unexpected but welcome events. It had to be changed – at once.

He would use the cavalry to go forward, probe the redoubts and test the Russians' reaction to an advance; if they wavered or withdrew so much the better; the cavalry had the mobility to seize a chance. Infantry could well still be needed to mount an attack on the redoubts later if the enemy was not intimidated by the cavalry, but they would not arrive in time to take up the fleeting opportunity that Raglan felt might now have presented itself. His intention, if the infantry was needed, was to have an attack (or an advance) on two fronts. Cathcart's 4th Division would advance along the line of the Causeway Heights, rolling up the redoubts one after another from the east. That was the first front. The second was to be the Duke of Cambridge with the 1st (Guards)

Division. He would proceed across the South Valley towards Balaclava and, if necessary, support Cathcart with an attack from that direction. This was what Raglan wanted to happen; these were his intentions that had to be translated into action.

He lowered his telescope, turned to Airey and issued his third order to the cavalry. As the Commander-in-Chief was crippled, Airey wrote all his orders. It was approaching half past nine when an unknown ADC turned his horse's head to the lip of the escarpment and gingerly started the steep descent to find Lucan 400 feet below. Next, Wetherall was called over to be given the task of locating Cathcart and ensuring the 4th Division understood its new role. Five or ten minutes passed while Raglan got increasingly concerned at the non-appearance of the infantry. Why were they so late? Where exactly were they? Would Wetherall be able to chivvy them up and ensure they implemented the changed plan? Raglan decided somebody more senior might be needed; he would send his quartermaster-general. Airey cantered off along the top of the escarpment towards The Col.

The 1st Division had overtaken the 4th in the march for The Col. The Guards Brigade of three battalions (Grenadier, Coldstream and Scots Fusilier), and the two (42nd and 79th) of the Highland Brigade had crossed the Woronzoff road and headed across country over the Sapoune Heights. They came close to the edge of the escarpment, following far behind, but in the tracks of, 'C' Troop RHA. Weeks without proper washing had long since destroyed any parade-ground polish. The once-bright scarlet jackets were dark with dirt and sweat stains. Morale, however, was given a welcome boost when many were treated to a grandstand view of the Heavy Brigade action. Colonel Tower of the Guards noted in his diary:

'The Heavy cavalry charge was just going on as we came in sight of the Turkish redoubts; we could distinctly see the grey horses and bearskin caps [the Scots Greys] swallowed up in a dense mass of grey-coated Russians, their sabres flashing in the sun.'[2]

A short while afterwards, possibly as the division reached the top of The Col, the exasperated Nolan had ridden by and had his brief, angry words with the Grenadiers' adjutant (Higginson). By ten o'clock the Duke of Cambridge's division was starting out across the South Valley towards Kadikoi. In a little under two hours it had covered four miles – nothing to boast about, but quite good going in comparison with Cathcart.

Ewart, the young ADC who had so courageously refused to be browbeaten, albeit politely, by the divisional commander at the outset, was still with the 4th Division. His continued presence, however, had failed to add momentum to the marching pace of the seven weak battalions that made up the division. By ten o'clock it was half an hour behind the 1st, despite having a far shorter distance to go. This was particularly regrettable as all along Raglan had in mind that the 4th should play the leading role in securing Balaclava, and now

119

in spearheading any infantry attack necessary to retake the redoubts. It was intended to be the first British infantry on the scene, not the last – a not unreasonable calculation considering its camp was a mere two miles from The Col. It was Cathcart's lack of enthusiasm that had wrecked this estimation.

By about 10.20 am the leading battalions of the 4th Division were at last debouching into the South Valley. A mile ahead, but as yet out of sight, the Guards and Highlanders were deploying north of Kadikoi village. Immediately in front, on the line of march, was a large vineyard at which the track forked. Ewart's original task had been to guide Cathcart to Campbell and, as this instruction had not been countermanded, the right-hand fork was chosen as this route led to Kadikoi and then to Balaclava. At this moment 'Captain Wetherall, the DAQMG, galloped up calling out, "You are going the wrong way!" ' The column halted. At first Ewart was inclined to believe that they should push on in accordance with Raglan's original orders, but Wetherall insisted that he carried fresh orders for the division to swing left of the vineyard in the direction of the nearest redoubts. As they debated this new prospect Airey appeared. He emphatically confirmed the change of plan. Speaking to Cathcart he said, 'Lord Raglan wishes you to advance immediately and recapture the redoubts from the Russians.' Turning to Ewart he added, 'Captain Ewart, you are acquainted with the position of each redoubt. Remain with Sir George and show him where they are.'[3] Airey rode back the way he had come. By 10.30 the 4th Division was moving, in no great haste, towards the Causeway Heights.

Cathcart still refused to be rushed. He was decidedly hostile to the idea that his division, straight from the trenches and exhausted, should be chosen to undertake the leading role in a potentially hard fight to regain the redoubts while the Guards (in his view) hung around and watched. His men were being overloaded. He would continue to be difficult. When Ewart pointed out the line of redoubts Cathcart immediately countered, 'You must be mistaken.' There was no error, so Ewart tried again. This time, after Canrobert's Hill had been carefully indicated, Sir George muttered, 'It's impossible that there can be one as far away as that.' Ewart patiently explained that he knew them all, that he had surveyed them and been inside every one. The truculent Divisional Commander was determined to have the last word. 'Well,' he said, 'it's the most extraordinary thing I ever saw, for the position is more extensive than that occupied by the Duke of Wellington's army at Waterloo!', at which battle he had been present.[4]

By eleven o'clock, or thereabouts, the 4th Division was moving slowly towards No. 6 Redoubt behind a line of skirmishers. It was found to be empty so Cathcart left some 'men of a red regiment' in it, and looked towards No. 5 which had, by this time, been occupied by some enterprising Turks. To his right front, and facing east, was the 1st Division drawn up in two lines; to his right rear, back the way he had come, were the two French infantry brigades that had preceded him into the South Valley. The Chasseurs d'Afrique had disappeared along the foot of the escarpment into the North

Valley (see Map 12). From close to ten o'clock until eleven o'clock, while the infantry had, at long last, been deploying in the South Valley, the cavalry had achieved nothing.

It was about 9.50 when Lucan was handed Raglan's order. It read:

> 'Cavalry to advance and take advantage of any opportunity to recover the Heights. They will be supported by infantry which have been ordered. Advance on two fronts.'

This version was supposedly the one Lucan actually received, but as, in the circumstances, the last two sentences verge on gibberish this writer will give Airey the credit for at least getting the punctuation right and consider the second version. This was marginally more intelligible:

> 'Cavalry to advance and take advantage of any opportunity to recover the Heights. They will be supported by infantry which have been ordered to advance on two fronts.'

Lucan's interpretation of this was quite simple, and precisely the opposite of what Raglan intended. He was to wait for the infantry; when they arrived there would be a combined advance to retake the redoubts. No doubt somebody would elaborate on what was meant by 'two fronts' in due course. Meanwhile he would mount the division, bring forward the Light Brigade into the North Valley – and wait. As he was to put in his official report to the Commander-in-Chief two days later, 'The division took up a position with a view of supporting an attack upon the heights.' Lucan's detractors and Raglan apologists at the time, and ever since, have roundly condemned Lucan for not rushing off with his brigades towards the line of redoubts – straight at the very infantry and guns that had forced his earlier retirement. That was what Raglan wanted because he felt that a purposeful advance at once might induce the abandonment of the redoubts. There was little he could see or know at the time that would indicate this would happen. Nevertheless he had thought it worth trying.

The ADC, whoever he was, had not been briefed. In response to any queries from Lucan no attempt was made to alter Lucan's interpretation of the third order. Perhaps he had not been given verbal instructions, or he had failed to ask for them; either way the staff system was not working well. Solely from what he read how could the Cavalry Divisional Commander be expected to realize that he was to advance against the enemy without infantry support? Two additional words would have made all the difference. Insert 'immediately' after 'advance' in the first sentence and 'later' after 'advance' in the second, and Lucan would have found it impossible to turn the meaning upside-down.

To Lucan the message, as written, could never have conveyed what Raglan really wanted. Cavalry never advanced uphill, unsupported against fixed positions defended by infantry and guns; shellfire from the redoubts had

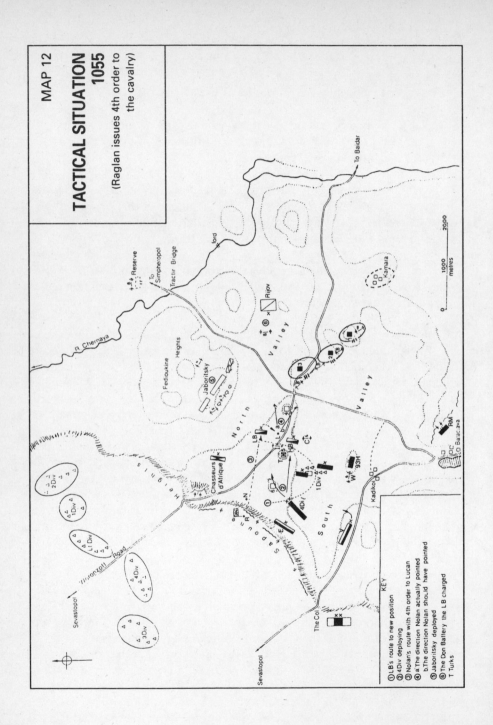

MAP 12

TACTICAL SITUATION 1055

(Raglan issues 4th order to the cavalry)

KEY

1 LB's route to new position
2 4Div deploying
3 Nolan's route with 4th order to Lucan
  a The direction Nolan actually pointed
  b The direction Nolan should have pointed
4 Jaboritsky deployed
5 The Don Battery the LB charged
T Turks

0   1000   2000
metres

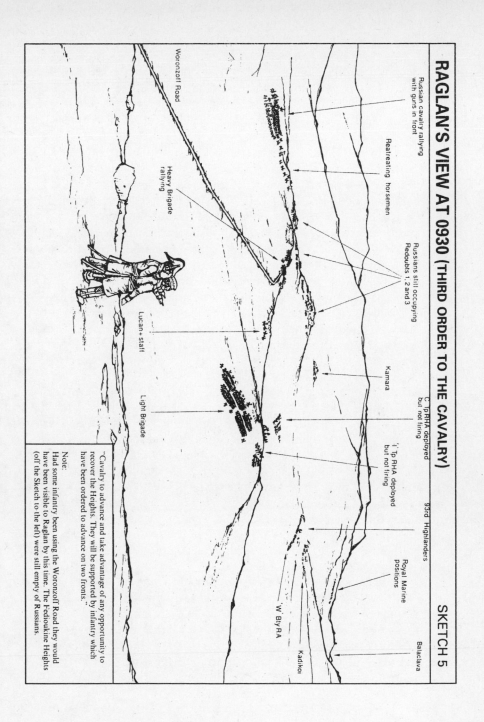

RAGLAN'S VIEW AT 0930 (THIRD ORDER TO THE CAVALRY)

SKETCH 5

Russian cavalry rallying
with guns in front

Woronzoff Road

Retreating horsemen

Heavy Brigade
rallying

Russians still occupying
Redoubts 1, 2 and 3

Lucan + staff

Light Brigade

Kamara

'C' Tp RHA deployed
but not firing

'I' Tp RHA deployed
but not firing

93rd Highlanders

Royal Marine
positions

'W' Bty RA

Kadikoi

Balaclava

"Cavalry to advance and take advantage of any opportunity to
recover the Heights. They will be supported by infantry which
have been ordered to advance on two fronts."

Note:
Had some infantry been using the Woronzoff Road they would
have been visible to Raglan by this time. The Fedioukine Heights
(off the Sketch to the left) were still empty of Russians.

123

already proved too much for the Cavalry Division; there was not even any urgency in the wording. To advance in these circumstances without infantry support would be totally counter to Raglan's policy of reining in the cavalry since the landings in the Crimea. Time and again, at the Bulganek, at and after the Alma, and as recently as that very morning, Raglan's concern to husband his horsemen had led to Lucan being restrained, recalled, pulled back and forbidden to engage the enemy.

The second part of the order was certainly open to the interpretation that the advance should not start without infantry. If it did how could the cavalry get the infantry's support? The phrase, 'they *will* be supported by infantry' implied that without it the advance should not go ahead. Logically from this, if there was no infantry around but they had been ordered to participate, the answer must be to wait for them. Three days later, in his despatch on the battle to the Duke of Newcastle, Raglan admitted that any retaking of the redoubts was to be a combined cavalry/infantry operation when he wrote, 'I directed the cavalry, supported by the fourth division, under Lieutenant-General Sir George Cathcart, to move forward and take advantage of any opportunity to regain the heights.' The message was fatally flawed. It had been issued by a man unable to convey his intentions accurately, written by a man without staff training or field experience, delivered by a man without knowledge of its implications and received by a man who had never before been permitted the slightest degree of initiative with his division.

When he first got the order Lucan was probably sitting somewhere near No. 5 Redoubt. By ten o'clock, if he looked towards The Col as he must surely have done countless times, he would have seen the 1st Division beginning to arrive in the South Valley. The infantry were coming. What more natural than to wait for them? After all they were supposed to support him, but they could not do that until they had joined him. It was as simple as that. As time passed, and more and more infantry arrived, with the 4th Division moving in his direction, Lucan saw less and less reason to do anything other than wait for them. He had no idea that with every passing minute his Commander-in-Chief's mood was deteriorating, and that what little confidence he now had in Lucan was all but gone. Those near Raglan noticed that his empty sleeve was twitching, a rare but ominous sign.

With an advance in the offing the Light Brigade had to be brought forward in line with the Heavies. The division would form up astride the Causeway Heights with the Light Brigade in the North Valley facing east, and the Heavy Brigade on the ridge near Redoubt No. 4 and in the South Valley (see Map 12). From this position they would be ready to advance once Lucan had coordinated details with the infantry commanders. Private Mitchell described how it was done:

'We were soon moved by "Threes from the right" down to the bottom of the vineyard and through an opening to the left which brought us through our camp, and bearing up towards the ridge, crossed it, and were

124

formed facing down the next plain, or valley. . . . We could see along the ridge several dead Russians lying, and here and there a Cossack horse, all of which had been killed in the attack on No. 4 Redoubt. . . . We were now dismounted, and soon we could see the enemy had placed a number of guns across the lower part of the valley nearly a mile and a half from us. At the same time a field battery ascended a hill on our left front [the Fedioukine Heights], where it was placed in a position facing us. They also placed a field battery on the slope between the redoubts and the valley on our right.'[5]

After a while the Light Brigade relaxed. It was a bright sunny morning and those able to do so ate some breakfast, while many smoked pipes. The infantry were a long time coming.

*　　*　　*

It was the fourth order that triggered the Light Brigade's charge. During the 15 minutes from the moment it was written, at a few minutes before eleven o'clock, until the Light Brigade moved off at about ten past, circumstances and personalities were fused together in such a way as to make a tragic and costly error almost inevitable. It is important to try to place the reader in the position of each horseman at the time it was issued in order to understand his mood and his thinking.

Raglan had been chafing at the non-appearance of the infantry, Cathcart especially, all morning. Now he turned his impatience on the cavalry. He had watched as Lucan mounted the division before moving the Light Brigade forward and positioning it in the North Valley in two lines. This was encouraging. Then, to his amazement, he could see the men dismounting; some sat down, some actually started eating! What was that fool Lucan doing? He was living up to his nickname with a vengeance. It was imperative for the cavalry to advance. Ten minutes passed, then twenty, half an hour, then, unbelievably, 45 minutes and still the cavalry were idle. During this time the infantry finally emerged into view and with painful slowness moved eastwards into the South Valley. The French Chasseurs positioned themselves down below him near the head of the North Valley. Even some Turks emerged from Balaclava and took possession of No. 5 Redoubt. There also was a significant change in the Russian deployment. Over on the Fedioukine Heights Raglan could pick out infantry and guns positioning themselves within what looked to be artillery range of the Light Brigade. But Lucan did nothing.

The brief opportunity that Raglan considered he had instructed the cavalry to seize had all but vanished. The scene below him had changed for the worse while he watched. The Russian cavalry's defeat had not been followed up (which must surely boost their morale); the enemy infantry and guns on the eastern end of the Causeway Heights were in no way threatened or even under fire; there was artillery across the North Valley, while Rijov had rallied and been given time to reorganize; now more enemy had appeared in substantial numbers on the Fedioukine Heights (see Sketch 6 and Map 12). Lucan had a lot to answer for.

Close by was a familiar and distinctive figure. Always conspicuous, ostentatious even, with his red and gold forage cap and tiger-skin saddle cloth, Nolan was waiting for his next duty. Yet again he had been a mere spectator in a battle that had seen the cavalry in action; yet again he had found cause to curse Lucan and give vent to his opinion that he was an incompetent, dangerous ass. Why had Lucan not sent the Light Brigade into the flank or rear of the Russian cavalry? Why had Cardigan just stood and watched? These were certainly the questions that repeated themselves over and over again as Nolan mentally raged against 'Lord Look-on' and the 'Noble Yachtsman'. This Irish-Italian, reckless, foolhardy but courageous officer was in an agony of frustration and eagerness. Paget's description of him has a realistic ring to it:

'An officer named Captain Nolan, who writes books, and was a great man in his own estimation, and who had already been talking very loud against the cavalry, his own branch of the service, and especially Lucan.'[6]

This man was to deliver the order.

Lucan was still waiting to implement the third order. He and his staff were positioned between the two brigades, probably on the ridge near No. 5 Redoubt. From there they could follow most of the progress of the infantry as they advanced towards them. The closer the infantry came the more convinced Lucan was that he should not move forward without them. Their very approach confirmed his interpretation of Raglan's instructions. By 10.30 the 1st Division was beginning to form up in two lines south of No. 5 Redoubt, the kilted Highlanders being only a few hundred metres away. Shortly after this some of Cathcart's men occupied No. 6 Redoubt; others seemed to be heading in his direction. By eleven o'clock Lucan must have been considering meeting his infantry opposite numbers to coordinate the 'advance on two fronts'. But he was about to receive another order.

Then there was Cardigan, sitting aloof and silent in front of the Light Brigade in ignorance of what had passed between the Sapoune Heights and his divisional commander. It had not been a good morning for him so far, and his brigade looked as though it would see no action. He had been late joining his command; he had been told to defend a piece of ground that had no significance and from which he could see little; he had been forced to watch and wait while the Heavies got the glory; he had felt obliged to stay put even when an insubordinate captain had embarrassed him in front of his staff and nearby officers; then he had had an impertinent and uncalled-for reprimand for doing as he was told by his witless brother-in-law. From where he sat on his chestnut charger, Ronald, he could see Russian guns well over a mile away down the North Valley to his front. He had watched as more guns and infantry deployed on the Fedioukine Heights, uncomfortably close, on the left front of his position. If they were to fire roundshot they could hit the Light Brigade at the second or third bounce. For the moment, however, they showed no sign of trying. Reluctantly, for he had had no chance to show off his flair and flamboyance, Cardigan considered the fighting finished. He was wrong. The fourth order's journey would end with him. He would have to execute it.

The situation as it might have appeared to Raglan when he issued the fourth order is depicted on Sketch 6 and Map 12. He was almost bursting with impatience at the cavalry's inaction below him when a member of his staff shouted out that the Russians were removing the guns from the redoubts. By this he meant the British guns lost by the Turks when they abandoned their positions earlier that morning. It was claimed that, through a glass, teams of horses could be seen coming forward with lasso tackle to drag the cannons away. With a distance of 4,000 metres between the closest redoubt (No. 3) and Raglan the accuracy of this observation must be highly questionable. The British pieces were inside the redoubts; if teams of horses were seen nearby they could just as easily be being used to adjust the positions of an enemy field battery. Nevertheless, the suggestion had been made. An unnamed officer's warning had lit a fuse at precisely the right moment in a highly charged atmosphere.

The thought of permanently losing guns to the enemy galvanized Raglan into action. A deep-rooted military tradition had it that to bring home enemy artillery pieces was proof of victory; it had been so since guns were invented. Conversely, to lose some was a serious stain on a general's reputation. It smacked of defeat.

The situation as seen through Raglan's eyes had become intolerable. Not only had Lucan failed to carry out his orders, but now, because he had done nothing for so long, the chance of recapturing the guns was likely to be lost for ever. A moment's thought and Raglan would have realized that there was absolutely nothing he could do in the time to save the guns if they were about to be removed. But his patience had snapped; he was not thinking rationally, only of spurring that blockhead Lucan into action. He spoke rapidly to Airey, who, resting a nondescript piece of paper on his sabretache, scribbled away in pencil, translating his Commander-in-Chief's words into the fourth order. When he had finished Raglan took it, read it and added the word 'immediate' at the bottom under Airey's signature. It said:

'Lord Raglan wishes the cavalry to advance rapidly to the front – follow the enemy and try to prevent the enemy carrying away the guns – Troop Horse Artillery may accompany – French cavalry is on your left – R Airey
    Immediate–'

It was just as sloppily constructed as the previous one, again reflecting Airey's inability to write accurate and unambiguous messages and Raglan's dislike of giving orders that sounded like orders instead of suggestions. He much preferred to let his wishes be known. The use of words such as 'wishes', 'try'

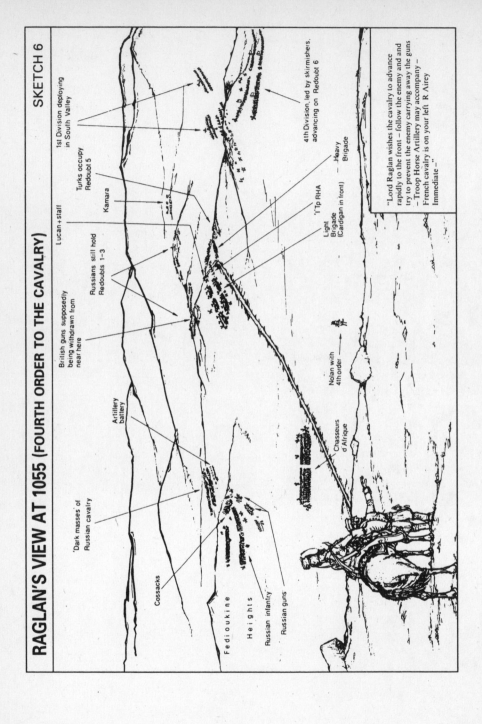

RAGLAN'S VIEW AT 1055 (FOURTH ORDER TO THE CAVALRY)

SKETCH 6

'Dark masses of Russian cavalry

Cossacks

Fedioukine Heights

Russian infantry

Russian guns

Artillery battery

British guns supposedly being withdrawn from near here

Russians still hold Redoubts 1–3

Lucan + staff

Turks occupy Redoubt 5

Kamara

1st Division deploying in South Valley

4th Division, led by skirmishers, advancing on Redoubt 6

Heavy Brigade

'I' Tp RHA

Light Brigade (Cardigan in front)

Nolan with 4th order

Chasseurs d'Afrique

"Lord Raglan wishes the cavalry to advance rapidly to the front – follow the enemy and and try to prevent the enemy carrying away the guns – Troop Horse Artillery may accompany – French cavalry is on your left R Airey

Immediate –"

128

and 'may' are seldom conducive to decisive military action. It remained to be seen what the recipient would make of it.

Calthorpe was the next ADC for duty and he made a move to take the paper from Airey but was waved aside by Raglan with the words, 'No, send Nolan!' Raglan was thinking of speed when he chose Nolan. He wanted the order delivered and implemented at once and Nolan was one of the finest, fastest and most accomplished riders in the Army. Whether he was temperamentally suited to taking such an order to Lucan, the man he had been openly reviling for so long, never entered Raglan's head. Everything was forgotten except the need for haste.

As the Commander-in-Chief's messenger Nolan was expected to understand the order he carried. He was supposed to be able to explain it, answer questions on it and to do his utmost to ensure it was complied with as his Commander intended. That was the theory. It needs to be emphasized again that the recipient of an ADC's order was obliged by military law to obey the instructions of the ADC as if the general himself was speaking. If in doubt he could, indeed should, question the ADC to clear up any possible misunderstandings, but in the end he had to obey to the best of his ability. On the battlefield there could be no other way.

Whether Nolan was briefed on the fourth order is a crucial question that most writers have avoided. Almost certainly he read the message, if only hurriedly, as this was normal routine procedure, but did he fully understand the thinking behind it, the objectives of the cavalry and the reason for mentioning the horse artillery and French cavalry? He would probably not have known the precise contents of the third order as he was not involved in its delivery, although he should have picked up the gist of it from observations, general conversation and activity at headquarters. Nolan was with Airey (and therefore close to Raglan) for much of the time; he heard many of the discussions of the senior officers as well as other ADCs. His view was exactly the same as Raglan's and everybody else up on the escarpment. In other words he undoubtedly made it his business to be well versed in what was happening at headquarters and why. It would be entirely out of character had he not done so.

If Nolan was properly briefed then it follows that there is a strong possibility that he deliberately misled Lucan, or at least failed to correct his obvious misunderstanding, after delivering the order. There is strong evidence that Nolan *was* well informed on the order, apart from what he would have heard by just being close by when it was conceived. (It is likely he heard Raglan issuing it to Airey for example.) Calthorpe, the officer who almost took the order himself and was on the spot watching and listening as Nolan took the piece of paper, in his *Letters from Headquarters*, written four years later, wrote:

'This order was entrusted to Captain Nolan, aide-de-camp to General Airey, a cavalry officer of great experience. Previous to his departure he received careful instructions from both Lord Raglan and the Quartermaster-General.'[7]

The picture emerges of Raglan insisting Nolan take the order, Airey handing it to him, Nolan hurriedly reading it while Airey makes sure he understands what is written. Then Raglan joins in, quickly emphasizing what he wants Lucan to do, i.e. advance at once and try to stop the guns in the redoubts from being removed. He may have added something along the lines that Lucan should take his own horse artillery and not to worry about his left flank as the French Chasseurs would protect it. Nolan is itching to be off and appears to understand what is required. All this takes but a minute or so and, at about five minutes short of eleven o'clock, the second horseman is released. As he spurs away Raglan ruins both his written and verbal instructions by calling out the words that seal the fate of the Light Brigade: 'Tell Lord Lucan the cavalry is to attack immediately.'[8] What wonderful words! Nolan has waited months to hear them. Now they have come from the mouth of the Commander-in-Chief and he, Nolan, is the instrument that will vitalize Lord 'Look-on' into action. Fate at last has rewarded him. The cavalry is to attack and he has been sent to ensure they do. And, by God, he will ride with them! Such must have been the thoughts that flew through Nolan's head as he forced his horse over the edge of the escarpment.

Nolan spurred his mount at breakneck speed down the almost sheer slope. Perhaps he followed a goat track, perhaps not, but his horse slithered on its haunches, stumbled over the rocks and crashed through the scrub. At any moment the watchers above expected to see man and horse fall and tumble to the bottom. After an endless minute Nolan reached the foot of the escarpment, his horse trembling with the effort. There followed a frantic gallop of a little over a mile to where the cavalry was resting. After careering through the Light Brigade lines, getting shouted directions as to Lucan's whereabouts and excitedly yelling, 'You'll see, you'll see,' in response to Morris' eager inquiry of 'What's going to happen?', Nolan skidded to a halt in front of the man he despised above all others. Lucan, with his knot of staff officers, was waiting on the Causeway Heights about halfway between Nos. 4 and 5 Redoubts and between his two brigades. It was a little after eleven.

Because Nolan was shortly to die, and was therefore never able to give his account of what he said and did, it is important to identify witnesses other than Lucan who saw or heard what happened in the next two minutes. Most of his staff were there at the time but the present writer has only been able to find two who actually recorded something of what occurred. They were Captain Walker, an ADC, and John Blunt, the civilian consular officer who was Lucan's Turkish interpreter. The latter's letters in particular give some fascinating details.[9] Nolan thrust Airey's note into Lucan's outstretched hand. It took but a few seconds to read. His face clouded with doubt and disbelief. Turning to Trumpet-Major Joy he quietly told him to 'mount the Division'. The call was sharp, clear, and insistent.

What was he being asked to do? This order seemed totally unconnected with the last and raised numerous queries. It appeared he must take his entire division and 'advance rapidly to the front'. But what about the infantry he was supposed to be waiting for? The 4th Division was within a few hundred metres of his position, clearly visible. Lucan had waited for 45 minutes or more for them; was he now being asked to dash off without them? And what was meant by front? Which front? Whose front? The previous instructions had spoken of two fronts, but now the latest order seemed to refer to the Cavalry's front. Well, from where he stood, if he looked towards the east (in the general direction of the enemy), his view was mostly blocked by the empty No. 4 Redoubt. That was directly to his 'front'. In front of the Light Brigade was the North Valley, stretching away down to a Russian battery with cavalry massed behind it. If the Heavy Brigade advanced to its front it would head into the South Valley towards the redoubts. It was confusing. It needed explanation.

So did 'follow the enemy.' What enemy? To 'follow' implied the Russians were withdrawing, but Lucan could see no indication of it. 'Try to prevent the enemy carrying away the guns.' Which guns? From where Lucan sat on his horse the only enemy guns he could possibly see were those over a mile away in the North Valley or across it on the Fedioukine Heights. Of course he knew that there were enemy guns on the Causeway Heights close to, and between, Redoubts 1–3 as they had engaged him periodically, causing some casualties, all morning. There were also the captured British guns inside those three redoubts. The message was less than specific as to which, but he was almost certain it must be the ones the Turks had lost. In his official report to Raglan two days afterwards Lucan confirmed this when he wrote that the fourth order was 'to prevent the enemy carrying away the guns lost by the Turkish troops in the morning'. Nevertheless, in either case he could see none of them. There was now no mention of the Heights being regained with infantry support; instead this part of the order seemed to mean the cavalry must try to stop guns in the redoubts being removed without it. Therefore his division was to advance uphill, alone, against entrenched infantry and artillery. Could Raglan really mean this? How does cavalry stop an enemy removing guns other than by attacking? Once again the whole idea seemed to run absolutely counter to every instruction he had received from Raglan since the landings. All these months he had been prevented from doing anything worthwhile with the cavalry; now, suddenly, Raglan seemed to want him to undertake a high-risk venture. Lucan was baffled; he needed clarification.

'Troop Horse Artillery may accompany.' Even this simple phrase gave rise to uncertainty. Shakespear's (formerly Maude's) RHA guns were under his command, in direct support of the Cavalry Division, so why was the word 'may' inserted. Perhaps Raglan was giving Lucan permission to use his own guns? Or did this mysterious 'may' mean that artillery support was dependent on the whim of Captain Shakespear, or Lucan – or chance?

Then, perhaps the most puzzling of all, the bald statement of fact, something that Lucan was well aware of as he could see them plainly – 'French cavalry is on your left'. So what? Are the Chasseurs d'Afrique involved in some way? Have they been told to cooperate in this advance? Am I supposed to ride over half a mile to them to find out? Am I supposed to coordinate my movements with theirs? These are some of the questions that must have raced through Lucan's mind as he considered the final part of the order. Then there was the last word, 'immediate'. If Lucan was to get French support it would take time and 'immediate' seemed to decisively rule this out. What a muddle!

Lucan looked up, bewildered, from the piece of paper in his hand to the staff officer beside him. In his own words he began to 'urge the uselessness of such an attack and the dangers attending it'. According to Blunt, who was standing nearby, Lucan did not lose his temper, and that what occurred at this stage was the beginning of a 'discussion' rather than an argument.[10] However, he had no time to put many questions as Nolan's response cut him short with startling abruptness. Lucan's querying the order was exactly what Nolan had both expected and wanted. It was typical of the Divisional Commander – always mishandling the cavalry, always neglecting an attacking opportunity. But now he, Nolan, had the whip hand. He would force this incompetent aristocrat into action. Burned into his brain were those parting words of Raglan's, words that even Lord 'Look-on' could not ignore. 'In a most authoritative tone', Nolan repeated those words, 'Lord Raglan's orders are that the cavalry are to attack immediately'. No writer describing this crucial encounter makes the additional point that Nolan, in all probability, also told Lucan he was to attack the guns. It is implicit in the general's startled response.

This made matters infinitely more perplexing for Lucan. There was nothing in the written message about 'attacking' anything, although he realized that to advance on the redoubts could only have an attack as an outcome, unless the enemy were so craven as to retreat at the mere sight of his moving on them. Now he was being told to attack at once. There was obviously not going to be any infantry support and no time to check on French involvement. The cavalry was to be thrown at the Russians unsupported. But attack what? He had to be certain of his objective. The only enemy he could actually see were on the Fedioukine Heights, and those at the eastern end of the North Valley, and the latter had never entered his head as having anything to do with the piece of paper still in his hand. With anger and bewilderment lending sharpness to his tone Lucan demanded, 'Attack, sir! Attack what? What guns?'

Nolan's insubordinate response and insolent gesture to these questions launched the Light Brigade into immortality. Throwing back his head, all thoughts of his briefing forgotten, Nolan flung out his arm and pointed, tauntingly, insultingly, not at the redoubts (which would have meant pointing into the South Valley), but down the North Valley at the battery drawn up across it over a mile distant. 'There, my lord, is your enemy, there are your guns!' It may have been at about this time that Cardigan's ADC rode up for instructions, having got his Brigade mounted. Lucan brusquely told him the Light Brigade was to attack down the North Valley. The staff officer raised the obvious problems such a move would involve. He was waved aside.

Many writers have since claimed that Nolan's gesture was general, vague, sweeping, mocking certainly, but never intended to be specific, never intended actually to point out an objective. These claimants can never have stood where Nolan and Lucan stood that day. Up on the Causeway, anywhere between Nos. 4 and 5 Redoubts, if you want to indicate the guns in the redoubts then occupied by the Russians you must point into the South Valley. If you want to indicate the battery that the Light Brigade was soon to charge then you must point to the left of No. 4 Redoubt, down the North Valley. Even if you are in one or other of the valleys, rather than on the dividing ridge, you must point either down it or out of it. You cannot indicate one direction and mean the other. Not that there was any doubt at the time in Lucan's mind or amongst his staff. They were aghast. Walker, watching nearby, wrote in a letter five days later;

'An order was brought [to Lucan] by an officer personally hostile to him, and received without the discretion fitting in an officer of high rank. . . . Lord Lucan, instead of taking the order and exercising his own judgement as to how he carried it out, asked Captain Nolan what he was to attack, and was answered by his pointing to the Russians drawn up across the valley, with the words: "There, my Lord, is your enemy, there are the guns".'[11]

This was the nub of all later contemporary criticism of Lucan, including that from Raglan as he wriggled to get off the uncomfortable hook of responsibility himself. It was not that Lucan had misunderstood Nolan's outstretched arm, but rather that he had failed to use his discretion, failed to get from the arrogant ADC a rational explanation as to what was really required. He was after all a lieutenant-general confronting a captain. If the order seemed impossibly rash, obviously unsound, he was entitled, indeed it was his duty, to use his own judgement. But was he? Lucan's predicament was not that simple.

The first point was made by Paget and reiterates the long-established principle that a messenger, of any rank, represents the general who sends him. Quoting the Duke of Wellington Paget wrote:

' "that a private Dragoon can be the bearer of a written order as well as a staff officer, but for this consideration, that a staff officer is supposed to know somewhat the views of the commander who sends the order" . . . [in this instance] not only was the officer on the headquarter staff selected to convey the order, but that the mode in which he delivered it was in such a tone – of manner, as well as of assumed authority . . . as to render it impossible for Lord Lucan to disobey.'[12]

Lucan had been put on the spot. His Commander-in-Chief had sent him a written order that he did not properly understand and that appeared to contradict the accepted norms of cavalry warfare. He started to remonstrate with the

ADC, only to be cut short with the peremptory order, repeating Raglan's exact words, to attack at once. Still more confused and angry at the manner in which he was being addressed by a junior officer, Lucan tried again to find out the correct objective for his division. This time the response is even more impertinent and emphatic. There was no mistaking the guns pointed out so insultingly. There was no doubt in anybody's mind at the time that Nolan had indicated the Russian artillery in the North Valley. We know that Walker was convinced, as was Blunt, who for some time was apprehensive that he, an unarmed civilian, might have to accompany Lucan down the valley. 'There can be no doubt,' wrote Captain Tremayne of the 13th Light Dragoons who was close to Lucan when the order was brought to Cardigan, 'that Nolan gave the order to go where we did go. Cardigan told me this repeatedly afterwards.'[13]

As will be seen below Nolan himself did nothing to disabuse anybody that the guns actually charged were not the correct ones, and he had ample opportunity to do so beforehand. Nolan was almost certainly within earshot when Cardigan got his orders. He himself spoke to Cardigan and then chatted to Morris while awaiting the order to advance. It is stretching the bounds of human nature to believe that Nolan never mentioned to his friend the target down the valley ahead. Morris was always certain the charge went in the direction intended by Nolan.

So, having been put so rudely in his place, what could Lucan do other than obey? Not much. He could have forced a showdown. Some writers later suggested he should have had Nolan arrested for insubordination. This was out of the question at the time. A general arrests the Commander-in-Chief's ADC when he brings him an urgent message to attack at once because he does not like his manner. What would that achieve, other than the general's own demise shortly afterwards? Lucan knew that sometimes circumstances arise in a battle when the Commander-in-Chief has to make painful decisions, perhaps to surrender one unit to save the situation elsewhere. Raglan could see virtually the entire battlefield; Lucan could not. For all Lucan knew there might be some overriding reason why Raglan seemed about to sacrifice the cavalry. There is an old Army tenet on the question of orders that soldiers break at their peril: obey first, ask questions afterwards. If this rule is not followed no Army can function effectively.

There was also the time factor. There was no mistaking the words 'advance rapidly' or 'immediate'. Nolan had repeated them verbally. Lucan was required to take action at once. If he disbelieved the messenger it would take for ever to refer back to Raglan up on the escarpment. Could he risk his neck by causing all that delay? Lucan wanted to keep his command, not throw it away in the middle of a crisis on the battlefield. The shame and disgrace of delaying, or even worse disobeying what could easily turn out to be a proper order, was unthinkable to a man like Lucan.

Nevertheless, had there been less bottled-up emotion, less animosity and less contempt on the part of Nolan he would not have provoked Lucan so blatantly. Had Lucan been more even-tempered he might have calmed Nolan, asked more pertinent questions, and quietly insisted on fully understanding what was required of him. That was where he was at fault. Nolan had indicated the wrong objective although he must have known the right one. The fact that he remained silent about the error gives rise to the possibility, mentioned above, that it was deliberate rather than inadvertent. Did his feelings and frustrations so consume him that he knowingly launched the cavalry on a death or glory charge in which he had every intention of participating? We will never know. But we do know that he considered such an operation perfectly feasible. Years before, as an instructor at Maidstone barracks, had he not sketched such an action as the one the Light Brigade was about to undertake on the wall of the quartermaster's store? It was just a question of being sufficiently daring and enterprising. Fortune always favoured the brave.

Lucan was not so sure. Blunt described his mood as:

'surprised and irritated at the impetuous and disrespectful attitude and tone of Captain Nolan and looked at him sternly but made no [further] answer.'[14]

Instead he turned his horse towards the Light Brigade. Followed by his staff and Nolan, Lucan trotted across to find his brother-in-law. It was the Light Brigade's misfortune to be in exactly the right place, facing in precisely the right direction. Anyway, it was their turn to do something.

Perhaps not surprisingly the key words used by the brothers-in-law when Lucan passed on Raglan's order are disputed by them both. Lucan's version was that he told Cardigan to '*advance* in two lines' and that he was to do so 'very steadily and keep his men well in hand'.[15] According to Cardigan he was ordered 'to *attack* the Russians in the valley about three-quarters of a mile [it was a mile and a quarter] distant.'[16] Whether he gave Airey's piece of paper to Cardigan to read is disputed and the evidence inconclusive either way. The balance of probability is that he did not, as neither ever subsequently claimed it happened, and Cardigan's response makes it clear he had just been told to advance (or attack) straight down the North Valley.

'Certainly, sir; but allow me to point out to you that the Russians have a battery in the valley in our front, and batteries and riflemen on each flank.'

Whereupon Lucan resorted to the time-honoured ploy of commanders having to give unpopular orders, that of blaming his superior.

'I know it,' he said, 'but Lord Raglan will have it. We have no choice but to obey.'[17]

Cardigan formally saluted his divisional commander with his sword. Nolan looked on, making no attempt to change the direction of the advance or its objective. Neither senior officer questioned him; they now seemed in no doubt as to what was expected.

Lucan then decided to alter the formation of the Light Brigade. He considered that three regiments in the first line was too many. Perhaps he felt he should not expose so many men to the carnage the leading line could hardly

135

avoid; perhaps he thought the brigade's frontage too wide, or that there should be greater depth to the advance. Probably it was a combination of them all. For whatever reason, he told Cardigan that the 11th Hussars should drop back to the second line. This would leave the 13th Light Dragoons and the 17th Lancers (Lucan's Regiment) in the place of honour, while relegating Cardigan's former Regiment to a supporting role. It is highly unlikely that Lucan selected the 11th Hussars deliberately; it just so happened that they were on the left of the line so their dropping back would cause the least confusion. Cardigan did not see it that way. It was insufferable that *his* regiment should be slighted in this manner. It is quite possible he refused to order the change as, when Cardigan rode off to find Paget, Lucan went to tell the Commanding Officer of the 11th (Douglas) personally, with the words, 'Douglas, withdraw your Regiment and support the 17th Lancers.'[18]

It is of interest to read the statements that both Lucan and Cardigan later gave to Kinglake on their meeting that morning in front of the Light Brigade. First Lucan's:

'With General Airey's order in my hand I trotted up to Lord Cardigan, and gave him distinctly its contents so far as they concerned him. I would not say on my oath that I did not read the order to him. He at once objected, on the ground that he would be exposed to a flanking battery. [This rings true, as the battery near No. 3 Redoubt, which would fire into the Brigade's right flank, was invisible from where they stood.] When ordered to take up his then position [confusing English, but probably it meant when Cardigan had received the order to 'mount'] he had expressed, through his aide-de-camp, [unidentified, but possibly Maxse] the same apprehensions. I told him that I was aware of it. "I know it," but that "Lord Raglan would have it," and that we had no choice but to obey. I then said that I wished him to advance very steadily and quietly, and that I would narrow his front by removing the 11th Hussars from the first to the second line. This he strenuously opposed; but I moved across his front and directed Colonel Douglas not to advance with the rest of the line, but to form a second line with the 4th Light Dragoons.'[19]

Then, Cardigan's version:

'The brigade was suddenly ordered to mount, upon which I sent one of my aides-de-camp to reconnoitre the ground. [A puzzling remark, but in the light of Lucan's version probably meant the ADC went to Lucan to find out what was happening and, when told, had voiced his misgivings. Lucan then rode over to confront Cardigan personally.]

Lucan then came in front of my brigade and said, "Lord Cardigan, you will attack the Russians in the valley." I said, "Certainly, my lord," dropping my sword at the same time; "but allow me to point out to you that there is a battery in front, a battery on each flank, [this was almost

136

certainly made with the benefit of hindsight, as the Russian guns on his right could only just be seen from the escarpment, and not at all from the North Valley] and the ground is covered with Russian riflemen."

Lord Lucan answered; "I cannot help that; it is Lord Raglan's positive order that the Light Brigade [inaccurate, as it was the Cavalry Division the Commander-in-Chief wanted to advance] is to attack the enemy;" upon which he ordered the 11th Hussars back to support the 17th Lancers.'[20]

Scores of anxious eyes had been following these hurried and ominous consultations carried out in full view of the 13th Light Dragoons and 17th Lancers. Captain Tremayne of the former recalled:
'I saw Lord Lucan's and then Lord Cardigan's evident astonishment at the message; Nolan pointed right down the valley.'[21]
Mitchell, of the same Regiment later said much the same:

'Captain Nolan . . . came galloping down and handed a paper to Lord Lucan. We now felt certain there was something cut out for us, but Lords Lucan and Cardigan appeared to demur, when Captain Nolan, pointing to the guns, said something which caused us to get the order to mount [they had already mounted by this time].'[22]

After making this final adjustment with Douglas Lucan rode back towards the Heavy Brigade. He had been ordered to attack with the cavalry, that is the whole division, so he must now coordinate the support for the Light Brigade. As Lucan fully appreciated the enormity of what was likely to happen as the result of his instructions to Cardigan, he contemplated how best to secure his own reputation. He was aware that, whatever the consequence, the price in casualties for this foolhardy undertaking would be high. If, as seemed probable, the advance was repulsed with heavy loss then a scapegoat would be required to take any suggestion of blame off the shoulders of the high command. That was the military system. It appeared to Lucan that he would be a prime candidate for this role. His face would fit the frame perfectly and he was determined to dodge the allegation if it came. He would need evidence that he was only doing what he was told. With this in mind Lucan handed the fourth order to Blunt for safe-keeping. As things turned out it was a wise precaution as Nolan was killed and, typically, Airey had not made a copy for headquarters.

III
THE CHARGE

CHAPTER SEVEN
Running The Gauntlet – I

Cannon to right of them,
Cannon to left of them,
Cannon in front of them
Volleyed and thunder'd;
Lord Tennyson

STRELAI! The order was high-pitched, sharp and imperative. Five guns flashed and roared out, virtually simultaneously. As Corporal Thomas Morley (17th Lancers) was later to write:

'The Russian gunners were well drilled. There was none of that crackling sound I have often heard in the American [Civil] War and other places, where one gun goes a little ahead and the others follow, having the effect of a bunch of fire-crackers popping in quick succession. In such cases the smoke of the first gun obscures the aim of the rest. The Russian Artillery at Balaclava went off at the word of command, all together.'

Number 1 Battery of Position of the 16th Artillery Brigade had at last opened up on Cardigan and his men. The Russian officer who shouted the command had been itching to fire since his battery had deployed on the south-west edge of the Fedioukine Heights some 45 minutes earlier. Although the British cavalry was 1,000 metres away that was comfortably within range of his 12-pounder guns and 18-pounder howitzers. It had been exasperating to watch them through his glasses. They were right out in the open, relaxing, as though on some leisurely field day. He had felt sure he could drop a few shells on top of them, shake them up a bit, but his battery commander had forbidden it. Then he had watched excitedly as a bugle sounded and the enemy horsemen mounted. Within a few minutes they were on the move, six, seven, maybe eight hundred cavalry were advancing east directly across his front. Within seconds they began to trot. The young officer had permission to open fire.

There were ten guns in the battery, which had been split into two divisions of five each. This had facilitated finding good, reasonably flat gun positions among the rough, uneven slopes of the Fedioukine Hills and enabled the battery to cover a wider arc. The battery occupied an important position guarding the extreme right flank of Jaboritsky's force, indeed of the entire Army. It had the close protection of a company of the Black Sea Foot Cossacks, two companies of riflemen, and with the Vladimir and Sousdal Infantry Regiments in support close behind. There was no obvious threat to its position; even the enemy guns that overlooked it from the Sapoune escarpment were out of effective range, and the Woronzoff road where it descended from the heights was deserted. The only enemy troops clearly visible were two bodies of cavalry (the Light Brigade and the French Chasseurs). The Russian gunners felt secure, delighted to relieve the boredom with a textbook, but testing, shoot on to a target that was not threatening them.

The battery had a magnificent field of fire to its front. The enemy cavalry was packed tightly together in two lines and was obligingly presenting its left flank to enfilade fire. Enfilade fire could be particularly lethal as a miss on one man would almost certainly hit his comrade behind. Artillerymen always strove to find positions from which they could fire into the flank of enemy troops. With roundshot capable of ploughing through 15–20 men and more, sweeping from one end of a line to another, it was possible to obliterate a unit with ease with enfilade fire. It was the six Russian 12-pounders that were using roundshot. The great thing about cannon balls was that they bounced and took a lot of stopping. If the gun was fired with the barrel horizontal, or very slightly elevated, the shot would not travel at much above a man's height and would bounce two or perhaps three times within a thousand metres. The shot literally bowled over men and horses like so many skittles. The fact that the troops could sometimes see them coming was deceptive. Even on the third bounce the weight and velocity of the ball would still take off an arm or foot.

The four howitzers never fired roundshot. Howitzers' lower muzzle velocity and higher trajectory made them unsuitable. Their ammunition was the shell rather than the shot. They fired at a higher elevation than the gun, lobbing their projectile, which was designed to explode over, or on, the target. This 'common shell', as it was called, was basically an iron ball filled with gunpowder that was exploded after a predetermined time of flight by a fuze. The fuze was a train of powder whose length was adjusted at the time the fuze was set. It was ignited by the flash of the propellant charge. It could be damaging, but was not always accurate. The problem was in getting the right burning time for the fuze for the range, and the fact that there was only the fragments of the ball to fly around when it exploded. Spherical case (or shrapnel) – a ball filled with bullets – which was much more effective and was in use by the British – had, fortunately, not yet reached the Russian artillery. If the fuze in a common shell was too short a high air burst, with the consequent waste of lethality, resulted. If it was too long the shell lay spluttering on the ground while all nearby got out of the way.

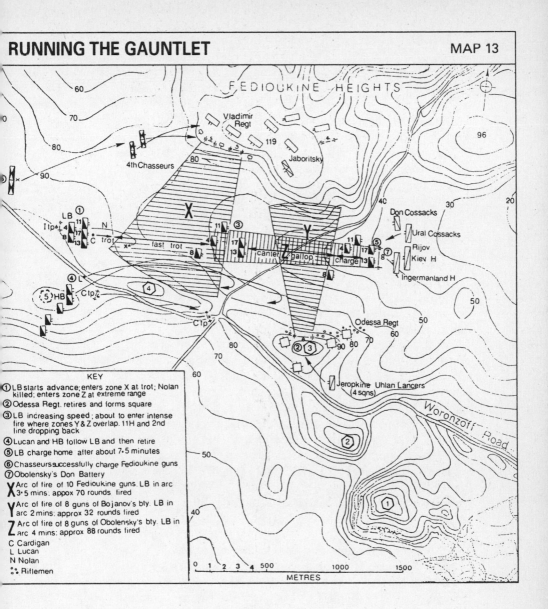

FEDIOUKINE HEIGHTS

Vladimir Regt

4th Chasseurs

Jaboritsky

Don Cossacks
Ural Cossacks
Rijov
Kiev H

Ingermanland H

Odessa Regt

Jeropkine Uhlan Lancers (4 sqns)

Waronzoff Road

KEY

① LB starts advance; enters zone X at trot; Nolan killed; enters zone Z at extreme range

② Odessa Regt. retires and forms square

③ LB increasing speed; about to enter intense fire where zones Y & Z overlap. 11H and 2nd line dropping back

④ Lucan and HB follow LB and then retire

⑤ LB charge home after about 7·5 minutes

⑥ Chasseurs successfully charge Fedioukine guns

⑦ Obolensky's Don Battery

X Arc of fire of 10 Fedioukine guns. LB in arc 3·5 mins; appox 70 rounds fired

Y Arc of fire of 8 guns of Bojanov's bty. LB in arc 2 mins; approx 32 rounds fired

Z Arc of fire of 8 guns of Obolensky's bty. LB in arc 4 mins; approx 88 rounds fired

C Cardigan
L Lucan
N Nolan
∴ Riflemen

0 1 2 3 4 500   1000   1500

METRES

141

The Russian artillery was their élite arm, well trained, with the best personnel selected as gun detachments. The battery on the Fedioukine Heights was being offered a splendid opportunity to show off its skills. As the firing became general, however, things became more difficult for the gunners. The first hindrance to accuracy was, as Morley so rightly emphasized, the smoke. A cloud of white smoke drifted across the position after every shot, often obscuring what had just been a perfect view. As the gun had to be manhandled forward several feet to be realigned (aimed) after the recoil of each firing, smoke screening the target did not make for meticulous shooting.

Then there was the problem of movement, rapid movement. The Light Brigade was starting to gather speed, its initial parade ground trot quickly becoming brisk. A moving target is always hard to hit, especially if you are firing heavy cannons. With a musket or rifle the firer could swing his weapon in his shoulder, following the target as it moved, and fire when he had the best aim. Not so for the artilleryman. He must point his gun ahead of the target and fire just before it passed the end of the muzzle – if he could see through the smoke when the right moment had arrived. If the enemy were moving fast the gun might only be able to fire once or twice before its target was out of view or range.

This lack of time was the Fedioukine Battery's other difficulty. The Light Brigade was only in the Russians' arc of fire for about three to three and a half minutes (see zone X on Map 13). After that it became impossible to swing the guns sufficiently to their left to make continued firing worthwhile. So the ten guns had to do their damage in a limited period. The usual rate of fire was about two rounds per minute as reloading and realignment procedures were tiring and time-consuming. This meant that the Fedioukine guns were able to fire about seven times each before their target disappeared down the valley. Perhaps seventy shots, around 30 shells and 40 roundshots, were fired at the Light Brigade from this battery over the first 800 metres of its advance. The Russian gunners would never know it, but one of their first, possibly their very first shell slammed into the chest of Nolan as he attempted to spur ahead of the Brigade.

No sooner had the Light Brigade left their arc of fire than they were replaced by another target. More cavalry appeared from over the Causeway Heights and trotted boldly out into the North Valley. This was Lucan and the Heavy Brigade starting to follow their compatriots. The Russian battery continued to fire and inflict casualties. But not for long. Possibly the detachments toiling at their guns had failed to notice that the body of cavalry that had been stationary under the escarpment had moved. These French Chasseurs had been ordered forward by their Cavalry Division Commander, General Morris. Their objective was the Fedioukine Battery. Four squadrons of the 4th Chasseurs, in two lines, led by Major Abdelal, advanced half-left to come in on the gunners' right. They broke through the Russian skirmishers and advanced rapidly on the right-hand division of the battery, which was only able to limber up and retreat with difficulty and confusion. The other five guns followed suit. Only the later arrival of the Vladimir Regiment caused the French to sound the retire. At the minimal cost of ten killed and twenty-eight wounded the Chasseurs had completely neutralized the Russian artillery on the north side of the valley. The attack had been audacious, skilful and effective. The Light Brigade's survivors would have reason to be grateful ten minutes later as they straggled back over the same ground.

*　　*　　*

Captain Bojanov commanded Number 7 Light Battery of the 12th Artillery Brigade. He had six 6-pounder guns and two 9-pounder howitzers. Since about 8.30 he had been in position close to No. 3 Redoubt with the four battalions of the 24th Odessa Jaeger (Rifle) Infantry Regiment deployed in and around the entrenchments. His guns were facing west along the Causeway Heights and into the South Valley. This was where all the action had been. His battery had fired a few rounds at the British cavalry and artillery when he had thought them within range, but he had been a spectator for most of the morning. Bojanov had been disappointed at the failure of his cavalry comrades to charge home on the Highlanders, and visibly shocked to see the recoil of Rijov's entire brigade shortly afterwards. The morning had started so well. Now there seemed to be a stalemate.

At eleven o'clock there was no change in Bojanov's position. About ten minutes later, however, he heard the thunder of his compatriots' guns across the North Valley on the Fedioukine Heights. Standing on the higher ground near the redoubt he turned his glasses north-west to scan for the gunners' target. Within seconds he had seen it – a large body of British cavalry moving quickly down the North Valley. If they continued they could either turn up on to the Causeway or sweep past him to attack the Russian cavalry and guns at the eastern end of the valley. Either way his battery was poorly placed to engage them. It was at this moment that the realization came to him that the Odessa Regiment was forming squares. This was the standard tactic for infantry threatened by cavalry. Four tightly packed walls of muskets bristling with bayonets was almost certain to see off any mounted charge, whereas if caught in line infantry were highly vulnerable to being ridden down. Like Bojanov, the infantry thought the enemy were heading in their direction.

Bojanov then made his decision. He yelled his orders to limber up the battery and redeploy it facing into the North Valley. It says much for the training of the Russian gunners that they were able to get into the new fire positions within about five minutes.[1] By this time their target was almost opposite them, a mere 500 metres away and cantering down the middle of the North Valley. As the enemy cavalry passed No. 3 Redoubt a volley of fire rang out from the nearest Odessa battalion square, and the riflemen began a continuous popping as individual skirmishers loaded and fired at will. With the limited range of the Russian musket this was a waste of ammunition as far as the squares were

143

concerned, and it only served to obscure the field of fire with smoke. The riflemen, however, stood a better chance with their much longer-range weapons. As the guns arrived they opened up immediately.

A slight digression is in order here. Kinglake, and other historians since, have described how, when the Light Brigade was seen to start advancing down the North Valley the Odessa Regiment retired to positions strung out along the low ridge that juts out from the main Causeway ridge immediately east of No. 3 Redoubt (see Map 13). Kinglake's map shows four battalion squares along this ridge, the furthest being at least 800 metres from its original position near No. 3 Redoubt. It also depicts Bojanov's battery moving along this ridge and deploying in three separate divisions, with some guns moving 600 metres. According to Kinglake, and the others who have followed him, these troops were able to fire on the Light Brigade as it charged the guns at the end of the valley. In the present writer's view this is nonsensical.

The Light Brigade's charge was all over (i.e. it had reached the guns) in about 7½ minutes. Assuming the Russian infantry and gunners had reacted instantly the moment they saw the advance starting they would have had five minutes at the most to get into Kinglake's new locations if they were to have any hope of joining the firing. It was a physical impossibility. Neither Bojanov's gunners nor the Odessa infantrymen were supermen. They could not cover the distances in time and adopt their new fire positions or formations and be ready to fire before the Light Brigade finished its charge. Kinglake's account conjures up visions of infantry and artillery sprinting madly along the ridge in a race to keep ahead of the Light Brigade. This would never have been contemplated, let alone attempted, by infantry threatened by cavalry. To be caught on the move by charging horsemen was not just dangerous, it was certain death. The Odessa battalions had no reason to move any distance. They were supporting each other; they were on high ground and their gunners were nearby. Their immediate reaction to the Light Brigade's advance was to form square at once.

Russian sources do not support Kinglake's version either. The Russian author M.N. Boganovitch, author of *The Eastern War of 1854–6* wrote:

> 'As soon as the advance of the English cavalry was observed by the Russians, the Odessa Regiment of Rifles *retired to No. 2 height* and formed square.'

Albert Seaton, an acknowledged expert on the Russian military and author of *The Crimean War – A Russian Chronicle*, states of this incident:

> 'The Odessky, holding the high ground on the southern heights, as soon as they sighted the enemy, hurriedly *fell back over the ridge of the Causeway on to No. 2 Redoubt* and formed square.'

This, however, would have all four battalions off the Causeway ridge to the south and quite unable to fire into, or be seen from, the North Valley. It is possible that one or two battalions felt too exposed on the slope of No. 3 Redoubt and pulled back a comparatively short distance before forming square. We know for sure from several Light Brigade survivors that infantry squares were shooting at them towards the end of the charge. Map 13 indicates the most probable movements of the Odessa Regiment at this time.

As the Russian detachments got their cannons pointing north the Light Brigade was probably just entering zone Y on Map 13. It was about two-thirds of the way through its charge and was moving at a canter, which soon became a gallop. So, although the enemy were presenting their flank to the gunners and the range was only 500–600 metres, they were far from easy to hit. By this time the first line had shrunk in size due to casualties and the 11th Hussars dropping back, while the second line had, in reality, become three quite separate units.

While still having the smoke problem to contend with from the infantry and their own firing, Bojanov's men were acutely disadvantaged by the speed of their target's movement. The Light Brigade were in the Causeway gunners' arc for 600 metres at most, and they covered that distance in around two minutes. Each detachment had to aim his loaded gun across the path of the approaching horsemen, then fire just before they reached the end of the muzzle. A split second too early and the round would strike in front, a split second too late and there would be a miss behind. Once a gun fired it is doubtful if it fired again at the first line. More likely it waited for the second, as that would not entail swivelling the gun to take a fresh aim – much simpler to wait for the second line. Three or four shots were probably the most Bojanov's guns could get in before it became dangerous to swing further east for fear of hitting the Don Battery in the valley. If each gun and howitzer fired four times Bojanov's battery would have fired about 32 shots (perhaps 8 common shells and 24 roundshot). Not a great deal. Certainly nothing like the numbers implied by many British accounts (contemporary and modern) and in Tennyson's famous poem quoted at the start of this chapter.

*   *   *

Only moments before Cardigan gave the order for his Brigade to advance General Rijov had ridden forward to speak to the commander of No. 3 Battery of the Don Cossacks. Colonel Prince Obolensky had over 200 men with four 6-pounder guns and four 9-pounder howitzers deployed facing west up the North Valley. Unbeknown to him his battery had been erroneously pointed out as the objective of the entire British Cavalry Division. His guns were deployed for 'action front' in a line. At 20-metre intervals between guns the battery frontage was 150 metres, perhaps a little more. The ground was open, flattish, and with an excellent field of fire right up the shallow valley. The enemy, however, was a long way off (the Light Brigade was 2,000 metres away), so the Russian gunners on foot could barely make out anything remotely hostile. The *Artillerist's Manual and British Soldier's Companion*

145

(1839–59) had this to say on the subject of visibility in clear conditions with the naked eye:

> 'Good eyesight recognises masses of troops at 1,700 yards: beyond this distance the glitter of arms may be observed. At 1,300 yards infantry may be distinguished from cavalry, and the movement of troops may be seen; the horses of cavalry are not, however, quite distinct but that the men are on horseback is clear.'

Nevertheless, through glasses, the officers and Obolensky had been able to identify the British cavalry.

The Colonel was proud of his battery. Don Cossacks made good soldiers and his gunners were handpicked. They were in effect horse artillery and as such had been attached to the 6th Hussar Brigade. Until now they had seen no action that morning. They had been kept to the rear during the Russian cavalry's half-hearted attempt to advance on Balaclava and for almost two hours had been doing nothing. The battle so far had left them, like the British Light Brigade, virtually untouched. To some it may have seemed a dull day.

The guns were not yet loaded. There was no target, no threat, so Obolensky preferred to keep his options of ammunition type open. Piled beside each gun were a number of rounds of roundshot and canister, beside the howitzers common shell and canister. Some five yards to the rear of each cannon were the limbers – large rectangular boxes mounted on two big wheels containing the first line ammunition and various tools and equipment essential for keeping the guns in action. When on the move the limbers were attached to the gun and the whole pulled by a team of six (in the case of light guns) horses. It was the duty of an NCO and several soldiers to keep the guns supplied with ammunition of the right type, and with the correct fuzes, from the limbers. The teams of horses were kept further back, in this case with little threat, probably 100 metres to the rear. They would be called forward to move the guns when necessary. Further back still would be the reserve ammunition wagons and teams.

Behind each gun a linstock was planted into the ground supporting a slow-burning match. For some 250 years the most reliable method of firing guns had been by the use of linstock and portfire. The linstock was really a forked stick holding a slow match that was burning at both ends. 'Slow' was something of an understatement as the match burned at the rate of a metre in eight or nine hours. The means by which fire was transferred from the linstock to the vent of the gun was by the use of a portfire. This was a 16½ inch tube made of a composition of saltpetre, sulphur and mealed powder that burnt at the rate of one inch per minute. It was remarkably similar to the modern sparkler variety of firework. Immediately before firing the firer ignited his portfire from the linstock and stood by his gun awaiting the order to fire.

In Rijov's own account of the battle he stated that, while talking to Obolensky, 'the sharp eyes of the Don gunners noted the far distant cloud of dust raised by enemy cavalry coming down the slopes into the valley. Two minutes later it was clear that the enemy was coming along the valley in to the attack.'[2]

The Colonel's command to load was shouted down the line – guns with round-shot, howitzers with shell, maximum elevation. All eyes strained towards the target. The enemy was a long way off, still out of effective range – but closing. Obolensky was sitting on his horse in the centre of the position, a few metres back from the gunline. His glasses were glued to his face. He could see shells bursting over the enemy horsemen. The guns on the Fedioukine Hills were in action. He was judging distance, anxious to fire as soon as possible but not so soon as to fall short. After a few seconds he glanced to left and right to check each detachment commander had his arm raised to indicate his gun was ready. They had. Another look and, without taking his gaze off the target, Obolensky yelled, '*Strelai!*' The Russian battery commander was not to know it but the Light Brigade was some 1,200 metres away and almost exactly four minutes from his position.

Sergeant Sulina commanded one of the 6-pounders near the centre of the position.[3] He was what the British call the 'No. 1' on the gun, responsible for supervision and laying. He had to align the gun, traverse it, elevate it and make final adjustments with the 'tangent sight'. To do this he needed to see the target. This was always the problem. The ear-splitting crash of the first volley was instantly followed by the gun leaping about four feet backwards and dense clouds of billowing smoke. With the entire battery continuously in action there would be no respite from the smoke. With the first shots away the familiar characteristics of a gun position in action were instantly present. First, the difficulty in seeing the enemy; second the awful ringing in the ears, the tempo-rary deafness that meant all orders had to be bellowed from a few feet away, and third the acrid smell of burnt gunpowder.

Sulina helped the other four members of his detachment push the gun back into position and yelled, 'Roundshot – Reload!' He peered along the barrel. Aiming accurately was impossible and he had really no idea where his first shot had gone. As the attackers seemed to be coming at him head on, and at speed, he would merely reduce the elevation and fire again – there was little need to worry about traversing. As he was doing this the reloading drill went ahead automatically. It was a drill that had been repeated a thousand times in training. Every movement had to be done in precisely the right order, every action must be exact. Short cuts cost lives; gunpowder was highly inflammable; one tiny spark could lead to oblivion. This was the reason why all implements used by the gunners were made of wood or copper.

Gunner Popov was the spongeman. His position was in front of the wheel on the right of the gun. His equipment was a spongestaff and a bucket of water. The wooden staff had a soft sponge (often of fleece) nailed round one end and a harder rammer on the other. As the gun fired Popov dipped the sponge into the water and thrust it down the bore. The object was to extinguish any burning fragments before a fresh charge was loaded. This duty was not the most popular.

Next, the loader, Gunner Remchenkov, who was standing by the muzzle on the left, placed a bag of propellant charge down the bore, and Popov, quickly reversing his spongestaff, rammed it home. At the same time Gunner Sazonov, the 'ventsman', standing behind the wheel on the right, was supposed to put his thumb (covered by a leather thumbstall) over the vent. The passage of the charge and rammer down the bore created an air current out of the vent if it was not blocked. This could ignite any small hot embers left after the swabbing – with nasty consequences for Popov. It was an easy thing to forget altogether, or not do at precisely the right moment in the heat of battle. Failures by ventsmen in this respect were not uncommon. Often they would get away with it and nothing happened. In the British service tradition had it that the spongeman was entitled to hit the ventsman over the head with his spongestaff if he failed to 'serve his vent'. The penalty, if things went badly wrong and there was a premature explosion, was not that the barrel burst and the detachment were all killed, but that the spongeman usually lost his hands.[4]

Next Remchenkov put the roundshot down the bore and again Popov rammed it down. Time was sometimes saved by not bothering to ram the charge. Meanwhile Sazonov had inserted his 'pricker' in the vent to puncture the bag with the charge in it to make firing more certain. This was immediately followed by his 'priming the gun'. It entailed placing a small copper (formerly a quill) tube filled with powder in the vent. It was the fifth member of the detachment, Gunner Teplov, who actually fired the gun. On the order being given he applied his portfire to the vent. Teplov's portfire tended to fizz and sparkle with obvious risks, so if the firing rate was slow he had to extinguish the sparks after every shot. He did this by cutting off the burning end with the portfire cutter which was permanently attached to the trail of the gun. In a situation demanding continuous and rapid firing, such as now confronted the battery, this was not possible.

Speed, with safety, had been drummed into every man throughout training. Practice, practice and more practice, coupled with brutal punishment for errors, had been the daily drudgery of barracks. The manuals stipulated a minimum of two shots a minute, which required slick, smooth drills and no mistakes, no fumbling from any member of the detachment. As this was only the second shot safety had not been compromised. The drills were second nature to all and, although the adrenalin was beginning to circulate, nobody was actually firing at them and so, as yet, there was no sense of urgency. Sulina put his arm up to signal he had reloaded before looking over his left shoulder at his battery commander, awaiting the next order to fire. '*Strelai!*' Again the flaming muzzles and thundering crash. Again the rolling banks of smoke. Again the reloading drills. About a minute had elapsed since first opening fire. In that time the Light Brigade had advanced 300 metres towards them.

During the next minute the battery got off another two rounds per gun. In that time the enemy broke into a canter and covered another 300 metres, which put them only 600 metres from the gunline. This was a serious attack which, despite heavy losses, showed no sign of slowing, let alone stopping or turning back. Obolensky's field glasses had never left his eyes since the first volley was fired. After the third volley he had ordered all guns to fire on the horizontal, with no elevation. Although the target was larger now, it was still infuriatingly obscured by the smoke. When it thinned, he noticed that the Fedioukine guns had ceased firing but that the battery on the Causeway Heights had taken over. The leading line had thinned appreciably. There were empty saddles – men and horses crumpled and fell as he watched – but the British cavalry kept closing in on the centre, still in formation, still under control. As a professional soldier Obolensky admired them for that. He could now make out that the first line consisted of lancers and hussars. One moment the lances had been upright, clearly visible, the next they had disappeared, or rather been lowered to the 'engage'. If they were not stopped it looked as if the lancers would hit the right half of his battery and the sword-waving hussars the left. The speed of the attackers was most disquieting. He shouted for the guns to increase the rate of fire. As a precaution he sent an orderly back to the limber teams for them to come up behind the gunline as the battery was not under artillery fire so there was no risk to his horses.[5] Rijov left the battery to send forward some of his cavalry to support the gunners.

Increase the rate of fire! Sulina and his detachment were already sweating hard from their exertions. Sulina was screaming at them to load faster, while cursing their supposed clumsiness. In fact they were doing remarkably well, apprehension, not yet fear, enhancing their labours. Sulina had given up all pretence of aiming or aligning the gun; his right arm was up almost before Popov had withdrawn his rammer. He was now ramming down both the charge and the ball at the same time and his sponging had become one quick thrust with a sopping sponge. The result of this fresh urgency was the firing of three rounds during the next minute. This undoubtedly caused more casualties but it also brought the enemy to within 300 metres of the gun position. The next minute would surely see the climax. Either the cavalry would be blown away at point-blank range or they would charge home and overrun the battery.

Obolensky was about to play his last card – canister. This giant shotgun cartridge was devastating, capable of carrying away huge swathes of an attacking formation in one blast. It could even be fired 'double-shotted', that is two canister rounds or one of canister and one of roundshot. The 'spread' of canister (or case shot as it is sometimes called) was recorded as being a circle of 32 feet diameter at 100 metres. It was difficult to miss. The battery commander was incredulous that the enemy was still advancing, that they had got so close under such galling fire from three batteries, plus rifle fire from the flank, as they came east of No. 3 Redoubt. He had seen, but not heard, the scores of little puffs of smoke all along the slopes of the Causeway Ridge which indicated that the infantry had joined in. And he had briefly glimpsed at least one infantry square discharge a volley with their muskets. The next sixty seconds would decide it. 'Load canister! Load canister! Fire at will! Fire at will!' There was more than a hint of desperation in the prince's voice as he bawled his last fire order.

As the Light Brigade closed in at the gallop there was no panic on the gunline. Each of the eight detachments stood their ground and served their guns. That does not mean they were not now frightened, merely that their discipline was holding and that most appreciated that their best chance of surviving lay, not in running, but in blasting the attackers away from the very muzzles of the guns. Up to the last second the battery could win.

Sulina, being a senior artillery NCO and a soldier of some service, had anticipated his battery commander's final order and already loaded canister with a roundshot after the seventh shot. He now got authority to fire at will, which meant that he gave the actual order to fire as soon as he was ready and as often as he could without waiting for anybody. To further speed things up Sulina decided to take a risk with safety: he ordered Popov to cease swabbing the bore after every shot. No more sponging, only ramming. If any burning fragments were still in the bore from the previous shot then the fresh charge could explode when it was rammed home. If that happened it was doubtful if the two gunners at the front of the gun would know anything about it. For the next few shots Sulina would stand by Sazonov's ear screaming 'Vent! Vent!' to ensure that in the excitement he did not forget to put his thumb over it before the charge was rammed in. Sulina knew the odds were heavily against an accident happening. In a tight spot sponging had successfully been abandoned before, and this was the tightest spot he had ever been in. By doing this it should be possible to fire four, or at a pinch five, rounds of canister in the last minute of the charge. In the circumstances Sulina considered it well worth the gamble.

Within 30 seconds his gun had flamed twice, each time belching out enough bullets to tear apart a dozen horsemen. Now all eight guns were firing the instant they had reloaded. Every detachment worked like demons and the firing of each gun was so quick that it still seemed that the battery was firing volleys. There is evidence that some detachments fired double-shotted, and that not all guns fired canister at the end. Neither Cardigan nor Captain Morgan, who both survived being fired on at the mouth of the muzzles, would have escaped unscathed had all those final shots been of canister. The noise was indescribable, each crash splitting heads with sledge-hammer blows. The smoke swirled around like a dense bank of fog through which it was often impossible to see the gun to the left or right.

At the muzzle of the gun Popov and Remchenkov had no time to so much as glance towards the rapidly approaching enemy. Their attention was riveted on getting ammunition down the bore, on ducking away as the cannon roared. Half-seen through sweat and smoke were their comrades: Teplov, desperately applying his portfire every time Sulina thumped him on the shoulder,[6] and Sulina's blackened face silently mouthing in Sazonov's ear as he frantically 'served his vent'.

From behind the gun, however, frightening glimpses of the Light Brigade's approach were now possible through the haze. With 150 metres to go the charge had not been broken. Teplov looked along the top of the barrel as the third canister round was loaded. Lancers! There was no mistaking them, even through the smoke. Many were falling, horses and riders crashing down in a tangle of flying legs, arms and bodies – but they kept coming. Until now the enemy had been remote, distant, something to shoot at but not really a threat, and mostly unseen. Now they were hurtling forward with lances levelled, only seconds away. Why was the battery not limbering up to retire? Where were the Russian cavalry? Surely they had not been abandoned? How could un-supported cavalry charge guns over such a distance and not be destroyed? Popov applied his portfire and the awful vision was momentarily blanketed out again. Perhaps there was just time for another shot. There was. Sulina's detachment managed to fire for the last time with the 17th Lancers a few feet from the muzzle. It was their eleventh shot in four minutes,[7] a very creditable performance, well above anything they had achieved in training – but it was not good enough.

Before the smoke of the final shot had cleared the enemy horsemen swirled around the guns, led by a single officer in hussar's uniform, riding erect with sword arm raised. Teplov had caught sight of him to his left for a split second before he dived under the trail and crouched under the gun feigning death. Popov met the fate so dreaded by foot soldiers: he was skewered through the back as he tried to run. Remchenkov and Sazonov got away, or at least succeeded in disappearing. Sulina had drawn his sword. Through the spokes of a wheel Teplov could see the legs of a horse and those of his sergeant turning and twisting.

The Light Brigade had ridden 2000 metres in seven and a half minutes. They had been under fire from 26 guns and howitzers which had fired, if Sulina's gun was average in the Don Battery, a total of 190–200 shot, shell and canister at them. Statistically this represents an overall total of 7.7 shots per gun or 2.25 shots per second of the charge. From the Russian point of view this was in-sufficient and is meaningless until related to the casualties inflicted, which are better discussed after running the gauntlet with the Brigade.

CHAPTER EIGHT
Running the Gauntlet II

'For the hand of my rider felt strange on my bit,
  He breathed once or twice like one partially choked,
And sway'd in his seat, then I knew he was hit –
  He must have bled fast, for my withers were soak'd'
Adam Linsday Gordon.

The dark blue lines of the Light Brigade moved forward together – but only for a short distance. Within a few metres the 17th Lancers and 13th Light Dragoons pulled ahead at the trot, the remaining regiments continuing at a walk. This was to allow a proper interval between the lines, and for the 11th Hussars to drop back. The finest light cavalry in the world was being asked to do what most would have said was impossible. To the front was the long, gently shelving, shallow valley. First a patch of plough, then rough turf stretching evenly down for well over a mile. In peaceful circumstances an inviting gallop.

Many survivors recall the hush that seemed to come at that moment, as if everybody held their breath in the realization of what lay ahead: the sudden recognition of mortality, an awful awareness that mutilation and death had never been closer.

Joy's and Britten's bugle calls rose up sharply, to be heard with ease on the Fedioukine Heights, on the Causeway Ridge and even high up on the Sapoune escarpment where the generals were gathered a mile to the rear. Within the ranks there was little or no talking. Then the creak of leather as scores of horsemen leant forward in their saddles, stood in their stirrups and pushed their mounts into the stiff, jigging trot of the cavalry. Quickly the thudding of hundreds of hooves was deadened by soft soil. Within moments of entering the plough the Russian guns on the Fedioukine flank opened up with the first salvo. The Light Brigade had begun to run the gauntlet. Like the man suddenly caught in a shower of rain who begins to walk briskly, the first line involuntarily increased the pace.

Cardigan was a single, solitary rider, well out ahead of his Brigade. He looked the part of a cavalry leader, resplendent in a Hussar uniform that fitted him perfectly, as, despite his 57 years, he had kept a youthfully slim figure. His chest was a blaze of gold lace as he wore his pelisse like a jacket, not slung from the shoulder. He was an accomplished horseman and had long since perfected the 'military seat' with the legs almost straight in the stirrups. His charger, Ronald, was a magnificent thoroughbred chestnut. He had been broken in specially for Cardigan by his old regimental riding master.

All eyes in the front ranks were riveted on the brigade commander. Cardigan took great satisfaction from that. Nevertheless, it was infuriating that Lucan, at the last minute, had ordered *his* 11th Hussars out of the first line. It was insulting, unnecessary, an unwarranted, but typical, interference with his command. That sort of decision should be made by the brigade commander. Without doubt he had been given a stupid, probably suicidal, order by his brother-in-law, but he would carry it out with style and panache. Cardigan lacked many virtues but courage was not one of them. The prospect of the next few minutes was daunting, but it was also the apogee of any cavalry career. The likelihood was that he would be blown to pieces, but his death would be a soldier's death at the head of his men; it would be honourable – magnificent even. There was nothing complicated in what he had to do. He was certain as to his objective – he could see it far ahead. No tricky manoeuvres were required; there was no need to change formation. All he had to do was press on and control the pace. He had practised his Brigade often enough during those dreadful days at Devna in much more intricate drills. If it had to be done he would do it well.

But, after advancing less than 200 metres, Cardigan's concentration on the task in hand was shattered by what was, for him, a truly appalling sight. From the corner of his eye he became aware of a rider cantering, almost galloping, up from behind and to his left. There was no mistaking the uniform, or that red and gold forage cap. It was Nolan. He had his sword raised and seemed to be trying to catch up with his brigade commander. Was he deliberately forcing the pace? Was he endeavouring to lead the Brigade? Whatever the reason it was inexcusable behaviour. The man was a menace and would have to be dealt with severely if they ever came through the next half hour.

It was at that precise moment that fate intervened. The Fedioukine battery opened fire and almost decapitated the Light Brigade. A shell burst within a few feet of Cardigan but left him untouched. Nolan, however, was not so fortunate. A jagged lump of hot metal tore open his chest. With a hideous scream that men remembered for the rest of their lives his horse veered sharply right across the front and disappeared through the squadron interval of the 13th Light Dragoons. Nolan had been mounted on a former troop horse of the 13th, so perhaps its instinct was to rejoin them when its rider ceased to guide. Nolan's body remained in the saddle until it finally toppled to the ground behind the first line. TSM Linkon, who at that moment had also just lost his horse, was able to grab Nolan's mount and catch up with his Regiment. For Cardigan the embarrassment had been removed, but he would not forget the incident either during or after the charge.

Private Wightman, who rode in the first rank of the 17th Lancers, almost directly behind Cardigan, described what he saw:

153

'We had ridden barely 200 yards, and were still at the trot, when poor Nolan's fate came to him. I did not see him across Cardigan's front, but I did see the shell explode of which a fragment struck him. From his raised sword-hand dropped the sword, but the arm remained erect. Kinglake writes that "what had once been Nolan maintained the strong military seat until the erect form dropped out of the saddle"; but this was not so. The sword-arm indeed remained upraised and rigid, but all the other limbs so curled in on the contorted trunk as by a spasm, that we wondered how for the moment the huddled form kept the saddle. It was the sudden convulsive twitch of the bridle hand inward on the chest, that caused the charge[r] to wheel rearward so abruptly. The weird shriek and the awful face as rider and horse disappeared haunt me now to this day, the first horror of that ride of horrors.'[1]

History will never know with absolute certainty why Nolan was dashing forward at the time he was hit. It must, however, have been for one of two reasons. Either he wanted to change the direction of the advance or his all-consuming eagerness got the better of him and he dug in his spurs prematurely. The first is the most intriguing theory, which did not become current until some time afterwards. The contention was that he was desperately trying to right the wrong he had so recently perpetrated and redirect the Brigade on to the correct objective – the Causeway Heights, or rather, Redoubt No. 3, the scenario being that Raglan's message had been misunderstood by Lucan, that Nolan had no intention of pointing out the guns at the end of the valley as the ones to be attacked. Then, when he realized the Brigade was in fact starting to advance towards them, he tried to prevent the disaster.

It was Kinglake, writing some 12 years later, who first introduced this speculation, it never having surfaced before in written form. It has since been frequently accepted by historians, if not as absolute fact, at least as the most likely explanation for Nolan's conduct. Only one or two modern authors have cast doubts on this interpretation. The belief that Nolan wanted to change the direction of the advance rests on Kinglake's surmise, and on the written reminiscences of two former soldiers in the 17th Lancers published over 40 years after the event. Kinglake states:

'Nolan audaciously riding across his [Cardigan's] front from left to right . . . Nolan turning round in his saddle was shouting, and waving his sword as though he would address the brigade . . . [what] he sought to express by voice and signs would apparently mean something like this – "You are going quite wrong! . . . Bring up the left shoulder, and incline to your right as you see me doing. This is the way to get at the enemy!" .'[2]

In a footnote Kinglake added, 'Supposing my interpretation to be correct,' thereby acknowledging that he had no real evidence of Nolan's intentions.

Old soldiers love to reminisce, to tell a good yarn, especially if it involves famous exploits in which they took part. They are often not above embellishing the story and, particularly if age has dimmed the memory, the results can sometimes seem far removed from the reality. Sergeant-Major James Nunnerley who had ridden some forty years earlier as a corporal with the 17th Lancers is quoted in the *Short Sketch of the 17th Lancers and Life of Sergeant-Major Nunnerley*:

> 'He saw Captain Nolan . . . ride up to Captain Morris, then commanding the 17th Lancers, to whom he said, "Now, Morris, for a bit of fun!" Scarcely had he uttered these words than he was shot . . . after giving a kind of yell which sounded very much like "Threes right", and throwing his sword hand above his head his horse wheeled to the right and he fell to the rear. As though obeying this death-like order, part of the Squadron wheeled "Threes right". . . . Nunnerley immediately gave the order "Front forward" and so brought them into line again.'[3]

Morley, writing in 1892, supported Nunnerley's version of Nolan's actions and the brief wheeling to the right by part of the 17th Lancers. A far stronger case can be made out that Nolan's strange behaviour had nothing to do with changing the objective of the advance.

All accounts by witnesses agree certain facts. Nolan spurred forward soon after the advance started, before the first line had gone more than 100–200 metres. He was moving from left to right and he had his sword held high. As he got out in front, about level with, and close to, Cardigan, the shell exploded and hit him. His mortal wound caused him to scream in agony as his horse wheeled to the right and carried him back through the 13th Light Dragoons. Kinglake's interpretation is beguiling, which is why it has been perpetuated, but does not, in the present writer's opinion, stand up to close scrutiny.

Firstly, there were no witnesses who came up with this suggestion immediately afterwards. No less than fourteen officers on Cardigan's staff, or who rode in front of the first line, survived. Not one of them seriously put this interpretation on Nolan's actions. Perhaps it was discussed as a possibility but with no one did it carry sufficient conviction for it to occur in correspondence or diaries. One, Lieutenant Fritz Maxse, who rode as Cardigan's ADC, and whose horse almost collided with Nolan's, was incensed when he read Kinglake's account. He went so far as to write to *The Times* that he had 'no recollection of Nolan having attempted to create a divergence in the line of advance either by deed or gesture'. Again, in a letter to his brother, Maxse was even more emphatic, 'Nolan was killed close to me and Kinglake's account is . . . *absurd* [the word was underlined] as to Nolan wanting to charge any other guns but those which he did.'[4]

Cardigan himself had not the slightest suspicion that he might be trying to indicate that he was going the wrong way. Captain Tremayne (13th LD) wrote a few days later that:

155

'I saw Lucan's and then Cardigan's evident astonishment at the message; Nolan pointed right down the valley.'[5]

Later, when he read the suggestion that Nolan might have been pointing to the redoubts on the Causeway as the correct objective, Tremayne's comment was terse and emphatic – 'I think not'.

Captain Morgan, who as a squadron commander in the 17th Lancers was well able to watch what happened, afterwards told Assistant Surgeon Cattell of the 5th Dragoon Guards:

' "We had not gone many yards before we were under fire, I think from a heavy battery on our left [Fedioukine], the first shot from which killed poor Nolan, a splinter going through his heart and his horse carried him back to us." '[6]

He made no mention of anything untoward in Nolan's actions.

Perhaps most important of all, his close friend Morris, with whom he must have chatted just before the advance began, never indicated that Nolan had mentioned the objective was other than the guns straight ahead. Morris's impression when Nolan spurred away was that he had lost his head with the elation of the moment. He yelled out after him, 'That won't do, Nolan. We've a long way to go and must be steady'. His moving across the front of the Brigade is more likely to have happened *after* he was hit, when his horse wheeled right and through the ranks in the centre of the 13th Light Dragoons' line. Before this he had no reason to ride other than almost straight ahead, even if he had wanted to catch Cardigan up and impart urgent information to him.

The belated recollection of Nolan supposedly shouting 'Threes right' was probably Nunnerley's imaginative mind years later thinking the long-drawn-out shriek sounded similar to 'Threeees riiiight'. Nobody else seems to have heard it that way apart from a, by then, thoroughly disgruntled Morley who latched on to Nunnerley's idea. If a part of the 17th Lancers inclined to the right and then were ordered to face the front again it is strange that none of the troop or squadron officers recalled it. In those circumstances, at the start of the advance, the rank and file would be watching their own officers ahead of them for directions, pace and control. They would comply with their movements, signals or orders, not one from an ADC from a different regiment who happened to be present.

More cogent still is the fact, until now totally ignored, that Nolan could not have thought Cardigan was heading for the wrong objective when he suddenly darted forward. The Light Brigade had advanced barely 200 metres when they came under fire and Nolan was killed. That was far too short a distance for anybody who thought the objective was the Causeway Heights (No. 3 Redoubt) to appreciate that the Brigade was not heading the right way. From where the Brigade started, if its destination was No. 3 Redoubt area or the guns at the end of the valley, the route to either would be the same for much further than 200 metres.

A glance at Map 13 makes this clear. From the start point to No. 3 Redoubt was a mile in a straight line. Such a route, however, would be difficult. It would mean passing over No. 4 Redoubt and then traversing broken ground with the narrow ridge dividing the Brigade so that some regiments would have been in the North and others in the South Valleys. It would have involved attacking along the line of the Causeway Heights, which would have been difficult for cavalry with the ground giving considerable control problems in getting the regiment to arrive on the objective together. Paget confirmed this when he wrote:

'If it be correct that the object of Lord Raglan's last order was that the cavalry should, instead of going down the valley, have made an advance *along* the Causeway Heights, to recapture the guns on the redoubts, then the nature of the ground must be considered; and I think I am right in saying that such an advance would have been attended with much difficulty, the ground being broken and uneven, and of such configuration that the cavalry would have acted on it at a great disadvantage.'[7]

Given a start in the North Valley, a cavalry commander whose objective was the area of No. 3 Redoubt would move down the valley and then wheel right when he got as nearly opposite his objective as possible. He could then advance square on to the redoubt ridge. He would not attempt to make this wheel until after at least 1,000 metres, possibly more. The Odessa Regiment anticipated this when they formed squares. When Nolan moved ahead at the outset he could not have yet realized Cardigan was taking the wrong route.

Interesting though this is, it does not answer the even more tantalizing question – had Nolan misunderstood or mistaken Raglan's intentions, or had he deliberately altered the objective? Supposing he had intentionally goaded Lucan into attacking the wrong guns, then supposing, at the very last moment, after sitting quietly waiting for the Brigade to form up, the enormity of what he had done became apparent, the horror of it might have been enough to bring about a frantic bid to put things right. This would explain the failure to discuss the objective with Morris while they waited. This would explain the sudden darting forward as the line moved off. In these circumstances there would be no waiting until it was obvious the advance was heading the wrong way. Nolan would know it was wrong, and would be out to catch up with Cardigan from the outset.

It is easy to blame a dead man, but the circumstantial evidence against it being an error is weighty. Nolan had spent the morning, apart from one known absence, up on the Sapoune escarpment. He had been in close attendance on Airey and in close proximity to Raglan. He had been surrounded by generals and other staff officers. He had seen all that they saw, and he had heard all the various discussions and observations made as events unfolded. He would have to have been an exceptional dullard not to have known what was happening and why. Nolan was not dull. It is safe to assume he was fully aware of the concern at headquarters that the Russians were thought to be removing the captured British cannons, and that the object of Raglan's fourth order was to prevent them completing this task. Calthorpe, who was expecting to take the order, is emphatic that Nolan was given a verbal briefing before departing. There seems little room for doubt. He knew exactly what was expected of the cavalry.

As he slithered down the slope he made up his mind to ride with the cavalry rather than return to Airey as was his proper duty. Raglan's parting words rang in his ears again and again, 'Tell Lord Lucan the cavalry is to attack immediately'. He would make sure they did so this time. Then came Lucan's hesitation, his questioning, his reluctance to take action. Here was the Commander-in-Chief's order being queried as, yet again, Lord 'Look-on' dithered. Then, the crass question, the ultimate proof to Nolan that Lucan was the incompetent dunderhead he had always thought. 'Attack, Sir? Attack what? What guns, Sir?' Nolan pointed to the only guns both could see – down the North Valley. He had made an instant decision to force Lucan to move. If that meant going directly for the guns then so be it.

* * *

As the 11th Hussars fell back the length of the first line was cut by one third. Now only some 275 officers and men formed the first line covering a frontage of not more than 150 metres, precisely the same as the battery they were to attack. As yet they had only come under fire from the left, on the Fedioukine Heights, but the gauntlet would get progressively more deadly the closer they came to their objective. Not only would more guns be brought to bear but the range would be shrinking. Over the final furlong of the charge they would expect canister to cut them down as a scythe reaps corn.

A mile and a quarter was a long way in the circumstances. If it was done too fast the horses would arrive blown and exhausted at the very moment they needed momentum and energy. Cardigan was mindful of taking it steadily, of keeping the pace down. Being a perfectionist over details of drill, formations and parades, he would have been well versed in the *Regulations for the Instruction, Formations and Movements of the Cavalry* which had been issued by the Horse Guards in May, 1851. It was the cavalry's bible. He would surely have known the contents of Sections VIII and IX. They covered the theoretical aspects of a charge. On the speed of the attack the *Regulations* gave the following guidance:

> 'Whatever distance a Line has to go over, it is desirable, if the nature of the ground will permit, that it should move at a brisk trot, till within two hundred and fifty yards of the enemy, and then gallop [the present writer has reason to believe that in practise this meant canter], making a progressive increase [of speed to the gallop], till within forty or fifty yards

of the point of attack, when the word "Charge" will be given, and the gallop made with as much rapidity as the body [of horsemen] can bear in good order.'

That was how Cardigan intended it to be, but it was not how it happened.

Kinglake's account gives the impression of Cardigan riding the length of the valley without so much as a glance to the left or right, always staring fixedly towards the guns in front, oblivious of what was happening behind him. This picture is an exaggerated one. Cardigan was continually trying to keep the pace down to a trot. The Brigade took their speed and direction from him. Specifically, the directing squadron of the 17th Lancers under Captain White was responsible for keeping the proper distance behind the brigade commander. The rest of the first line took their dressing (alignment) from that squadron. From the outset, however, the tendency was to push on and, as the shelling increased, to go faster and faster. Eventually Cardigan was forced to give way to this pressure behind him if he wanted to stay in front.

Wightman was well positioned to observe Cardigan's efforts to check the speed:

'Lord Cardigan, almost directly behind whom I rode, turned his head leftward toward Captain Morris and shouted hoarsely, "Steady, steady Captain Morris!" . . . [The 17th Lancer squadron] was held in a manner responsible to the brigade commander for the pace and direction of the whole line. Later when we were in the midst of our torture and mad to be out of it . . . I heard again, high above the turmoil and din, Cardigan's sonorous command, "Steady, steady the 17th Lancers!" '8

The fast trot quickly developed into what equestrians like to call a 'collected canter' which meant the horses were cantering slowly, held in check by their riders. As Map 13 shows the Brigade was by then half-way down the valley and about to enter the arc of fire of Bojanov's battery. For the next 500 metres or so they were under a devastating hail of fire from two directions simultaneously. As the map shows it is doubtful if, at any time, the Light Brigade was under effective artillery fire from three directions at once. This is one of the myths that has refused to go away for 140 years.

Nevertheless, it was a hellish ride. As Morley later commented:

'If we had been moving over uneven ground we should have had some slight protection in the necessary uncertainty of aim of the guns, but moving as we did in compact bodies on smooth ground directly in range, the gunners had an admirable target and every volley came with terrible effect.'9

The urge to gallop ahead was becoming increasingly hard to contain. Morley again:

159

'There is a natural instinct to dodge cannon balls. In such fire as we were under it changed to an impulse to hurry. There was no time to look left or right, and the guns in front were what I looked out for. They were visible as streaks of fire about two feet long, and a foot thick in the centre of a gush of thick white smoke, marking about every three hundred yards of the way, as they would reload in 30 or 40 seconds.'[10]

As gaps appeared in the line the riders on either side of the falling horses concertinaed as they strove to close in again. Shouts such as, 'Close in! Close in on the centre,' 'Back on the right!' 'Come up on the left!' 'Steady, steady!' were the only orders given by officers and NCOs alike. At one stage White, the 'directing' squadron commander, allowed himself to break into a gallop and actually came abreast of Cardigan. Wightman saw Cardigan 'check him with his voice and outstretched sword', the blade of which was held across White's chest.

Like the man caught in the shower that becomes a downpour who sees his house ahead and begins to run, so, as the intensity of the fire became intolerable, the Light Brigade leant low over their horses' necks and broke into a gallop. When Britten sounded the 'Gallop' it merely endorsed what was already happening. Men were giving up all pretence at reining in their horses. Individuals were darting ahead. The Brigade trumpeter's call opened the leaky flood gates and the final mad rush for the guns began. The 'Charge' was never sounded. Not only had it become unnecessary, but Britten was grievously wounded within a few seconds of letting the bugle fall from his lips. Cardigan was being ignored.

No longer was the leading line covering 150 metres. It had been horribly thinned. As it rapidly closed the distance, and as the close-range volleys of double-shotted canister and roundshot crashed out, it seemed that the 13th Light Dragoons, or what was left of them, would strike the right (southern) half of Obolensky's battery, while the 17th Lancers took on the left (northern) part. Cardigan, miraculously, was still out in front – but only just. He was still aiming to hit the centre of the objective. The last few moments are best described by survivors.

Charging with the 13th Light Dragoons was Private Mitchell who had cause to pray aloud (successfully) twice.

'As we drew near the guns in our front supplied us liberally with grape and canister, which brought down men and horses in heaps. Up to this time I was going on all right, but missed my left hand man from my side, and thinking it might soon be my turn, I offered up a short prayer. "O Lord protect me; and watch over my poor mother." We were now very close to the guns, for we were entering the smoke which hung in clouds in front. I could see some of the gunners running from the guns to the rear, when just at that moment a shell from the battery on the right struck my horse. . . . I found my horse was lying on his near side; my left leg was

160

beneath him. . . . I tried to move . . . at that moment I heard the second line come galloping towards where I lay, and fully expecting to be trampled on, I looked up and saw the 4th Light Dragoons quite close. I called out, "For God's sake don't ride over me".'[11]

Captain Tremayne (13th LD) recalled: 'The last thing I heard before I went down [his horse was hit] was one man saying to his neighbour, "Come on; don't let those bastards [meaning the 17th Lancers] get ahead of us".'[12] Still pounding along with the Lancers was Wightman. He was later to record the death of his comrade, a soldier with eight years' service, Lee:

'Poor old John Lee, my right-hand man on the flank of the regiment, was all but smashed by a shell; he gave my arm a twitch, as with a strange smile on his worn old face he quietly said, "Domino, chum," and fell out of the saddle. His old grey mare kept alongside me for some distance, treading on and tearing out her entrails as she galloped, till at length she dropped with a strange shriek.'[13]

When his friend Private Dudley exclaimed, 'What a hole that bloody shell made,' Wightman heard Private Marsh shout at him, 'Hold your foul-mouthed tongue. Swearing like a blackguard when you may be knocked into eternity next minute.' Which is what happened to Wightman's sergeant a few moments later:

'Sergeant Talbot had his head clean carried off by a round shot, yet for about thirty yards further the headless body kept the saddle, the lance at the charge firmly gripped under the right arm.'[14]

Morgan, out in front of the 17th Lancers, shut his eyes and waited to be blown to oblivion:

'When about a hundred yards from the guns I noticed just in front of me a gunner apply his fuze [portfire] to the gun at which I appeared to be riding. I shut my eyes for I thought that settled the question as far as I was concerned, but the shot missed me and struck the man on my right full on the chest.'[15]

Then, finally, as the guns were neared Wightman again:

'Cardigan was still straight in front of me, steady as a church, but now his sword was in the air; he turned in his saddle for an instant, and shouted his final command, "Steady! Steady! Close in!" Immediately afterwards there crashed into us a regular volley from the Russian cannon. I saw Captain White go down and Cardigan disappear into the smoke.'[16]

Morley afterwards considered that:

'The ride down the valley could not have occupied five minutes [it took 7½]. I could have run it on foot in seven minutes, as I was a champion runner in those days, but the formation of the brigade with which we started had been destroyed, probably within three minutes.'[17]

Of the moments before reaching the battery he said:

'The last volley went off as we were close on them. The flame, the smoke, the roar were in our faces. It is not an exaggeration to compare the sensation to that of riding into the mouth of a volcano.'[18]

The first man into the enemy battery was Cardigan. According to his Regimental ADC, young Cornet Wombwell, who was only a metre or so behind:

'Just as he [Cardigan] got close up to a gun, it went off, luckily without touching him, and not being able to see from the smoke he rode right up against the gun.'[19]

Cardigan has described his arrival at the objective somewhat less dramatically and less accurately than those immediately behind him:

'We reached the battery in a very good line, and at the regular charging pace; and here many officers and men were killed. On leading into the battery a gun was fired close by my horse's head. I rode straight forward at the same pace.'[20]

*   *   *

Paget was surprised when he noticed the 11th Hussars holding back. Cardigan had made no mention of any changes in the Brigade's formation when he had stressed that he wanted 'best support' from the second line. Paget commanded the supports, and now he was uncertain whether the 11th was to drop back to join his line or if there were now supposed to be three lines. Should he wait for them, should he try to catch them up or should he leave things as they were? Lack of communication was the cause of this confusion. Paget was unable to solve the quandary. During the next few minutes he was to lose control over the second line so that he ended the charge commanding only his own Regiment, the 4th Light Dragoons.

The 4th Light Dragoons and 8th Hussars together numbered some 241 officers and men. In two ranks their frontage was around 120 metres and they were formed up, according to Paget, 100 metres in the rear of the first

162

line. This was too close for supports. From the start Paget began to have problems:

'The first line started off . . . at a brisk trot, the second line following, though at rather a decreased pace, to rectify the proper distance of 200 yards [the manuals specified 400]. When I gave the command to my line to advance, I added the caution, "The 4th Light Dragoons will direct". Before we had proceeded very far, however, I found it necessary to increase the pace, to keep up with what appeared to me to be the increasing pace of the first line, and after the first 300 yards [i.e. soon after they came under fire] my whole energies were exerted in their directions, my shouts of "Keep up; come on," etc., being rendered the more necessary by the stoical coolness . . . of my two squadron leaders.'[21]

Poor Paget. The first line was drawing away from him; the 11th Hussars had dropped back; he was struggling to keep up in supporting distance; the Brigade now came under effective shell fire and he began to lose the 8th Hussars. In his *Journal* he describes his difficulties:

'After we had continued our advance some 300 or 400 yards' distance, I began to observe that the 8th were inclining away from us, and consequently losing their interval. At the top of my voice I kept shouting, "8th Hussars, close in to your left. Colonel Shewell, you are losing your interval," etc.; but all to no purpose. Gradually – my attention being equally occupied with what was going on in my front . . . I lost sight of the 8th.'[22]

The 22-year-old Cornet Martyn, who was acting as adjutant, was riding on the right of the 4th and was thus quite close to Shewell. Believing Shewell unable to hear Paget's shouts, he rode across to the Commanding Officer of the 8th and said, "Lord George is holloaing to you to close in to the 4th." To which Shewell responded, "I know it, I hear him, and am doing my best." Try as he might Shewell never succeeded. The 8th Hussars were destined to slip further and further away from the 4th, miss the guns, and do very much their own thing. By the time the Brigade had got half way down the valley the second line had separated out, approximately as shown on Map 13.

Despite energetic attempts, Paget could not keep his line together. Having passed the half-way mark and given up on the 8th Hussars he attempted to catch up the 11th and thus form a new line of support. By the time the 4th reached the guns their first rank had almost come into line with the rear rank of the 11th. Nevertheless, there was no effective second line as the bulk of the 11th Hussars missed the battery to the north and the whole of the 8th Hussars missed it to the south. Although the 4th Light Dragoons went through the gunline they were too late to support the first line. As Paget himself said,

'When I got to the guns and saw all their host [Russian cavalry] advancing, I looked in vain for the first line, and never could account for them till I came back.'[23]

That the first line had suffered grievously was brought home to those in the second by the number of bodies of dead or wounded men and horses strewn in their path. The number of riderless horses and their behaviour impressed many, including Paget. Of them he wrote:

'Bewildered horses from the first line, riderless, rushed in upon our ranks in every state of mutilation . . . one was guiding one's own horse . . . so as to avoid trampling on the bleeding objects in one's path – sometimes a man, sometimes a horse . . . the poor dumb brutes, who by this time were galloping about in numbers, like mad wild beasts. They consequently made dashes at me, some advancing with me a considerable distance, at one time as many as five on my right and two on my left, cringing in on me, and positively squeezing me, as the round shot came bounding by them, tearing up the earth under their noses, my overalls being a mass of blood from their gory flanks (they nearly upset me several times, and I had several times to use my sword to rid myself of them).'[24]

Except for the final 500 metres or so the second line ran the same gauntlet as the first, the difference being that they were spared the last volleys of canister from Obolensky's guns. On the left the 11th Hussars suffered particularly from the Fedioukine guns during the early stage of the advance. They did not directly cover the 17th after dropping back, and as they swept down the valley they were the closest regiment to the Fedioukine Heights. Round shots and shells caused great havoc. Smith describes seeing the first man in his troop being hit:

'[He] was Private Young, a cannon ball taking off his right arm. I, being close on his right rear, fancied I felt the wind from it as it passed me. I afterwards found I was bespattered with his flesh. . . . Private Turner's left arm was also struck off close to the shoulder and Private Ward was struck full in the chest . . . When Private Young lost his arm, he coolly fell back and asked me what he was to do. I replied, "Turn your horse about and get to the rear as fast as you can".'[25]

Riding behind the rear rank, it was Smith's job to check why men were dropping out from his troop. After also telling Turner to get back he galloped after his regiment. At this time he chanced to look over his shoulder. In doing so he became one of the few actually to see the French Chasseurs attacking towards the Fedioukine guns. 'I saw the Chasseurs d'Afrique charging up the Fedioukine Hills at the battery that was taking us in flank.'[26] This meant the French had decided to attack in support of the Light Brigade within five minutes of the British advance beginning. It puts the time of the French intervention at around 11.15 am.

It was the 4th Chasseurs that Smith saw attacking, a unit with colourful uniforms and a colourful history in Algeria. The bright red kepis, pale blue jackets and baggy red trousers were unmistakable. The Regiment had acquired the nickname of 'The Travellers' due to the frequency of its moves around Algeria and the mountains of Morocco. The advance was under the direction of the brigade commander, General d'Allonville, with the first line commanded by Major Abdelal supported by a second under Colonel Champeron. It was a skilful flanking attack outside the arc of the Russian guns, so that the casualties suffered came from musket and rifle fire. As the French horsemen came pushing up through the scrub and bushes first the right half of the battery, then the left, were forced to pull out in confusion. Only when the Vladimir Regiment advanced against them was the recall sounded. The Fedioukine guns had been silenced.

Smith was not the only witness to the French support. An anonymous soldier of the 8th Hussars recorded:

'We pelted like mad; and looking back . . . I saw behind us the French coming down the hill. This was something to count on, far away as they were.'[27]

Riding with the 11th was Private Pennington, the future Shakespearean actor. His was an almost unique experience as he started the charge with his own Regiment, was wounded and dismounted, was rescued and given a fresh horse by the 8th Hussars, and completed the charge with them. He wrote:

'To see a forearm torn by shot or shell, bleeding and dangling by the tendons which still held it to the upper joint, or brains protruding from a shattered skull, would in cool moments have been a soul-moving and a sickening sight.'[28]

Not far from the battery Pennington was hit, he thought from the rifle fire from his right flank:

'A musket ball struck my mare's hind leg and lamed her so badly . . . that she became quite useless. I can recall the sinking of my heart as I beheld the 11th ride on, while I was left alone far from the British lines. . . . But I felt a strange reluctance to dismount. "Black Bess" had been the fastest mare in all the troop; high-bred and hardy, she had borne campaigning well. With no soul in sight, for the regiments in front had passed away obscured by dust and smoke, I made a good mark for musketry fire. A ball passed through my right leg, a shot from the left tilted my busby over my right ear, while "Bess" received the coup de grace which brought us both to earth, though I was still astride the mare. . . . I had some

165

half-formed plan of hobbling to the rear, when to my great surprise I heard the thud of hooves behind. They were the 8th Hussars! Great was my relief and joy as Sergeant-Major Harrison, seeing my plight, halted and bade me mount a grey mare he led as he rode serafile.'[29]

Harrison had the regimental nickname of 'Old Bags' because, as Pennington tells us, he wore his overalls 'loose and easy'. Pennington, bleeding profusely, rode on 'steadily at a moderate trot' with his new regiment.

According to the *History of the VIII Hussars 1693–1927* by Robert H. Murray that regiment deliberately kept the pace down (for which Pennington had reason to be grateful) even though the 4th Light Dragoons on their left gradually pulled ahead. They were kept in check by their officers as they erroneously believed the 8th was the directing regiment. It is probable, as Murray's account suggests (and Pennington endorses), that the 8th Hussars maintained a sharp trot throughout, and so never actually charged anything (there is no disputing they missed the guns). Writing home the day following the charge the adjutant, Lieutenant Edward Seager, states much the same thing:

'The fire was tremendous, shells bursting amongst us. Cannon balls tearing the earth up and Minié balls coming like hail, still we went on never altering our pace.'[30]

After seeing Fitzgibbon, Captain Tomkinson and Lieutenant Clutterbuck go down Seager positioned himself in front of the centre of the line as a squadron commander. He made no reference in his letter to a charge or even of attacking anything at that stage.

Clearly the second line had suffered severely but achieved little. Paget had not been made aware that the 11th Hussars were to be part of the supports. The pace of the first line was far faster than was anticipated; some officers in the 8th Hussars thought their Regiment was controlling the direction and speed of the line, and the entire line soon became split up into three small packets. The result was that the much reduced first line arrived at the guns well ahead, too far ahead, of its supports. By the time Paget got to the gun line the 13th Light Dragoons and 17th Lancers (and Cardigan) had disappeared. On the flanks the 8th Hussars and most of the 11th Hussars had hit thin air.

*   *   *

By eleven o'clock a large assembly of high-ranking French and English gentlemen, together with a sprinkling of ladies, had gathered high up on the Sapoune escarpment. While a few had responsibility for what was about to happen below them, the majority had come to see the show. It was a marvellous way to watch a battle. The view was stunningly clear, the troops below little blocks of colour, mostly red or blue, like so many tiny toys. Through glasses it was possible to pick out some of the enemy, especially the gun positions and the infantry scrambling about on the southern edge of the Fedioukine Heights. The chatter was that the Cavalry Division had just received orders to advance and try to prevent the Russians from taking away the British guns captured earlier in the redoubts. There were plenty of red-jacketed infantry in the plain but they seemed somewhat apathetic, standing around, moving a bit but seeming to get nowhere in particular. A cavalry advance promised to be much more dashing and spectacular. All eyes were on the horsemen as the nostalgic notes of the bugle floated up to the audience.

It is difficult to fathom what exactly Raglan anticipated seeing in the next few minutes. He had, for once, left Lucan to decide how he was to implement his orders. The guns were supposedly being taken from No. 3 Redoubt so he must have anticipated Lucan advancing to threaten that redoubt. The deployment of the cavalry at that time is shown on Map 12. The Light Brigade was in the North Valley, with the Heavy Brigade up on the Causeway ridge near No. 5 Redoubt and slightly into the South Valley. From Raglan's lofty position the most direct way of threatening No. 3 was to send the Heavies more or less straight ahead along the line of the ridge, using the Woronzoff road as an axis. The Light Brigade could then come up astride the Causeway Heights and follow as supports. Alternatively, possibly the Light Brigade would lead along the road. Perhaps one brigade would advance along the ridge while the second moved south-east across the South Valley. In any event the French Chasseurs would protect the exposed left flank. The least likely plan that Lucan might adopt was to approach the redoubt via the North Valley with his entire division. It would mean being exposed to flanking fire for some distance before inclining or wheeling towards the SE to take the objective.

The problem was that Raglan was sending cavalry to do infantry work. He hoped that the mere sight of cavalry advancing frontally against infantry and artillery on higher ground, with some in entrenchments, would be enough to get the Russians to abandon his precious guns and pull back. Wishful thinking? These same positions had already seen off the cavalry division and the Horse Artillery earlier that morning. The Heavy Brigade's (and artillery's) triumph in the South Valley against cavalry, the clamour among his entourage that the guns must be saved, had uncharacteristically emboldened Raglan to the point of rashness. It was worth a try he thought.

When the Light Brigade trotted off down the North Valley, followed shortly afterwards by the Heavy Brigade, and both came under a fearful bombardment almost immediately, the watchers, and Raglan in particular, were stunned. Gone was the cheerful atmosphere. As the enemy guns crashed out, as the pretty toy soldiers were knocked over, there came a realization that an awful error was being made below them. What are they doing? Where are they going? These were the comments and queries that flashed through a hundred heads. As the advance gathered momentum, as the barrage of shot and shell

167

took its terrible toll from either side and to the front, as the attack showed no sign of deviating from the valley, Raglan was beside himself with fury. He did not rant and rave, but the sight of what, to him, was the throwing away of his cavalry division, caused his face to contort and his empty sleeve to twitch. Lucan was going to have much to answer for. Calthorpe, sitting close by, was later to write:

> 'We all saw at once that a lamentable mistake had been made – by whose fault it was then impossible to say. Lord Raglan sent down two of his staff to ascertain the cause of all this, so little was it his intention that an attack of this nature should take place.'[31]

The French General Bosquet summed up the feelings of many when he exclaimed, *'C'est magnifique, mais ce n'est pas la guerre!'*

Lucan, however, was doing what he did with considerable reluctance, and only because he was convinced it was what his Commander-in-Chief was insisting on. Such is the friction of war. So how did Lucan propose to achieve the impossible and attack a battery so far away, frontally, and exposed to fire from three sides? He had been told to use the whole of his division. His intention seems to have been an advance, at a steady pace, by his entire command straight down the North Valley. The Light Brigade had done nothing so far, so they would lead (and anyway they were in the right position), with the Heavy Brigade following up as reserve. He wanted the attack to be in six lines, or waves – three in each brigade. It was for this reason that he told the 11th Hussars to drop back. He wanted to attack on a narrow front with plenty of depth, plenty of support, plenty of back-up. He felt, rightly, that the first line would bear the bulk of the casualties, that it might be blown away. It was a forlorn hope, a sacrifice to enable others to get to the guns. It must therefore be kept to a minimum. His inclination was to husband the Heavy Brigade. If he had to send the Lights on a mad venture Lucan was not about to embroil the Heavies too soon.

Assistant-Surgeon William Cattell, who was attached to the 5th Dragoon Guards, has given posterity an interesting insight into Lucan's thoughts as the Light Brigade moved off. The Heavy Brigade, or the two leading regiments of it, having as yet had no instructions, automatically started to follow down into the North Valley. Lucan having spotted this galloped up. Cattell wrote:

> 'Royals drawn up in line on left of Greys on N.E. slope of causeway Light Brigade on left and a little to our rear [see Map 13]. Ordered to advance we broke into a trot down the valley towards the Russian battery when Lucan galloped up shouting, "No! No! Halt the Heavy Brigade. They have done their duty. Let the Lights go!" We were accordingly halted and the Light Brigade trotted down on our left. As soon as they were some five or six hundred yards in advance of us they increased their pace to a gallop and we got an order to trot.'[32]

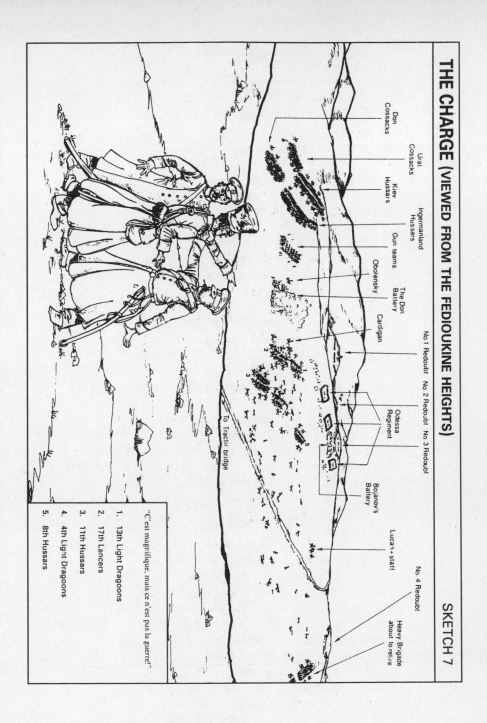

THE CHARGE (VIEWED FROM THE FEDIOUKINE HEIGHTS)

SKETCH 7

Don
Cossacks

Ural
Cossacks

Kiev
Hussars

Ingermanland
Hussars

Gun teams

Obolensky

The Don
Battery

Cardigan

No 1 Redoubt No 2 Redoubt No 3 Redoubt

Odessa
Regiment

Bojanov's
Battery

Lucan+staff

No. 4 Redoubt

Heavy Brigade
about to retire

To Tractir bridge

"C'est magnifique, mais ce n'est pas la guerre!"

1. 13th Light Dragoons
2. 17th Lancers
3. 11th Hussars
4. 4th Light Dragoons
5. 8th Hussars

After moving off, the Heavy Brigade mirrored precisely Cardigan's formation. There were three lines of two, one, and two regiments respectively. The first line (fourth in the division) was composed of the Greys and the Royals. Lucan positioned himself and his staff neatly between his two formations. We know all this because Lucan later explained his thinking:

> 'Be it remembered that I had carefully divided the Light Brigade into three lines, to expose as few men as possible in the first line, and that the first line should be efficiently supported. So soon as they had moved off, I instructed my aide-de-camp to have me followed by the Heavy Brigade formed in the same order of three lines.'[33]

Things rapidly went the way Lucan had feared rather than the way he had hoped. Once the Russian gunners opened up Cardigan rapidly seemed to lose control of the pace and allowed himself to be pushed, impelled almost, down the valley at way above regulation speed. An enormous gap began to appear between the two Brigades. Lucan attempted to bridge this with himself and his staff. To do so he had to gallop ahead into the maelstrom of flying metal that had by then descended on the North Valley. At a much more collected pace the Heavy Brigade followed.

Walker, on Lucan's staff, thought the situation decidedly hot. He had examined the enemy battery on the Fedioukine Heights just before moving off:

> 'Before they opened fire I saw these [Fedioukine] guns – or rather saw the horses – pulled out my glass, and in a moment saw what they were, and how completely they swept the whole [sic] length of our advance. I would not live over that moment for a kingdom . . . I hope I shall not soon again get such a pelting. Luckily a great many of their shells burst too high.'[34]

In a matter of minutes Lucan had lost touch with the Light Brigade. Even the much more steady 8th Hussars were ahead of him. He and his staff came almost level with No. 3 Redoubt before he began to have doubts as to the wisdom of proceeding further. His divisional headquarters was in danger of being wiped out. His young nephew and ADC, Captain Charteris, who had had a premonition of death, was killed at his side. Paulet had his hat knocked off, McMahon's horse was hit and he himself got a musket ball in his leg and his horse two hits in the body.

Some distance to the rear the Heavy Brigade was now at the receiving end of the Fedioukine guns. Their discomfiture was not to be prolonged, as the French Chasseurs were already advancing on the offending battery, but it was doing serious damage among the smart, orderly lines of the Heavies as they trotted down the valley. Colonel Yorke, the Commanding Officer of the Royals, who was about to be crippled for life, later had this to say:

'In a few moments we were in the hottest fire that was probably ever witnessed. The Regts were beautifully steady. I never had a better line in a Field Day, the only swerving was to let through the ranks the wounded & dead men & horses of the Light Brigade. . . . It was a fearful sight I assure you, and the appearance of all who retired was as if they had passed through a heavy shower of blood, positively dripping and saturated, and shattered arms blowing back like empty sleeves as the poor fellows ran to the rear. During this time there was a constant squibbing noise around me, proving even in these improved days of gunnery what numbers of shot do not take effect. However, another moment & my horse was shot on the right flank. A few fatal paces further & my left leg was shattered.'[35]

Further back the remainder of the Heavies did not seem to suffer so much. Major Forrest, riding with the 4th Dragoon Guards, thought his regiment got off quite lightly considering the amount of shot flying around:

'We, the Heavies, were taken down the valley as a support to the Light Brigade and we had batteries playing upon us, upon both flanks. We the 4th escaped in the most providential manner. The round shot were flying over us, both in front and rear and occasionally thro' the ranks; however, we had but one man severely wounded [later reported dead] and four men and three horses slightly wounded. I saw the Greys drop to the number of about 12 or 15 but I think they were principally horses that suffered.'[36]

Lucan had a hard decision to make. Should he press on? Should he support his leading brigade, or should he pull back the Heavies so that they did not suffer the same fate as Cardigan's command? To him Raglan's order had always been incomprehensible, if not downright stupid. Nolan had insisted he attack the guns at the end of the North Valley, and in complying it looked as though he had lost the Light Brigade. Should he reinforce what to him seemed a ghastly failure? It had never been a sound military principle. He made up his mind. Turning to Paulet he said, 'They have sacrificed the Light Brigade. They shall not have the Heavy, if I can help it.'[37] The Greys and the Royals were turned about, but Scarlett, who was not with Lucan and was therefore unaware what was happening, ordered his trumpeter to halt them. Not until Lucan came up and insisted did the withdrawal continue.

The Heavy Brigade came back to No. 4 Redoubt, out of the firing. Lucan was being somewhat economical with the truth when he later stated, 'they could only be useful in protecting the retreat of the Light Brigade; and I am confident that from their position they materially did so.'[38]

The Heavy Brigade performed no such service. They were almost a mile from the enemy battery and, when the remnants of their light cavalry comrades straggled back, were intercepted and attacked by Russian lancers, the Heavy

Brigade remained stationary, too far away even to see properly what was happening, let alone to assist. Lucan was not to know it, and nobody outside of the Light Brigade was to censure him for withdrawing the Heavies, but there was a chance that had he kept going he might have achieved the most spectacular of victories.

Throughout the entire day the Heavy Brigade suffered ten killed and eighty-seven wounded – virtually 100 total casualties. These included those inflicted during the early morning shelling and their own attack on Rijov's force. During their time in the North Valley they might have sustained seventy of them. Out of an overall strength of not much short of 800 it was worrying but hardly devastating. The Light Brigade had silenced the battery in front and the Chasseurs had done the same to the one on the left flank. Had they continued, the volume of fire they would have faced would have been cut by about two-thirds. It would have had nothing like the ferocity of a few minutes earlier. The next chapter will detail how the Light Brigade drove many of Rijov's cavalry in flight to the river; had another comparatively fresh brigade arrived in close support the story could well have had a different ending. A number of Light Brigade survivors roundly condemned Lucan for not giving them any worthwhile support, their argument being that if they could take the battery and push back the cavalry behind how much more would have been achieved with another brigade, and one that would have run a much less lethal gauntlet. This speculation has the benefit of hindsight. From Lucan's viewpoint at the time the position was grim indeed. He made his decision, probably the same one that nine out of ten commanders would have made. It remains one of those fascinating 'ifs' of military history.

Lucan forgot the artillery. This is not an uncommon accusation levelled against the Cavalry Division commander for failing to use the two troops of RHA available. Raglan's garbled message had contained a reminder that guns might accompany him. There is nothing quite like counter-battery fire to spoil the aim of enemy gunners. Cavalry were provided with light (6-pounder) Horse Artillery, able to manoeuvre at speed, for just this purpose. They were under the command of the cavalry general. Tactics manuals all emphasized the need to combine a mounted attack with gunfire. Smith (11th Hussars) was forthright on the subject:

'Strange as it may seem, although 12 Horse Artillery guns were at this moment close at hand, the divisional general made no use of them before he ordered the attack. It had been the custom on brigade field days for artillery fire to precede a cavalry attack. Had this simple rule been followed out on the Causeway Heights for a few minutes before we moved off it would have been of the greatest advantage to us.'[39]

The guns Smith refers to were I Troop RHA, then under Shakespear, and a part of Lucan's command, plus Brandling's C Troop RHA which was available due to the initiative of its commander, and was then waiting close by.

It is certainly true that Lucan gave his gunners no orders. When the Light Brigade moved off I Troop dutifully trotted along behind. Nobody had told Shakespear what to do or what was happening. As the Brigade gathered speed he found that he was about to take part in a cavalry charge under heavy fire. In no circumstances was this the role of Horse Artillery, although Shakespear followed until it became

'momentarily more and more apparent as the troop trotted steadily forward that, before it could render any efficient service, the Russian fire would entirely cripple it: accordingly the word was given to go about, and it retired to a position not far from the Heavy Brigade.'[40]

Their fellow Gunners in C Troop appeared even more neglected. They were deployed in the South Valley some distance in front of the Heavy Brigade (see Map 13). All had been quiet for a considerable time apart from the occasional shot from the area of No. 2 Redoubt. Brandling had gone up on to the Causeway ridge to see what was going on in the North Valley. His curiosity was well rewarded. The moment he saw the Light Brigade start down the valley he realized there ought to be a role for his guns. He pulled his horse off the crest

'[and] came galloping back, calling out "Mount, Mount", and when he came close in, he remarked in a loud tone to an officer that the Light Cavalry had begun an advance on the other side of the ridge.'[41]

The Troop moved off as quickly as it could up on to the ridge and advanced along it, 'keeping their right in the inner plain [South Valley], but their left gun was sometimes on the top of the ridge.'[42] Unfortunately the speed of events meant they only arrived in time to see, a long way off, the backs of the 4th Light Dragoons after they had swept through the battery.

Lucan, seemingly, never discussed his reasons for not trying to bring his guns into action, so it will for ever remain uncertain if he just forgot them, or whether he deliberately decided against using them. The latter is a possibility. There was some ploughed ground in the North Valley which would be difficult going for guns, slowing and tiring the teams. Lucan's main problem, however, was lack of time. He had been ordered to attack immediately. His objective had been pointed out and he was to advance on it at once. If artillery were to play a part, soften up the opposition or disrupt enemy positions while the cavalry approached, they must have time to get into firing positions. I Troop were immediately to hand but were at least 1300 metres from the Fedioukine guns. If they were to fire at that target they would have to take their 6-pounders much closer and deploy right out in the open under, literally under, the more powerful guns on the slope. Hardly a healthy option. C Troop needed time to get up on to, and move along, the Causeway ridge. To shoot at Obolensky's guns they would have to close the range considerably. In the event it took them at least eight minutes to get into position. It follows that Lucan would have needed to sanction a delay of some fifteen minutes to allow their fire to have some effect. With Nolan goading him on it is doubtful if Lucan gave the idea a second thought.

By 11.18 am the actual charge was over. What was left of the Light Brigade had disappeared into the smoke; the Heavies were back near No. 4 Redoubt and had been joined by 'I' Troop; 'C' Troop was watching impotently from close to the Causeway crossroads; the Chasseurs had silenced the Fedioukine guns; dozens of wounded men and horses were starting their agonizing struggle back up the valley and, unnoticed by the watchers, four squadrons of Russian Lancers were approaching the valley. They were looking for easy pickings.

CHAPTER NINE
Through The Guns

'My humble opinion is that it is quite sufficient for a
general of brigade to return with as well as lead the
attack of the front line, unless he should by chance
come in contact with his supports, in which case he
would remain with them.'
The Earl of Cardigan in a statement to Mr Kinglake.

Cardigan was the fourth horseman of calamity. It fell to him to execute the
infamous fourth order. As the brigade commander it was his duty to lead and
command his brigade personally, which in this instance involved a brigade
attack on a battery of guns. As the above quotation makes clear, Cardigan's
interpretation of his responsibilities was limited. Certainly, he accepted that a
brigadier-general must place himself ahead of the first line of an attack to
control the pace and direction. Cardigan did this admirably and without
flinching. But, having led the first line to the objective, that was, he felt, the
end of his obligations – unless by good luck he happened to contact his
supports! This was typical Cardigan. This was the cardboard soldier again. His
understanding of command in battle was minimal. Having got to the guns he
had launched his brigade with due impetus and considered his duty done.
What happened next? Where was his second line? Where was his second-in-
command, Paget? Should he rally his command and give fresh orders? Should
he locate his divisional commander? These were all questions which, if they
flashed through his mind, were never translated into any action. Cardigan,
after a few moments of half-hearted prodding from some Russian lances, extricated himself from behind the guns and rode smartly back up the valley, an
even more solitary figure than the one who had ridden down it.

Did Cardigan abandon the Light Brigade, or what was left of it, at the very
moment when it needed firm leadership most? To men like Paget it certainly
looked that way at the time. Pennington later related that, 'He [Cardigan]
certainly did not bring his brigade out of action, but he led them nobly in.' All
the manuals laid down the importance of rallying immediately after a charge.
Cardigan knew his drills and had laid much stress on the 'rally' being practised
again and again during his numerous field exercises in Turkey. So it can hardly
have been that he did not know what to do. Somehow he had to find his, or
any, trumpeter, some members of his staff, commanding officers, or

175

regimental officers and rally his brigade. This was the fundamental first step. His subordinates would try to regain control over their regiments, but he must make his presence known and endeavour to reassemble his command and give instructions on what to do next. Every regimental commander would be asking, 'Where is Cardigan?' 'What now?' 'Where next?' There was no answer from the 'Noble Yachtsman'.

Walker, Lucan's staff officer, was appalled by Cardigan's behaviour, writing a few days later:

'Don't believe any bosh you hear about Cardigan [i.e. that Cardigan had done splendidly]. He showed no head, and beyond riding *with* his brigade, no greater pluck than others. Old Scarlett is worth two of him.'[1]

It is time to examine what actually happened to Cardigan during the two or three minutes he spent behind the Russian guns.

Cardigan himself wrote that, 'During this affair [his period behind the guns] I was not accompanied by any aide-de-camp.' This was a fact. He had started off with four staff and a trumpeter, but by the time he hurtled through the smoke none of them was with him. Mayow, his brigade major, had become separated and embroiled with a group of the 17th Lancers; Maxse had retired hurt short of the battery; Lockwood was hit early on in the charge and disappeared (literally, as his body was never found and he was not taken prisoner); Wombwell had his horse shot from under him just as he reached the guns and the body of his trumpeter, Britten, was lying contorted and bleeding about a third of the way back up the valley.

That Cardigan and his horse reached the battery untouched, after being so conspicuously out in front for so long, was astonishing. The very first shell, the one that removed Nolan, burst closer to Cardigan than anyone else. Then, just a second before he reached the battery, the gun facing him fired. The flame almost scorched his horse's head and the blast blew him sideways. It was only with difficulty that Cardigan corrected the stumble, righting 'Ronald' as he swept past the gun. Once more neither rider nor mount were so much as scratched. It has been estimated that as Cardigan flashed through the enemy position waving his sword he was moving at seventeen miles per hour. Within moments he was clear of the guns and into the close-packed limbers and teams drawn up in rear, some of which were trying to withdraw or extract their guns. He almost crushed his leg against a limber as he plunged through the tangle. Behind him what was left of the first line hit the battery and began their brief struggle with those gunners who stood firm.

As Cardigan came clear of the smoke and confusion he was still alone. There, immediately to his front, not more than 100 metres away, was a large formed body of enemy cavalry. The momentum of the brigade commander's gallop carried him on towards the waiting Russians. By hauling viciously on the bit Cardigan managed to halt his wildly excited charger. He was then a mere 20 metres from the sword and lance points of his enemy. He faced certain death or capture. This time it was his being face to face with the Russians that saved him. By a miraculous coincidence, that no fiction writer would dare to invent, the senior officer opposite him recognized Cardigan, as he had met him on the London social circuit before the war. Prince Radziwill, being a man of honour and breeding himself, had no wish to see noble blood spilt needlessly so detached a troop of Cossacks to capture Cardigan alive. Equally important, the presentation of so prominent a prisoner to Liprandi could do his career no harm at all.

As the encircling Cossacks closed in, Cardigan sought to evade them. Not only was he alarmingly outnumbered, but the prospect of having to defend himself against several vulgar serfs simultaneously was not to his lordship's liking either. As he put it, quite rightly, it was 'no part of a general's duty to fight the enemy among private soldiers'. Cardigan had no way of knowing his adversaries were not about to kill him, so rather than defend himself he sought by superior horsemanship to dodge his opponents. At this time the adjutant of the 13th Light Dragoons, Lieutenant Smith, caught a brief glimpse of his brigadier. He noticed that Cardigan kept his 'sword at the slope and did not seem to take any trouble to defend himself'.

Fine equestrian though he undoubtedly was, Cardigan could not avoid all the lunging lances. He was poked, none too forcefully, in the ribs and received a minor wound in the thigh. The reluctant restraint of the Cossacks enabled him to wrench his horse's head around, swerve and dash back towards the gunline, away from the enemy who were trying to surround him. Cardigan's own account of this episode is surprisingly succinct, modest even:

'I led into the battery and through the Russian gun limber-carriages and ammunition wagons in the rear. I rode within twenty yards of the line of Russian cavalry. I was attacked by two Cossacks, slightly wounded by their lances, and with difficulty got away from them, they trying to surround me. On arriving at the battery through which I had led, I found no part of the brigade. I rode slowly up the hill, and met General Scarlett.'

'I found no part of the brigade.'[2] This was hardly unexpected as in the short time he had spent avoiding capture the second line had arrived at, and passed through or round, the objective. Apart from the dead, wounded and unhorsed men of his command, whom he could see straggling back up the valley, the bulk of his brigade was behind him engaging the enemy. Cardigan seems to have then assumed that 'the small broken parties retreating', as he later described them, were all that were left of his nearly 700-strong Brigade. Without attempting to find out what had happened, without so much as a backward glance, the cardboard commander closed his mind to what might be going on behind him and trotted back the way he had come.

Events had moved at high speed. The charge itself, including the arrival of the second line, lasted under eight minutes. Cardigan had spent two, perhaps three, minutes behind the battery, and the remnants of the Brigade had made their way back up the valley some twenty minutes after starting out. Although Paget's second line had become broken and separated it was never more than 300–400 metres behind the first. Thus the 4th Light Dragoons reached the objective at about the time the brigade commander was having his confrontation with the Cossacks. The 11th Hussars 'brushed' the battery to the north, while the 8th Hussars bypassed it to the south at almost the same time. Some members of the first line spotted Cardigan behind the guns. Smith had seen him being threatened by the Cossacks, while Morley (years later) alleged that he even suggested rallying on Cardigan. He wrote:

'My first thought after we were through the line was to look for an officer to see what we were to do. I saw Lord Cardigan at first but had no impulse to join him. I think no British soldier ever had. He led 670 and none relied on him. I saw troopers riding past him to the right and left. He was about 50 yards beyond the guns on their extreme left. I turned to look for some of my own regiment and mistook Lieutenant Jarvis [Jervis] of the 13th Light Dragoons, for one of my officers. . . . I galloped up to him and informed him that Lord Cardigan was about, pointing my sword to the place. . . . He replied "Never mind. Let's capture that gun!" '[3]

There can be no doubt that it was not lack of physical courage that made Cardigan retire prematurely. More likely was that he genuinely felt he had done his duty in leading the first line. That line had gone, perhaps been destroyed, and of the second he saw nothing. He did not consider it his job to go looking for them, so he turned his back and rode away. According to his thinking his presence was simply no longer necessary – it was the regimental officers' task to sort out the mess. A more accurate, if perhaps more modern, interpretation would be that Cardigan had abandoned his Brigade in battle.

*　　*　　*

As always happens in war both sides have their problems. With the Russians it was in getting Rijov's cavalry to face the terrible English horsemen again. After their ignominious retreat earlier, in the face of half their number, they were now required to turn back a wild charge that no amount of gunfire had been able to halt. Their response to Rijov's order to move forward to protect the guns was less than enthusiastic. The Russians, like most armies of the time, set great store by their artillery. In the circumstances pertaining, as the Light Brigade descended on Obolensky's battery, it was the duty of the nearby Russian cavalry to support their artillery. Rijov gave the necessary order for some of the Cossack squadrons to advance to save the guns.

On the slopes of the knoll near No. 3 Redoubt a group of Uhlan officers had watched the charge down the valley with a mixture of disbelief and admiration. One of them, a squadron commander in Colonel Jeropkine's regiment by

22. An unusual view of Balaclava and the Genoese towers. The photograph shows British hospital buildings in the foreground, and a glimpse of some of the ships in the harbour at centre right. Note the steepness and barrenness of the hills (Robertson – IWM)

23. A panoramic view from Raglan's viewpoint on the Sapoune escarpment taken in 1855. The diggings of Redoubts 5 and 6 are clearly visible while those of 2, 3 and 4 are just discernible. The tactical importance of the Woronzoff Road, which is unmistakeable, is immediately obvious. The complete lack of vegetation is in marked contrast with today. It is from this unique photograph that Sketches 2-6 have been compiled (Fenton - Royal Archives Windsor Castle)

24. A modern panoramic view of the Balaclava battlefield taken from near the same viewpoint as Fenton's. Modern roads, buildings, vineyards and the trees on the Sapoune make it difficult to recognize at first. Balaclava is on the extreme right, Canrobert's Hill is the rounded hill, partially covered in bushes, in the centre middle distance. The old Woronzoff Road is the track running diagonally from the botton left, across the vineyards of the North Valley, up on to the Causeway Heights (Farmer)

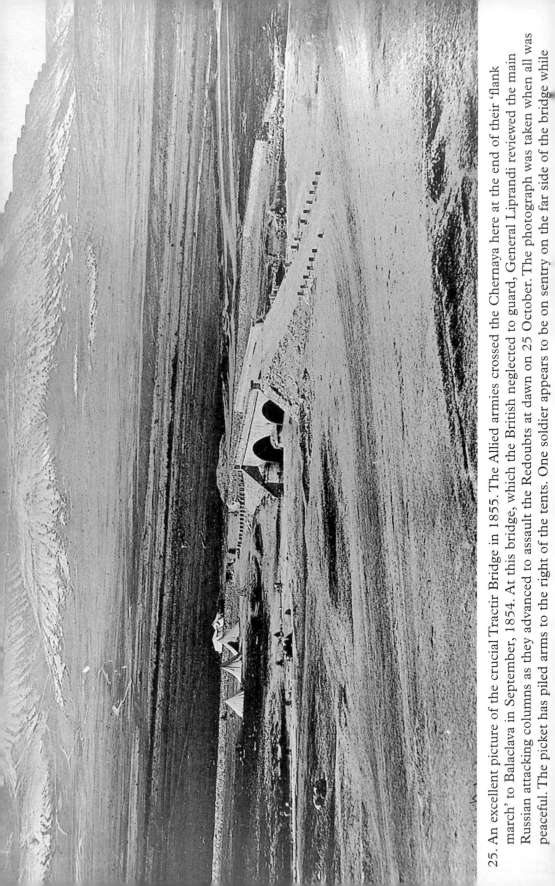

25. An excellent picture of the crucial Tractir Bridge in 1855. The Allied armies crossed the Chernaya here at the end of their 'flank march' to Balaclava in September, 1854. At this bridge, which the British neglected to guard, General Liprandi reviewed the main Russian attacking columns as they advanced to assault the Redoubts at dawn on 25 October. The photograph was taken when all was peaceful. The picket has piled arms to the right of the tents. One soldier appears to be on sentry on the far side of the bridge while

26. Typical Royal Artillery gun teams (IWM)

27. A group of the 8th Hussars, mostly senior NCOs, grouped around a makeshift field kitchen. Note the wife, who probably helped with the cooking, in the background (IWM)

28. A section of the Fedioukine Heights as it is today taken from the edge of the North Valley. There was little or no vegetation in 1854 (Farmer)

29. No. 1 Redoubt, Canrobert's Hill, today as viewed from the Causeway Heights. The bushes were not there in 1854 (Farmer)

30. The Sapoune escarpment down which Nolan plunged with the fourth order as it looks today. Raglan's actual viewpoint was a little to the left, off the photograph. The white building is a modern Russian military museum and observation post. The steep slope was bare and stony at the time of the battle (Gordon)

31. A modern view of the eastern end of the North Valley. Obolensky's battery was drawn up across the ploughed ground in the centre, with Rijov's cavalry massed forward to the line of bushes in the middle distance (Farmer)

32. A photograph of the act[ual]
4th order as written by
Airey. Not perhaps the [best]
handwriting to read in [a]
hurry under pressure. T[he]
reader will find the text [on]
page 127.

33. The old British
Balaclava memorial
that stands off the
Woronzoff Road just
south of the
Causeway Heights
near where the Heavy
Brigade attacked. The
scars were caused by
bullets fired in World
War 2. In the distance
at centre right is the
ouline of No. 1
Redoubt (Canrobert's
Hill) (Farmer)

the name of Kubitovitch, left an account of what he saw. As the Cossacks began, hesitantly, to close up towards the guns it soon became obvious to the watchers on the hill that Rijov had made a poor choice in selecting these irregular and ill-disciplined horsemen for such a crucial task. Kubitovitch described how

'the Cossacks, frightened by the disciplined order of the mass of cavalry bearing down on them, did not hold, but, wheeling to their left, began to fire on their own troops in their efforts to clear a route of escape; the Ingermanlandsky, who were covering the guns, were put into a fearful turmoil, being thrown back on the Kievsky; the whole Russian cavalry force in the valley made off, the good officers trying in vain to hold their men – some threw themselves forward against the enemy only to be cut down – General Rijov being one of the last to withdraw, seeking death, for he knew that he would be held responsible . . . the fault was indeed Rijov's for he should not have sent Cossacks ahead to meet the charge of line cavalry.'[4]

Map 14 shows how that end of the North Valley twists north-east and narrows slightly before gradually sloping down to the river. Before the banks of the river are reached, however, there is a man-made obstacle to be crossed – an aqueduct. This supposedly carried Sevastopol's water supply. There were two simple, flimsy footbridges over the aqueduct, neither of which had any side rails. They were both potential bottlenecks, particularly if large bodies of cavalry attempted to cross them in a hurry. Guns and wagons would have to cross slowly and one at a time. Once over the aqueduct retreating troops would head for the solitary ford, or north-west for the Tractir bridge some 1200 metres away. A defeated or withdrawing army has always been exceptionally vulnerable if it has fought with an obstacle at its back. Strike a man's hand with a hammer and it will be hurt, strike it when it is on a table and it will be crushed: the same principle applies to armies with barriers behind them. This was the situation confronting Rijov as the Light Brigade crashed through his guns. If his cavalry could not hold there was certain to be slaughter and chaos at the aqueduct, if not at the river.

According to Kubitovitch this is almost precisely what happened:

'The English then rode down the Don Battery sabring a number of gunners as they passed through; but the majority of the gun crews made off, mounted on horse teams and limbers. The enemy spiked some of the guns hoping to drag them off on his return, but most of the cavalry continued the headlong chase after the [Russian] hussars, slashing at them without mercy. The horse teams and limbers of the Don Battery together with 12 Horse Battery [the other battery of the 6th Hussar Brigade which had not engaged the British in the North Valley] and all the cavalry were soon milling about at the river, all trying to get over the

179

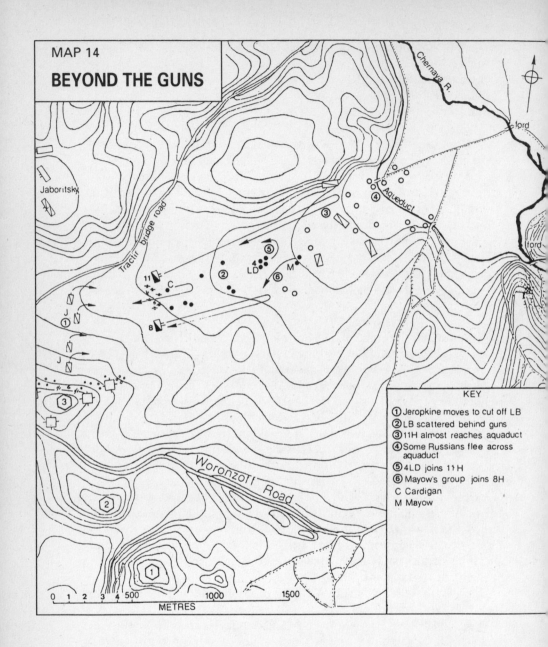

MAP 14

BEYOND THE GUNS

Chernaya R.

ford

Jaboritsky

Tractir bridge road

Aqueduct

ford

J
①

J

③

②

Woronzoff Road

①

METRES
0 1 2 3 4 500   1000   1500

11

C

8

LD

M

③

④

⑤

②

⑥

KEY

① Jeropkine moves to cut off LB
② LB scattered behind guns
③ 11H almost reaches aquaduct
④ Some Russians flee across aquaduct
⑤ 4LD joins 11H
⑥ Mayow's group joins 8H
C Cardigan
M Mayow

180

bridge, while the English chased them almost as far as the transport lines.'[5]

This account has some inaccuracies and has (unusually) exaggerated the panic in the Russian force, but does give a correct overview of the effect of the Light Brigade's charge on this part of Liprandi's line. No guns were spiked, as we know from Smith that only four spikes at most were carried at the start. Many, but not all, of Rijov's cavalry attempted to flee, and 'milled about' near the exits over the aqueduct rather than the river. It has not been established exactly where the transport lines were, but it is extremely doubtful if any of the Light Brigade got as far as the aqueduct itself which was a mile from the battery, let alone the river. Nevertheless, the Russian Hussars and Cossacks put up a poor performance against a greatly inferior force that had just been battered by sustained gunfire. When the charge ran out of steam, when the smallness of the attacking force was discovered, and when it was apparent that there was no support, no follow up, then the retiring Russians turned round and pushed back up the valley. At this stage the remains of the British regiments were in danger of being squeezed between Rijov's Hussars and Jeropkine's Uhlans.

When Liprandi, who was up on the Causeway Heights close to where the Uhlan officers were sitting on their horses, saw that no more British troops were coming down the valley, that the Heavy Brigade had been withdrawn and no attack had materialized along the Causeway ridge, he turned to Colonel Jeropkinc. The Light Brigade had overreached itself. A splendid opportunity was unfolding below to cut them off. The colonel was instructed to take his four squadrons of lancers, distinguishable by their green and white pennons, and move quickly into the valley to block their escape. Map 14 shows the route they took. Kinglake has caused some confusion for historians on this issue due to his knowledge of the Russian operations being understandably shaky. He marked two groups, of three squadrons each, of Jeropkine's Lancers on his maps. One was correctly sited near No. 3 Redoubt, but the other was a mile away across the North Valley on the Tractir bridge road. He had both these groups come out to cut off the Light Brigade.

Jeropkine had only four squadrons (not six) of Uhlans armed with the lance. His regiment formed the bulk of the cavalry element of Gribbe's left-hand column that had opened the battle from Kamara. When the redoubts fell he moved up just behind No. 3. At no stage did he split his command and send half across the valley to sit on the Tractir bridge road. To do so made no military sense at the time, that is before the Light Brigade attacked. Once they charged it was impossible, and too late.

Jeropkine led his Uhlans, including Kubitovitch's squadron, down the slope of No. 3 Redoubt and across the front of one of the Odessa Regiment's squares. The infantry were trigger-happy and the battalion commander nervous, still expecting a cavalry attack at any time. Unlike most of the Russian regiments which were mounted on horses of a uniform colour, the Uhlans, being a composite unit from various detachments, had horses of a

181

variety of colours. This may have confused the infantry. The square let go a volley which caused casualties. A minute or so was lost before matters were sorted out and Jeropkine could continue. He headed for the Tractir Bridge road which would put him well behind the Light Brigade. There he would turn to face the fleeing British.

*　　*　　*

Collins dictionary describes a mêlée as 'a noisy riotous fight or brawl'. Most participants in the few, frantic minutes of close-quarter combat behind the battery would feel this definition to be a serious understatement of their experience. Before highlighting some of the desperate individual exploits, however, it may be helpful to recap on the broader picture so that personal encounters are placed in context with what was happening within the Light Brigade. Map 14 shows the approximate movements of the regiments during this period.

The first line had been well shredded by the time it hit the battery. The 17th Lancers struck the left (northern) half of the gun position. A handful, under Morris, missed the guns completely and rode through to face the cavalry behind. The 17th were broken up into small groups and individuals each striving to keep together, strike down the gunners, capture guns even, and survive. Once beyond the guns the regiment was no longer functioning as such. It had become dispersed, with leadership depending on the nearest officer or NCO. On the right the same fate awaited the 13th Light Dragoons. They rode through the right (southern) half of the battery and immediately became scattered into small bodies of men hewing and hacking as best they could. They lost their commanding officer, Oldham, seconds before reaching the guns. His white mare, a notorious brute, had bolted, luckily straight ahead. This had resulted in Oldham seeming to rival Cardigan for the honour of being first at the objective. Captain Jenyns had seen him killed:

> 'Oldham I saw killed by a shell which burst under his horse and knocked over two or three others. It blew his mare's hind-legs off, and he jumped up himself not hit, when next moment he threw up his hands and fell dead on his face.'[6]

His body was never found.

A portion of the 17th swung involuntarily towards their left in order to avoid the clutter of limbers and horse teams immediately at the rear of the battery. While some of the first line attacked the Russian Hussars and Cossacks others swirled around in the thinning smoke until, within a minute, the 4th Light Dragoons came crashing through with Paget still at their head. They joined the turmoil and uproar. The 11th Hussars had all but missed the battery to the north, galloping past mere seconds before the 4th. After a brief pause to regroup they continued on, charging and dispersing bodies of cavalry until they came close to the aqueduct. Only then did they turn about and begin to fight their way out – still functioning as a Regiment.

Paget, with the stub of his cigar still clamped between his teeth (he had neglected to draw his sword), was looking desperately for Cardigan. His search took him to the left where he briefly met Colonel Douglas of the 11th. Douglas shouted, 'What are we to do now, Lord Paget?' He received no guidance. All Paget said as he spurred away was, 'Where is Lord Cardigan?' What was left of the 4th were gathered together by their colonel before pushing on towards the river, Paget still hoping to find the first line and his brigade commander, who was by this time behind him.

On the right of the second line, but arriving some minute or so later and wide of the guns, were the 8th Hussars. They had suffered like the rest during the advance but were well in hand, still moving at a fast trot or controlled canter. They advanced for some 300 metres and halted. Colonel Shewell was uncertain what to do. There was no sign of Cardigan or Paget so the 8th turned about to begin to fight their way back the way they had come. Within three or four minutes all those of the Light Brigade behind the gunline who were physically able to do so had wheeled around to begin the long ride back up the valley. The 17th Lancers' and 13th Light Dragoons' survivors rode in pairs or as individuals, if they could do so attaching themselves to larger bodies of the regiments of the second line who, by and large, had retained some cohesion as units.

A prerequisite for understanding what it was like to be in a Light Brigade saddle during those desperate but exhilarating minutes behind the enemy position is to read what a few of those cavalrymen endured.

*   *   *

Captain Morris, who only the day before had been on Lucan's staff, found himself in command of that part of the 17th Lancers that bypassed the northernmost gun. With not more than 15–20 men behind him he yelled over his shoulder, 'Remember what I have told you, and keep together,' and kicked hard into the flanks of 'Old Trumpeter'. He led straight into the Russian cavalry. He levelled his sword, edge outwards, arm thrust forward at shoulder height. He headed at full gallop for what appeared to be an Hussar officer. His point dipped neatly under the Russian's defence and within a split second Morris's sword had slid through the man's body. The momentum slammed the hilt home against his chest. The blade was sticking out of the Russian's back and had jammed tight in bone and muscle so that in falling the body had almost jerked Morris from his horse. He was effectively tied to the dead body, anchored to the spot and bent double out of his saddle trying frantically to withdraw the blade. He was reluctant to let go the hilt and disentangle his sword knot, so for some moments continued his tussle with the corpse. Seeing his predicament more enemy closed in.

Understandably, Morris was never again an advocate of the point over the cut. He would later say, 'I don't know how I came to use the point of my sword, but it's the last time I ever do.' As he wrenched and heaved Morris received two sabre cuts to his head, one of which sliced away a piece of bone above his ear. These blows knocked him from his horse and he lay for a few seconds unconscious. On coming to his senses he discovered that his sword had been freed, probably by his fall, but that he was now surrounded by Cossacks. A wounded man at bay was a favourite victim for a wolf pack. They lunged at him from several directions. Weakened, and partially blinded by his own blood, Morris resorted to the 'Moulinet', the whirling of his sword around his head. He slashed at least one thigh, but was struck again on his head by a lance point. Morris was sure he was about to die. At that moment an Hussar officer rode up slashing at the Cossack lances and calling out in English for Morris to surrender, and that if he did so his life would be spared. He handed over his sword and became a prisoner of war.

Morgan, a squadron commander of the 17th Lancers, later described to Assistant-Surgeon Cattell his brief encounter in the gunline.

> 'In another minute I was on the guns and the leading Russian horse [they were trying to hitch up and withdraw the cannon] shot I suppose with a pistol by someone on my right, fell across my horse, dragging it over with him and pinning me between the gun and himself. A Russian gunner on foot at once covered me with his carbine. He was just within reach of my sword and I struck at him, which disconcerted his aim. At the same moment a mounted gunner struck my horse with his sabre and the animal bolted with me right into the Russian lines.'[7]

Wightman described the smoke in the gun position as so thick that:

> 'I could not see my arm's length around me. Through the dense veil I heard noises of fighting and slaughter, but saw no obstacle, no adversary, no gun or gunner, and, in short, was through and beyond the Russian battery before I knew for certain that I had reached it.'[8]

Wightman was not far from Cardigan. As he rode forward he saw the brigade commander's ADC, Maxse, who was badly wounded, yelling at him, 'For God's sake, Lancer, don't ride over me!' and, pointing through the smoke, added, 'See where Lord Cardigan is. Rally on him.'

Wightman asserts that he tried to do exactly that. As he rode forward, however, a Cossack speared him in the thigh. Wightman's reaction was instantaneous and effective:

> 'I went for him, but he bolted; I overtook him, drove my lance into his back and unhorsed him. . . . When pursuing the Cossack I noticed Colonel Mayow deal very cleverly with a big Russian cavalry officer. He tipped off his shako with the point of his sword and then laid his head

right open with the old cut seven [a sword stroke straight from the drill book].[9]

Wightman pressed on, and was fleetingly involved in a fight that involved the brawny, 6 foot 2 inch ex-labourer and now old soldier of 23 years' service, Private Parkes (4th LD). Parkes was Paget's orderly cum bodyguard, but had become separated from him when his horse was shot. Now on foot Parkes came across Trumpeter Crawford of the 4th Light Dragoons (who by this time were part of the mêlée) who had been wounded, was dismounted and had lost his sword. He was being attacked by two Cossacks when Parkes intervened. With one arm supporting Crawford the other holding his sword Parkes kept his body between the enemy and his injured comrade, at the same time slashing out at his assailants and driving them off – with some assistance from Wightman. Parkes retired slowly, supporting Crawford as he did so. The pair were attacked again by six Russians but again Parkes kept them at bay until a heavy blow from a sword across his arm forced him to drop his weapon. Both were captured; both survived; Crawford to be promoted trumpet-major, Parkes to receive the Victoria Cross from the Queen at a parade in Hyde Park.

Meanwhile Wightman, joined by Privates Mustard (17th L) and Fletcher (4th LD), had come through the enemy cavalry that were milling about near the aqueduct and were actually able to see the river. They had gone far enough. On retracing their steps they were joined by Private Marsh (17th L) and together began the long journey back. Within a few moments they heard

'the familiar voice of Corporal Morley, of our regiment, a great rough, bellowing Nottingham man. He had lost his lance hat, and his long hair was flying out in the wind as he roared, "Coom ere! Coom ere! Fall in lads, fall in!" Well, with shouts and oaths he had collected some twenty troopers of various regiments. We fell in with the handful this man of the hour had rallied, and there joined us also under his leadership Sergeant-Major Ransom and Private John Penn, of the 17th. . . . [Penn] had killed a Russian officer, dismounted, and with great deliberation accoutred himself with the belt and sword of the defunct, in which he made a great show.'[10]

Earlier Morley had been involved in the attempt to capture a cannon with Lieutenant Jervis, who, it will be recalled, had shown little interest in rallying on Cardigan. Morley's account reads:

'We raced towards it. He [Jervis] said, "Cut down the gunners!" He shot one of the horses in the head bringing it to a sudden stop. The gunners disappeared between the horses and the gun-carriage as we slashed at them. We both dismounted and took out the dead horse while more of the Brigade gathered round to assist us. . . . We started back off the field at a gallop with the mounted cannon, and were near the place where I had seen Lord Cardigan, when a large body of Cossacks charged.'[11]

The captured gun was abandoned.

After Morris went down the 17th Lancers seemed to split into two small groups, one under Mayow, the brigade-major, and the other under Troop Sergeant-Major O'Hara. Mayow's party penetrated some 500 metres beyond the battery and within sight of the small bridges over the aqueduct. At this point they were looking around anxiously for support, being well mixed up with the enemy cavalry, when the cry went up, 'Look! Over there, the "Busby-bags" are coming!' They had spotted the headdress of the 8th Hussars with their scarlet bags hanging down the sides. They turned towards them. It was time to go.

*　　*　　*

On the left of what was supposed to be the second line were the 11th Hussars. They had allowed the first line to get ahead of them, but during the mad dash down the valley there never was a proper second line. Paget, with the 4th Light Dragoons and 8th Hussars, never quite caught up with the 11th. Viewed from the ranks of the 11th the first line seemed to dissolve. According to Douglas, 'this effect being caused by the dust, smoke, and the numerous casualties every moment taking place.' The 11th, as we have seen, 'brushed' the battery with its right-hand troop and rode through without difficulty. There were, however, a few casualties at this stage, one of whom was Private Pilkington who had enlisted under the false name of 'Tom Spring'. Spring later recounted to Pennington his astonishing escape.

Shortly after passing through the battery his horse fell, bringing him crashing down with one foot twisted and trapped in the stirrup. Seeing him struggling on the ground to extricate himself a Russian officer rode up and, at close range, emptied every chamber of his revolver into the helpless Hussar. Incredibly, although the bullets struck home, none penetrated sufficiently through his belt and jacket padding to cause a serious wound. Spring was taken prisoner but survived, living on well into his eighties, no doubt extricating maximum alcoholic benefit from his story on countless occasions.

Another Russian officer, 'decorated with several orders', came up to surrender personally to Douglas. He, however, like so many soldiers before and since in a similar situation, had no time to be bothered with prisoners, so ignored him. A hundred metres to the rear Douglas halted his men. He needed to rally and decide his next move.

It was at this moment that Paget had galloped up and, when Douglas had asked what to do next, had failed to respond other than to ride off asking, 'Where is Lord Cardigan?' Douglas has described what he did next:

'I here saw a body of Russian cavalry to my left front, and on the impulse of the moment I determined to attack them; my reasons were that I thought I could do it with very great advantage, being under the impression that I could jam them into the gorge of the valley, here forming a sort of cul-de-sac from which there would be difficulty in escaping, as the aqueduct and the River Chernaya barred any hasty egress.'[12]

Smith remembered these first moments behind the guns well. He later wrote:

'It now became my particular duty (the adjutant having remained behind when we received the order to attack, and the regimental sergeant-major's horse having been wounded) to note every word of command the colonel should give, as I now considered the adjutant's duty devolved on me. After halting, the word was, "Come in on the centre".... Colonel Douglas . . . having no order to retire and expecting any moment we should be charged by this body of cavalry to our front, called out, "Give them another charge, men. Hurrah!"'[13]

Douglas was, as he emphasized in his account, still thinking in terms of having supports coming up from behind. He expected 'that shortly both infantry and fresh cavalry would come up.' Like many others he had seen the Heavy Brigade and elements of the 4th Division in the vicinity long before the advance got under way. If the Light Brigade could get through others would surely follow – or so he reasoned.

The 11th Hussars surged forward again, swords held high. When within a short distance of the enemy the Russian cavalry suddenly wheeled about and fled. The 11th pursued them. Douglas recalled the thrill of the chase:

'As we came upon them they got into confusion and very loose order. My men got greatly excited, and we pursued at our best pace, they sweeping round the base of the hills to our left front, forming the end of the valley.'[14]

Meanwhile Smith, who was still riding at the rear of the Regiment, had been distracted by the prospect of capturing a cannon. On looking to his right he spotted a 'large brass gun with carriage painted green, drawn by six horses; there were only three men – drivers – with it.' He resolved to take it. Being a typical sergeant-major he immediately looked round for 'volunteers' to help him. Three men just ahead of him had been squeezed out of the tightly packed ranks and were thus riding independently. In his best parade-ground voice Smith bellowed, 'Follow me! Let's take this gun.' Turning his mount, Smith spurred towards the escaping gun team. When he had got within a short distance of it he noticed that a Hussar officer and three Cossacks had detached themselves from the main body and had advanced to protect the gun. These four horsemen had got between Smith and his regiment; three enemy with the gun, another four coming to assist and at the same time cutting him off from his comrades. Smith looked anxiously round for his 'volunteers'. There was no one in sight. He was alone, a good 300 metres from his regiment and 50 from the gun. Smith did not like the odds:

'Feeling it would be madness to attempt the capture singlehanded, I instantly halted, turned about and galloped off in the direction of my regiment. The first Cossack and the officer might have engaged me on my track, but they hesitated, calling out to the other two who were a little distance in their rear. This hesitation allowed me to rejoin my comrades.'[15]

By this time the 11th were well down towards the aqueduct (see Map 14), and had penetrated the furthest of any regiment. Smith arrived back in time to witness the incredible scene. His Regiment was completely isolated far behind the objective. Ahead and to the flanks were vastly superior bodies of the enemy. For a moment both sides sat a few metres apart staring at each other, wondering what to do, wondering who would blink first. Smith's description of this moment is impelling and vivid:

'We were now nearing the extreme end of the valley, about a mile and a half from our [original] position, still pursuing this body of cavalry. In their confusion, I saw one of the leading Cossacks fall from the bridge into the aqueduct, there being no parapet. Near the bridge was a moderately steep hill, which formed the end of the valley [indicated on Map 14], up which they rode a short distance; their rear being at the foot, close to us. They now halted, but remained for a few moments with their backs to us, looking over their shoulders. Seeing there were so few of us and without supports, they turned about and we sat face to face, our horses' heads close to theirs. As we looked up at them, they had all the appearance of a vast assemblage in the gallery of a theatre. The stillness and suspense during these moments was terrible. At last it was broken by their officers calling out to their men to follow them and break through us, which they themselves attempted to do by driving their horses at our front rank. But their men failed to show the same courage as their leaders, and our men showed a firm front, keeping close together and bringing their swords down to the right front guard, and so kept them at bay. Many of them now took out their pistols and fired into us, and the Cossacks began to double round our flanks and get in our rear. Our position became every moment more critical, for we were in danger of being surrounded, overwhelmed and killed to a man. But, had a few more of our squadrons [i.e. the Heavy Brigade] come up at this time, I am of the opinion that this body of cavalry would have surrendered to us, for we, numbering now not more than 80 sabres, held this Russian Hussar Brigade in a corner at bay for some minutes.'[16]

188

Of this moment Douglas said:

'They received us with a sort of irregular volley of carbines . . . my first
impulse was to charge; the word [of command] was almost out, but at
the instant I saw how fruitless such a proceeding would be. I halted the
regiment within forty yards of them, and gave the order to retire.'[17]

Paget never forgave Cardigan for his conduct after reaching the battery.
Throughout the charge Cardigan's words, 'Your best support', were burning
in his brain. Although Paget lost control of the 8th Hussars and never gained
control over the 11th, it was not for want of trying. Having got to the guns his
first thought was, 'Where is Cardigan?' As the brigade second-in-command he
had to find his Commander, or at least find out if he was a casualty. If he was
down then Paget would assume command. But Cardigan could not be found.
He had disappeared. Paget dashed about looking for him, asking for him to
no avail. His was a maddening predicament. Supposing his Brigade
Commander had pushed on into the enemy position he would expect his
supports to follow, indeed it was their duty to do so. If Cardigan had forged
ahead with the first line and he, Paget, did not back them up his military career
was over. On the other hand if Cardigan had been killed or wounded he would
need to make decisions, give orders and attempt to control the Brigade. The
fact that, as Paget was dashing around looking for his brigadier, that officer
had given up on the Brigade and was behind him about to retire was, at the
time, an unthinkable possibility.

As the 4th Light Dragoons charged through the battery and got caught up
in the mêlée Paget had tried to avoid involvement in swordplay although he
had become separated from his orderly (Parkes). In those few moments Paget
watched

'some fierce hand-to-hand encounters, and our fellows, in the excitement
of the moment, lost sight, I fear, of the chief power of their sabres, and
for the *point* . . . substituted the muscle of their arms, in the indiscrimi-
nate appliance of the cut, which generally fell harmlessly on the thick
greatcoats of the Russians.'[18]

Paget also witnessed, and countermanded, a gallant attempt by Lieutenant
Hunt (4th LD) to capture a gun. The Russian detachment had tried to pull it
out of the line with the team but in the process of the fight to get clear it had
overturned. Hunt, on seeing this, had sheathed his sword, dismounted and was
frantically trying to unhook the traces when Paget spotted him and shouted
for him to remount. Paget later explained his action:

'He [Hunt] thus disarmed himself in the mêlée . . . and the act which he
attempted would have been a most useful one, had support been near to
retain possession of the gun. . . . It was of course a useless attempt – but

none the less worthy of record and of a Victoria Cross, for which he would have been recommended, had the choice lain with me.'[19]

Paget thought highly of Hunt and recorded that, as far as he knew, he was the only subaltern in the Cavalry Division never to have missed a day's duty during the entire war, an astounding achievement in the appalling conditions of the Crimea.

The anonymous soldier in the 8th Hussars quoted earlier has left a lurid (and possibly exaggerated) account of the never-to-be-forgotten highlight of his military service:

'The first thing I did, once within the guns, . . . was to cut clean off the hand of a Russian gunner who was holding up his sponge against me. He fell across the gun-carriage glaring savagely, but I cared little for that. . . . I had seen too much. . . . Bodies and limbs scattered in fragments, or smashed and kneaded together, and blood splashed into my face.'[20]

He found at least one episode amusing:

'One incident that now occurred was almost comical. Corporal Taylor on my left, mounted on a horse either startled or mad with the noise of firing . . . couldn't hold his in at all; he had no control over his mouth. He shot past me like a rocket right bang into the heart of the Russian cavalry in the rear of the guns, which opened and then closed upon him. I heard after that he was wounded and then taken prisoner.'[20]

One informant claimed exciting involvement in the hand-to-hand fighting with the enemy cavalrymen:

'I had three Russians to deal with at once. . . . An Hussar made a desperate slap at my head which I parried, and with cut 'number two' gave him so tremendous a slash in the neck that it almost sickened me to look on. . . . I had now to wheel in order to meet a Polish [sic] Lancer who was just charging me full tilt. I saw that the butt was fixed against his thigh, and that he gave his lance a slight quiver, and that he seemed to know how to use it too. I bent down slightly on my saddle, received his lance on the back of my sword which passed over my shoulder, at the same instant that the point of my weapon, through the mere rush of the horses passing each other, entered his breast, and went clean through him, coming out at his back, so that I was forced to draw it out with a wrench as he rolled over the crupper.

'A Cossack was now upon me, but as I reined back in time his aim failed, and he shot by my horse's head, and I then rode after him, wounding him in the shoulder, and knocking man and horse over with my own, so that I was all but unseated.'[20]

190

Private Grigg, also of the 4th Light Dragoons, had some similar experiences and made two intriguing observations regarding the Cossack he fought and his lance. A mounted Russian driver

'cut me across the eyes with his whip, which almost blinded me, but as my horse flew past, I made a cut and caught him in the mouth, so that his teeth all rattled together as he fell. I can hear the horrible sound now. . . . Beyond the guns, the Russian cavalry. . . . We went at them with a rush. I selected a Cossack, who was making for me with his lance pointed at my breast. I knocked it upwards with my sword, pulled up quickly and cut him down across the face. I tried to get hold of his lance but he dropped it. As he was falling, I noticed that he was strapped on to the saddle, so that he did not come to the ground, and the horse rushed away with him. His lance, like all the others used by the Cossacks, had a black tuft of hair, about three inches from the blade, to hide a hook having a sharp edge, with which the reins of their enemies are cut when the lance is withdrawn after a thrust.'[21]

The 8th Hussars, reduced to one effective squadron, had found themselves isolated from the remainder of the Brigade by a substantial distance. When Sherwell pulled them up they were 300–400 metres behind and to the south of the gunline. The remnants of the first line had disappeared and the 11th Hussars were out of sight some 700–800 metres to the north. Sherwell was uncertain what to do. Should he continue and attack the Russians, large bodies of whom were to his front and his right flank, or should he withdraw? The enemy seemed to be in retreat, with numbers of them crowding around the aqueduct bridges trying to cross over. This was encouraging, so, after a pause of several minutes, Sherwell decided to resume his advance. He was worried that there was no formed body of British cavalry in sight and no sign of either Cardigan or Paget from whom he might expect directions. It was some relief, therefore, when a group of about fifteen men of the 17th Lancers emerged unexpectedly from the confusion on the left front. This party, under Mayow, seemed to Sherwell to be the only survivors of the first line. His first words were, 'Where is Lord Cardigan?' It was (relatively speaking) a long time since Mayow had seen his Brigade Commander.

As the two senior officers conferred as to what to do next someone spotted enemy Lancers moving across their rear. Pennington (11th H), who was now with the 8th, recalled these events:

'A small party of the 17th, who in the struggle at the guns had separated from that corps, added their strength to ours, and then we numbered only some seventy well-mounted men. It was now discovered that some squadrons of Russian lancers had ranged themselves across the valley to our rear, thus interposed between us and the British lines. "Cut off" was the excited cry.'[22]

191

According to Kinglake, Sherwell, in consultation with Mayow and de Salis, resolved the dilemma with a decision to turn about and fight their way out. However, the adjutant, Seager, told a slightly different version in his letter home the next day. This fascinating letter reveals the writer (an ex-ranker who rose to be a general) to be very much a sentimental soldier with a great love of his family. He had charged that morning carrying

'[in his sabretache] your's and the darling children's picture, my dear mother's present (prayer book and Testament), very small writing case with a lot of letters in it, and in the pocket of my jacket was your letter containing dear little Emily's hair which has been there since I received it. In my haversack was some biscuits and a bottle with some whisky and water in it and very useful I find it. In my pocket some sovereigns and also some Turkish coins, and round my neck was the dear locket you gave me in Exeter. All these I turn out with just the same as putting on my sword and revolver.'[23]

Now, with the rump of 8th Hussars plus a handful of hangers-on from other regiments caught between two enemy forces, the decision was made to come about. Seager's account hints strongly that it was himself who initiated the order. In his letter he wrote:

'They [afterwards] gave us great credit for wheeling about and attacking the Lancers; it enabled the other Regiments, who were previously broken, to get through them much more easily [in fact this was not the case]. The Colonel gets all the credit for it but Phillips [Lieutenant Phillips lived to see the opening of World War I, being then the last surviving 8th Hussars' officer], who was riding next to me, could tell you who it was that called to the Colonel to let us wheel about and attack them. But I will tell you all please God when I get home and it is well known to the Regiment.'[23]

According to Pennington Sherwell became flustered with the words of command necessary to get his men to face back the way they had come:

'Colonel Sherwell shouted, "Threes about!" There was some hesitation shown, for the withered ranks had kept together well, but lost their count by "threes". His able subordinate Seager interposed, "Excuse me, sir; 'tis right about wheel." The Colonel then cried, "8th right about wheel!" The 8th responded as if on home parade, and thus we faced the strong squadrons to our rear. This incident is vividly impressed upon my mind, but I have never seen a reference to it in any previous account.'[24]

By this stage there were only two formed bodies of the Light Brigade left. One was in the northern part of the valley, quite close to the Fedioukine Heights, the other to the south, much nearer the Causeway Heights. In the north were all that was left of the 11th Hussars who had been joined by Paget with the remains of the 4th Light Dragoons; in the south were the battered 8th Hussars whose numbers had been swelled by Mayow's group from the 17th Lancers. In between was a 500-metre gap, occupied mainly by casualties from both sides and riderless horses. The Light Brigade had done all that had been asked of it, and more. They had defied the rules of war and triumphed. Few if any of the observers, Allied or Russian, had thought any would reach the guns. The Brigade had taken its objective and then pushed vastly superior numbers of enemy cavalry back towards the river. Such was the reputation earned by the British cavalry that morning that the Russians could not be induced to face them. A number of Rijov's units were in a state of near panic. There were scenes of indecent haste to put the aqueduct and the river between them and their attackers.

Nevertheless, with the Brigade now a fraction of its strength a mere ten minutes earlier, with its Commander quitting the field, with no support materializing, with control gone, with a huge gap between the surviving bodies, a retreat was inevitable. The Russians turned and began to regain some fortitude when they realized how weak was the opposition. Some units even started to spur forward in pursuit. For the Light Brigade there lay ahead the awful prospect of retracing their steps all the way back up the valley – and still under fire. To make matters more difficult their way was barred by several squadrons of Jeropkine's Uhlans.

CHAPTER TEN
Back up the Valley

'God alone knows what has happened to my poor regiment.'

Lord George Paget, 4th Light Dragoons

'I contrived to slip in front of the grey mare that had done me such great service, and I kissed her on the nose, for to her I owed my life.'

Private Pennington, 11th Hussars

Cardigan was among the first of the unwounded (he was barely scratched by the lance) to reach the comparative safety of the Causeway Heights near No. 4 Redoubt. He had been the first through the enemy battery at the head of his Brigade; now he was, in effect, the first out, but with no Brigade. He had, literally, lost the Light Brigade. He had no clear idea of where they were. He seemed, by his actions, to believe that the majority of his command had either been destroyed or had retired before him back up the valley. He had seen nothing of them in the smoke of the gunline or beyond. He had surmised, as he gazed up the valley from the silent guns, that all that was left were the dozens of men and horses limping up the valley in front of him. Serenely satisfied that he had done his duty, he joined them. Most accounts have him walking his horse slowly back, ignoring the gunfire from Bojanov's battery which was still active to his left, and the rifle fire from both sides. This version is of questionable validity.

Firstly, the circumstances dictated more rapid progress. It was in Cardigan's interest to get back as quickly as possible. Not only might some degree of haste get him out of danger, but he would also be reunited more speedily with the remnants of his command, the bulk of which he erroneously believed to be ahead of him. Although Cardigan's physical courage was of a high order, human nature would have ensured he kicked 'Ronald' into a gait faster than a walk. Then there is the time and distance calculation. From the gunline he had to cover about a mile to get back to where the Heavy Brigade and 'C' Troop RHA were waiting. At a walk this would have taken perhaps 10–12 minutes of uninterrupted progress. However, it is certain he got back well in advance of the majority of his Brigade, all of whom had galloped or cantered at their best speed. If Cardigan had indeed walked he would have been overtaken, by the 8th Hussars for example, long before he reached the western end of the valley. It is also known that he took time to retrace his footsteps towards the guns for a short distance.

Neither was he caught by the Uhlan squadrons that quickly deployed across much of the valley to cut off the regiments as they withdrew. Clearly Cardigan escaped this movement by the enemy as he was already far enough back to avoid it. There are also at least two witnesses who claim to have seen the Brigade Commander hurrying back. The first was Private Ford of the 4th Light Dragoons who, in a brief statement wrote, 'My horse was killed about 150 yards from the guns. Whilst entangled with my dead horse, Lord Cardigan passed me galloping to the rear.'

The other, perhaps more reliable, was Mitchell of the 13th Light Dragoons. It was Mitchell whose horse had been killed some distance in front of the battery and who had survived being ridden over by the 4th as they dashed for the guns. He was actually spoken to by Cardigan. The circumstances of this happening are a not untypical vignette of the experiences of the men who never reached the enemy gun position. Mitchell described what occurred thus:

'After they [the 4th LD] had passed I tried to extricate my leg, which, after a short time I succeeded in doing. . . . I still had my sword in my hand, and soon found there were numberless bullets flying around me which came from the infantry on the flank of their battery, who fired at any of us who were dismounted. Just at this time a man from my troop, named Pollard [an old sweat of at least twenty years' service who knew how to look after himself], came to me, and throwing himself down beside the carcass of my horse for shelter from the bullets, called to me, saying, "Come here Mitchell; this is good cover." I said, "No; we had better make our way back as quickly as possible, or we shall soon be taken prisoners, if not killed if we remain here." Upon this he jumped up and we both started to get back. . . .

'The number of horses lying about was something fearful. As we went along we somehow got separated . . . and in another minute I found myself alone. Just then Lord Cardigan came galloping up from the direction of the guns, passing me at a short distance, when he turned about again, and meeting me pulled up and said: "Where is your horse?" I answered: "Killed, my Lord." He then said in his usually stern, hoarse voice: "You had better make the best of your way back as fast as you can, or you will be taken prisoner." I needed no telling for I was doing so as fast as I was able. He then rode back a little farther down, and in a few moments returned past me at a gallop.'[1]

Cardigan had galloped clear of the guns and then, for some inexplicable reason, turned and ridden back towards the battery for a short distance, during which time he stopped to ask his futile question of, and offer his patronizing and redundant advice to, Mitchell. Seconds later he reappeared, spurring away towards the rear once more. The schoolboy history-book image of him plodding sedately back does not bear much probing.

The next person to encounter Cardigan was Shakespear with the guns of I Troop RHA, who had followed the Brigade with his guns for some of the way before retiring. Even horse artillery could not participate in a charge, and he had been given no chance to support it by fire. With exaggerated composure Cardigan reined in for sufficient time to point to the hole in his pink pants, remarking, 'Damn nice thing this, Shakespear, and nothing to keep the cold out.' 'Pardon me, my lord,' replied Shakespear, 'the artillery are always prepared for an emergency.'[2] So saying he summoned his trumpeter who carried his flask and passed it up to his Brigade Commander. Nothing was said about what had just happened, Cardigan probably feeling the young gunner officer far too junior to engage in conversation on serious matters. After his drink he rode on at a walk.

Cardigan was by then back among British troops and had moved up the Causeway Heights near No. 4 Redoubt. It was there that he met up with the infantry, and General Cathcart in person, who was still accompanied by Raglan's ADC, Ewart. Neither of these gentlemen had witnessed the charge and were somewhat taken aback by Cardigan's opening, and only remark. 'I have lost my Brigade.' It was the truth. But whether Cardigan meant he had lost it in the sense of not knowing where it was, or that it had been destroyed, was not clear. The reality, at that moment, was that both interpretations were correct.

It was about this time that Cardigan, who until then had been entirely alone, was joined by the unloved adjutant of the 11th Hussars. Contrary to many accounts of the battle Yates did not ride in the charge. Sergeant-Major Smith makes a point in his diary entry dated 5 November, 1854, of stating with reference to the Light Brigade's charge, 'He [Yates], it will be remembered, remained behind when the brigade was ordered to attack.' Up until that morning he had been acting for the brigade-major, Mayow, who had been sick. Mayow, however, returned to assume his duties on the 25th at the last moment. Yates, whether by accident or design, was not present when the 11th Hussars had formed up a little before eleven o'clock that morning. His nickname was 'Joey' and, according to Smith was thought to lack 'bottle'. Eleven days later at Inkerman the rump of the 11th, with only four officers on parade, came under artillery fire. Smith's comment then was:

'We now came under fire, the cannonballs reaching us. It was now one of the men, who had turned round looking to the rear, called out, "There goes Joey". All within hearing turned to look when, in the distance, could be seen the adjutant galloping back towards the encampment. This caused great amusement and laughter – he had only been with us a month and had made himself thoroughly obnoxious to everyone.'[3]

Cardigan turned about and, accompanied by Yates, made his way back under the northern edge of the Causeway towards where 'C' Troop RHA were deployed.

By this time many wounded and unwounded stragglers had made their way up the valley watched by the gunners on the ridge, but, as yet, no formed regimental group had returned. Cardigan arrived at No. 6 gun, which was on the crest of the ridge, from the direction of the North Valley, with Yates a pace or two behind. There were a number of witnesses, including Brandling, the troop commander, as to what happened next:

'The horse seemed to have had enough of it, and his lordship appeared to have been knocked about, but was cool and collected. He returned his sword, undid a little of the front of his dress, and pulled down the under-clothing under his waist-belt. . . . He then, in a quiet way, as if rather talking to himself, said, "I tell you what it is – those instruments of theirs," alluding to the Russian weapons, "are deuced blunt; they tickle up one's ribs!" Having refastened his uniform, he pulled his revolver out of his saddle-holster as if the thought had only just struck him, and said, "And here's this damned thing I have never thought of till now." He then replaced it, drew his sword and said, "Well, we've done our share of the work;" and pointing up towards the Chasseurs d'Afrique, in our left rear, added, "It is time they gave those dappled gentry a chance" – this had reference to the colour of their horses. After this he asked, "Has anyone seen my regiment?" and the men, thinking it was the regiment dressed, like himself, in the crimson overalls, answered, "No sir." '[4]

Cardigan, who had been standing with his back to the valley, turned away and appeared to move in the direction from which the Light Brigade was returning. Within a moment or two the 4th Light Dragoons, and beyond them the 11th Hussars, came cantering up the North Valley, the former passing within 150 metres to the left of Brandling's guns. On seeing the gunners, Paget, who had come diagonally across the valley from the Fedioukine side, inclined more to his left and rode up to the Causeway ridge. Those who saw and heard him soon realized he was an exceedingly angry man.

'Lord George looked like a man who had ridden hard for a mile or two, and was heated and excited. He said, in a rather loud tone, "It is a damned shame; there we had a lot of their guns and carriages taken and received no support, and yet there is all this infantry about – it is a shame." '[5]

This was the overriding issue with many of the officers who returned at eleven-thirty that morning. Not so much that the Brigade had been set a seemingly impossible and immensely costly task, but that they had done their duty and more; they had captured the enemy battery, driven the Russian cavalry to the river – and received no support. They felt that their sacrifice had been for nothing. The French Chasseurs had done their part, but the 4th Division had sat on its hands, and even their comrades in the Heavy Brigade had just watched as the survivors limped home. Among the commanding officers, like Paget, there was also the nagging suspicion that the Brigade Commander's early disappearance was not entirely necessary.

As Paget was speaking Cardigan returned.

'Meantime, Cardigan had come back and was close behind Lord George while he was speaking, without the latter knowing it. Lord Cardigan stopped Lord George talking further by calling out his name, and, on the latter turning round, said to him in an undertone, "I am surprised," and tossing his head in the air, added some other remark that was not heard. Lord George lowered his sword to the salute, and without saying *anything*, turned his horse and rode on after his men.'[6]

At that moment Cardigan and Paget each regarded the other's behaviour as less than acceptable. The Brigade Commander was far from satisfied that Paget had provided his 'best support', and could not understand why he had seen nothing of the second line. Later he was to write, 'The supports ought to have followed me in the attack, instead of which they diverged to the right and left.' His second-in-command was speechless at seeing Cardigan unharmed behind him. Where had he been when he was needed? Why had he not rallied the Brigade? As will be seen later Paget felt so strongly on this point that it became the primary reason for his resigning his command and returning to England.

While all the officers and most of the men in the first line had seen Cardigan reach the battery this was not the case with the second. Virtually nobody in the 11th Hussars, the 4th Light Dragoons and 8th Hussars had any idea of whether their Brigade Commander had reached the enemy or, if he had, his subsequent movements. The men in the second line were too far back even to glimpse Cardigan as he was obscured by the first line and the dust and smoke of bursting shells. When Paget rode off, inwardly seething with indignation, Cardigan turned back once more to ride over to meet the 8th Hussars who were coming up the northern side of the Causeway Ridge. It was now, behind his back, that Cardigan was to be mocked, not only by his men but, almost inconceivably, by a senior officer as well. He was sufficiently insensitive to ride out and place himself at the head of the 8th Hussars. He intended to lead them in. There followed an incident that was both shameful and unique.

'He [Cardigan] had turned backward from the gun [No. 6] when the first cheer was given by some Heavies. He then halted, turned about as if to see what it was for, saw it was in compliment to the 8th returning, trotted towards them, turned about in front of Colonel Shewell, and took up the "walk". . . . And now occurred something rather painful to witness. As the 8th came into view of No. 6 Detachment . . . Colonel Shewell was in front, and Colonel Mayow behind on the left of the other officers. The moment Cardigan got his back turned round to the 8th, Colonel Mayow pointed towards him, shook his head, and made signs to the officers on the left of the Heavies, as much as to say, "See him; he has taken care of himself". Men here and there in the ranks of the 8th also pointed, and made signs. . . . Colonel Shewell neither saw this nor took any part in it. . . . Of course Cardigan did not know what was going on behind him while he was smiling and raising his sword to the cheers. . . . He was thus, in a way, held up to ridicule.'[7]

To the soldiers of the 8th Hussars it must have seemed as if Cardigan had never charged at all and was now coming forward to claim the glory. This was the first time many had seen him since Britten's trumpet had set the Brigade in motion. Now here he was coming out from the ridge near the artillery and Heavy Brigade and placing himself in front of their Commanding Officer. Little wonder there was muttered ill-feeling. It is possible that some soldiers may have mistakenly thought they saw Cardigan retiring before he reached the battery. This would have been Lieutenant Houghton of the 11th Hussars. Houghton was of a similar build to Cardigan, wore the same uniform and was likewise mounted on a chestnut horse with white stockings. He had been badly wounded before getting to the guns and thus forced to retire. Amid the noise, smoke and excitement of the charge some in the second line might have thought they had glimpsed the Brigade Commander pulling out.

Undoubtedly, for a time, Cardigan was wrongly blamed by many for what had happened during the previous half hour. Some thought he might not have charged at all, some suspected he withdrew before reaching the objective, while others thought he had failed in his duty by making no attempt to rally his command and retreating early. Among those who saw him come back there was, as Whinyates later wrote, a 'sort of tacit keeping away from him for a time; no congratulations on his escape were offered in the hearing of 'C' Troop.' Within the hour, however, it became clear to most where Raglan felt the blame lay.

It was typical Cardigan that the first words he uttered to Scarlett when they met were, 'What do you think of the aide-de-camp riding to the rear and screaming like a woman?'

Scarlett cut him short with, 'Say no more, my lord, for I have just ridden over his dead body.'[8]

Cardigan had not known that Nolan had been killed, and from this remark it is obvious that Nolan's behaviour was going to incur the Noble Yachtsman's wrath.

When he rode up to where the survivors were reforming, some wit called out, 'Hello my lord, were you not there?'

'Oh wasn't I though?' Cardigan replied. 'Here Jenyns, didn't you see me at the guns?'[9]

*   *   *

Although many of the Light Brigade made their way back as individuals, or perhaps as pairs, the majority returned as part of one or other of two organized groups. The first consisted primarily of the 11th Hussars under Douglas who were joined by the 4th Light Dragoons under Paget. This body started on the north side of the valley, under the Fedioukine Heights, and had to break its way out through the Uhlans. In doing so they inclined to their left across the valley, eventually reaching safety near No. 4 Redoubt. The second, or southerly group, was composed of Mayow's party of the 17th Lancers and Shewell's 8th Hussars. They got out, after a brush with the Uhlans, by following a route quite close to the Causeway Heights.

When Douglas decided to retire the 11th Hussars had some 3,000 metres to cover to reach No. 4 Redoubt. Douglas always hotly disputed, as did many other 11th Hussar survivors, that Paget took command of his Regiment after the 4th had joined them during the early part of the retreat. He did, however, accept that 'We had joined the 4th Light Dragoons or were close to them'. Then came the realization that there were Lancers drawn up in front of them, blocking their route. Pennons hung from their lanceheads so they could not be Cossacks. Douglas has described the moment:

> 'I saw in our rear [sic, at that stage the rear had become the front] two squadrons of Lancers drawn up. I instantly proclaimed, "They're the 17th. Let us rally on them." ... At this moment Lieutenant Roger Palmer rode up and said, "I beg your pardon, Colonel, that is not the 17th, that is the enemy." "Well," I exclaimed, "we must only retire and go through them." So with the 4th Light Dragoons we charged the Russians' Lancers and got past them with few casualties.'[10]

Close by, Paget confronted the same situation. Again there was doubt with some as to the identity of the Lancers.

> 'Hardly, however, had we thus rallied [after the charge], when a cry arose, "They are attacking us, my lord, in our rear!" I turned round, and ... saw ... a large body of Russian Lancers formed up, some 500 yards behind us ... and on the direct line of our retreat! On the impulse of the moment, I then holloaed out, "Threes about" – adding, "We must do the best we can for ourselves." '

At this stage Major Low said to Paget, 'I say, Colonel, are you sure those are not the 17th?' to which Paget replied, 'Look at the colour of their flags.'

Both the 11th Hussars and 4th Light Dragoons now summoned up their last vestiges of strength to fight their way out. It was hardly a charge as most horses were too exhausted, formation had gone and officers were at the rear as the bunch of horsemen spurred forward. As Paget wrote:

'Helter-skelter then we went at these Lancers as fast as our poor tired horses could carry us, rear rank of course in front . . . the officers . . . in rear, for it must be remembered that we still had pursuers behind us.'[12]

Then the Russians carried out a disquieting manoeuvre. As the British came towards them their right-hand squadron turned and wheeled right, to take up a position from which it looked as though they intended to charge into the British flank as they passed. The 11th and 4th pressed ahead, making for the gap that had now opened up but edging away to their left. Paget claims to have tried to get the 4th to wheel slightly right so as to face the enemy. It was impossible. The two Regiments were hell bent on getting away. Despite the risk, they would cross the enemy's front and expose their right to attack. All the Russians had to do was time their advance correctly and seize the splendid opportunity that was being offered. They did no such thing. Yet again the Russian cavalry were in the right position and were given the chance to inflict a serious reverse on their enemy. Yet again they failed. As the British fugitives, for they were almost that, rushed past, the Uhlans did virtually nothing.

Paget was incredulous:

'Well, as we neared them, down they came upon us at a sort of trot, . . . they stopped ("halted" is hardly the word) and evinced that same air of bewilderment (I know of no other word) that I had twice before remarked on this day.'[13]

Those on the right of the British group were poked at, but for the most part parried the feeble thrusts easily. Paget again:

'I can only say that if the point of my sword crossed the ends of three or four of their lances, it was as much as it did, and I judge of the rest by my own case, for there was not a man, at that moment, more disadvantageously placed than myself (being behind and on the right rear). . . . We got by them without, I believe, the loss of a single man. How I know not! It is a mystery to me! Had that force been composed of English ladies, I don't think one of us would have escaped.'[14]

Sergeant Bentley of the 11th Hussars very nearly did not. One of the few men to be seriously attacked by several opponents, Bentley soon found himself in trouble.

'On passing them I was attacked by an officer and several men, and received a slight wound from a lance. I was pursued by them, and cut the officer across the face. Lieutenant Dunn came to my assistance. I saw him cleave one almost to the saddle.'[15]

Seeing Bentley drop behind, Dunn had reined in and turned back to help him. He emptied his revolver at Bentley's assailants and then disposed of several others with huge swipes with his extra long sword. For this act of gallantry, and for afterwards cutting down a Russian who was attacking Private Levett in a vain effort to save that soldier as well, Dunn would receive the Victoria Cross.

To the south Shewell's 8th Hussars and Mayow's group of the 17th Lancers had also to smash through Jeropkine's line of Uhlans. With the cry, 'By God, they're Russians! Keep together, we'll ride them down,' the 8th galloped at them. The adjutant, Seager, whose horse had just been shot near the windpipe, was then leading the single squadron remaining. The following day he wrote home:

'We dashed at them. They were three deep with lances levelled. I parried the first fellow's lance, the one behind him I cut over the head which no doubt he will remember for some time, and as I was recovering my sword I found the third fellow making a tremendous point at my body. I just had time to receive his lance point with the hilt of my sword; it got through the bars, knocked off the skin of the top knuckle of my second finger, and the point entered between the second and top joint of my little finger, coming out the other side. I shall most likely be returned wounded in the Gazette but you see I have only got a slight scratch that might look interesting in a drawing room.

'After I found myself through the Russians, I saw the Colonel and the Major [De Salis] a long distance ahead going as fast as their horses would carry them, the batteries [sic, only one at that stage] and rifles peppering them in grand style. . . . On looking to see what had become of my men, I found they had got through and scattered to the left.'[16]

Pennington had started out with his regiment, the 11th Hussars, lost his horse, been 'rescued' and given another mount, and eventually found himself trying to fight his way out with the 8th Hussars. He was chased by Russian horsemen who endeavoured to catch him up on his left which was his vulnerable side. Pennington's desperate efforts to keep them to his right (sword arm side) well illustrate the tricks of cavalry combat.

'Colonel Shewell led the way. . . . The Russian squadrons, quite four hundred strong . . . to our amazement, their right fell back, and giving way assisted our design.

'I bore well to the left, quite losing touch with the 8th. Thus separated from all aid, lancers pursued me up the valley; but I kept them on my right and rear, my sword-arm free to sweep around. With many a feint at cut and thrust (for I feared to check the gallant grey) I kept them at arm's length, foiled their attempts to get upon my left, where they might strike across my bridle-hand, and the grey mare gradually drew ahead. . . . Balls from the causeway ridge raised up the dust around my grey mare's hoofs, but happily their force was spent.'[17]

It was at the end of this momentous ride that Pennington gratefully embraced and kissed his horse.

Perhaps understandably the Russian accounts of the Light Brigade's escape are somewhat different. The four squadrons had moved out from behind No. 3 Redoubt until they neared the Tractir Bridge road. They then turned east to face down the valley approximately as shown on Map 14. The Uhlan squadron commander Kubitovich's version of what happened gets high marks for interest, but less for accuracy:

'We saw the enemy coming back down the valley at a trot, and we marvelled how they came along steadily in faultless style as if on manoeuvres. No. 1 Squadron under Rotmistr [captain] Verkhkitsky, moving first to the right and then to the front [this seems to refer to the movement to get on to the flank of the British seen by Paget and others]. This left ample room for my squadron and I led it straight at the foe. . . . For then the desperate slashing started. Our artillery and infantry reopened fire, but I must say they caused as many casualties to us as to the enemy, particularly to our horses. The English fought with amazing bravery, even the dismounted and wounded fighting on until they dropped. . . . We pursued them almost as far as No. 4 Redoubt.'[18]

A Lieutenant Kozhukhov, who was serving with a light artillery battery not far from the Tractir Bridge witnessed the Uhlans in action but was less fulsome in his praise of their performance:

'General Ryzhov (Rijov) said that the Uhlan attack was brilliant. Well, we watched the Uhlans and we saw no brilliance. Indeed no real attack, as such, took place and the fighting which did occur was not pressed home. . . . How can one explain how the "disorganised groups of enemy cavalry, already defeated, and without leaders or direction", could fight their way through a mass of fresh cavalry [Uhlans], leaving in our hands only the dead and wounded?'[19]

*   *   *

The struggle back up the valley was the final ordeal. For every man that made it as part of a group or formed body another did it on his own. Each one had his own horrifying tale to tell. Some made it on foot, the luckier ones in a saddle. Strewn all over the valley were dead and dying men and horses, walking wounded, crawling wounded, and horses in every aspect of agony. Some animals had lost legs, yet struggled piteously to get up; others stumbled uncomprehendingly along on three; some had their legs but had been disembowelled. One officer later recalled the devotion of some riders for their mounts. Soldiers, whose love for their horses can only really be fully understood by a cavalryman, refused to leave their wounded or dying friend, preferring to risk capture or death in order to tend their wounds.

Many horses staggered along with their riders on their backs despite numerous wounds. Captain Hutton of the 4th Light Dragoons was shot in both thighs but was brought to safety on a charger with eleven wounds. Hutton's grief when his horse had to be destroyed was intense and unashamed. Captain Jenyns (13th LD), whose horse was hit four times wrote:

> 'I only brought nine mounted men back! Poor old "Moses" [his charger] was shot through his shoulder and through the hip into his guts, but *just* got me back.'[20]

Even unwounded animals were utterly spent by the time they got back. Most had covered three miles, much of it at the gallop, so it was not surprising that Paget was using both his spurs and the flat of his sword to keep 'Exquisite' moving at the end.

Those who were dismounted had to walk, trying to dodge the Cossacks who roamed the valley seeking out the wounded and defenceless to kill or capture. Lieutenant Chamberlayne (13th LD) was one of these. His horse 'Pimento' had fallen during the charge and he had sat down beside it. There he waited until Lieutenant Smith rode past on his way back. Chamberlayne called out to ask what to do. Smith advised him to take off the saddle and bridle and start walking, remarking, 'Another horse you can get, but you will not get another saddle and bridle so easily.' Chamberlayne followed this advice, and with the saddle over his shoulders and head plodded slowly back. Almost certainly he was not attacked by the Cossacks as they failed to recognize him under his burden or mistook him for a Russian looter.

Five more Victoria Crosses were to be won by soldiers trying to save the lives of wounded officers during this retreat. The two officers concerned belonged to the 17th Lancers. The first incident involved Captain Webb who had his ankle smashed by a shell. In excruciating pain and unable to ride, it seemed that he was certain to be captured and probably to die of loss of blood. When Sergeant Berryman (17th L) rode up, however, he determined to try to save his officer. Under fire and with the enemy pressing, Berryman carefully half-lifted, half-dragged Webb from his saddle. Then with the help of Sergeant Farrell (17th L), Corporal Malone and Private Lamb (both 17th L) they began the laborious business of carrying the wounded officer up the valley. Webb had a raging thirst so Lamb went back to search the dead bodies until he found a waterbottle. Several times Webb implored Berryman and the others to leave him and save themselves.

Webb was eventually evacuated to the base hospital at Scutari but died there twelve days later, being too weak and shocked to recover from the amputation of his leg. Berryman, Farrell and Malone were all awarded the VC but Lamb got nothing. Lamb always maintained that this was because he lost out in a draw for the award with Malone. His story was that after the war each commanding officer had one VC to allocate to his men and that Malone and he had to draw lots to see who got it.

The other wounded officer was Morris. That Morris was to live through his ordeal, had the stamina to stagger back up the valley despite three head wounds, a fractured arm and several broken ribs, was due to his enormous physical strength. Later, after his recovery, he returned to serve again in the Crimea. He was something of a fitness fanatic, being aptly nicknamed 'the pocket Hercules'. After surrendering, the officer to whom he had done so disappeared, leaving Morris glaring back at several hostile horsemen. As they moved in on him Morris broke away and half-ran, half-stumbled into the nearest patch of smoke. A riderless horse cantered past and Morris grabbed a loose rein, permitting himself to be dragged for some distance. The effort was too great, his wounds and loss of blood too severe, for him to hang on for long. He blacked out.

On recovering consciousness, and as he struggled to his feet, Morris was almost bowled over by another loose horse. This time he managed to slow the animal sufficiently to heave himself up, ignoring the intense shooting pain in his chest when he lay across the saddle. He hauled the horse's head round and began to ride up the valley. For some minutes all went well, with considerable progress being made. Then, just as horse and rider were almost out of the cross-fire, the horse was hit. It staggered, pitched forward, rolled over – and died. Morris lay with his leg trapped under the animal's body. After much agonized wriggling and twisting he extricated himself and once more lurched unsteadily to his feet. He tottered on towards safety. Eventually he collapsed, falling to the ground just short of where the body of his friend Nolan was lying. Whether this was pure coincidence or, as Kinglake stated, because he recognized Nolan and realized he must have almost reached the British lines, can never be known for certain.

Nevertheless, Kinglake's assertion that, 'being now quite exhausted, he laid himself down beside the body of his friend, and again became unconscious,'[21] is highly questionable and is more likely one of the minor myths of the battle. That Morris fell close to Nolan is almost certainly correct, but deliberately lying down alongside him unlikely. The account by the man who first discovered Morris (and Nolan) seems more plausible. This was Captain Ewart, Raglan's ADC, who was present near No. 4 Redoubt as the Light Brigade limped home.

'Released by Sir George [Cathcart], Ewart crossed over the ridge of the Causeway, and galloped down the North Valley a little way towards the enemy. At no very great distance, he found poor Nolan with a dreadful wound in his chest, quite dead; and a few yards farther on he came across Captain Morris. . . . Ewart dismounted and spoke to him but he was almost insensible. . . . A little farther on was a man from the Heavies with his jaw frightfully smashed. Ewart asked him his regiment, but he could not speak, and only pointed to his buttons, from which Ewart saw he belonged to the 5th Dragoon Guards. As he could not lift Morris or the others by himself, Ewart rode back towards No. 4 Redoubt [more likely No. 5.]'[22]

Ewart was looking for help to carry the bodies off the battlefield. Near the redoubt he found some Turks (some of whom, it will be remembered, had occupied No. 5 Redoubt before the charge). By gesticulating frantically he gathered together a small group to accompany him back down the valley. On reaching Nolan's body two Turks were detailed off to carry it in, while Ewart and the others moved on to Morris and the wounded dragoon. At this moment some Russian guns opened fire, bouncing several balls close by. This fire probably came from the Fedioukine Heights where, by this time, one or two detachments of the Russian battery may have come into action again after the Chasseurs' withdrawal. They were enough to frighten the Turks. Nolan was dropped as they fled back to the redoubt. Ewart yelled to a nearby horseman, instructing him to fetch help and stretchers from the 17th Lancers as their Commanding Officer was lying grievously wounded in the valley. The soldier spurred away.

According to Assistant-Surgeon Cattell, who was with the 5th Dragoon Guards, the soldier sent with this urgent message could have been a Private George Smith. In his account he describes what happened thus:

'Pte Geo. Smith informed Sgt. O'Hara of the spot where Morris lay and Scarlett sent the staff surgeon with Tr. Sergt. Major Wooden [sic, Sergeant Wooden, 17th Lancers] to bring him in. They found a trooper trying to arrest the bleeding from the scalp. Presently some Cossacks attacked the party and the doctor, Mouat, said he had to draw his sword, which he described as "a novel experience".'[23]

Some four years later the award of the VC to Mouat and Wooden was gazetted. They had gone out, dressed Morris's wounds under fire and under possible threat from Cossacks, and then had him carried back to safety. When last seen he was being carried away towards Kadikoi in a semi-conscious state, moaning in pain and calling out, 'Lord have mercy on my soul!'

There are a number of problems with reconciling the circumstances of this award of the VC. For the Cross to be given both Mouat and Wooden would have needed to be under heavy fire. Some guns on the Fedioukine Heights would have been the only ones able to bring fire to bear if Morris had got back as far as No. 4 Redoubt. Cattell claims Mouat said he had to draw his sword when the Cossacks threatened. Could the pursuing Russians have come as far up the valley as No. 4 Redoubt? It seems doubtful when the area was occupied by elements of the 4th Division and the Heavy Brigade was nearby. Then there is the difficulty of Nolan's body. All accounts agree that Nolan fell between

200–400 metres from the start line. That puts his position well to the west of, that is further up the valley than, No. 4 Redoubt. If Morris collapsed close to Nolan he would surely have been out of reach of any Cossacks and hardly under heavy fire. An interesting puzzle that will probably never be resolved satisfactorily.

After Cardigan had left Mitchell the latter continued up the valley, struggling laboriously over a soft patch of plough towards the Scots Greys whom he could see some distance away. He was clearly feeling the effects of the long morning and, like all dismounted cavalrymen, utterly lost without a horse:

'There were several riderless horses galloping about the plain. I tried very hard to get one but could not. I saw two officers' horses belonging to my own regiment. I could tell them from the bindings on the sheepskins and the saddles. They appeared almost mad. I would have given a trifle just then to have had my legs across one of them, for I was getting tired; for we had been out since 4 a.m., and had nothing to eat since the day before. And to make matters worse there was a piece of ground that lay in my way which . . . was very loose, which made it heavy travelling.'[24]

Then, to Mitchell's horror, the Scots Greys turned around and pulled yet further back.

'To my dismay I saw the Scots Greys . . . now about 500 yards from me, in the act of retiring at a trot. I thought there was no chance now, when our support was retiring. . . .

'I could now see some Cossacks showing themselves in swarms on our right. . . . As soon as I saw them approaching, I bore more away to my left front, and towards a party of Chasseurs d'Afrique. . . . These having shown themselves menacingly, it had the desired effect of turning the Cossacks from their purpose.'[24]

On reaching the spot where the Heavy Brigade had been formed Mitchell found a solitary wounded soldier standing alone and seemingly abandoned by his comrades. Mitchell continues:

'As I came along he heard me, and calling out, said: "Is that an Englishman?" I answered: "Yes;" and on going to him found he had been wounded by a piece of shell just between the eyes, which had blinded him. He had bled very much, and was still bleeding. I had a handkerchief in my breast, which I bound round his wound, and taking him by the arm, led him along.'[25]

Mitchell, with his blind comrade in tow, had not gone much further when he came upon the badly wounded form of his friend who had ridden down the valley on his left. This man was spreadeagled on the ground on his back, gurgling and gasping as he struggled not to choke on his own blood which was pouring from his mouth.

'I could see death in his countenance, but turned him over and placed his arm under his forehead, thinking he would be better able to relieve himself of the blood than by lying on his back.'[25]

His next comments reinforce the view that No. 4 Redoubt was sufficiently far from any effective fire as to render persons near it to be relatively safe. They also confirm that the redoubt was occupied by British infantry by then – that is by 11.30.

'By the time we had gone a little farther we were pretty well out of range of their fire, and coming near No. 4 Redoubt, which the enemy had left, and which was now held by our 68th Light Infantry [part of the 2nd Brigade of the 4th Division and which later became the Durham Light Infantry]. The men were lying down in the ditch, and the poor Scots Grey whom I was leading saying he felt faint from loss of blood, I led him up to the ditch of the redoubt and coming to an officer first, I said: "Have you a drop of water, sir?" . . . [the officer produced a pint mug of half rum and half water] The wounded man drank about half, and his moustache and mouth being covered with blood had dipped into it, but that did not matter then. I emptied the cup . . . we made another start. . . .
     'We met our commissary officer, Mr Cruickshank, mounted on a pony with saddle bags filled with bottles of rum. He was making his way to meet any of the men returning from the charge. He very kindly gave us a good drop each, which helped us along nicely [modern soldiers will recognize in Mr Cruickshank the makings of an excellent quarter-master]. . . . I could see a couple of ambulance wagons, and several surgeons hard at work dressing wounds. After seeing the poor fellow's wound dressed, and assisting him into the wagon, I bade him goodbye, and have never seen him since. I have heard that he was discharged blind, and was allowed a shilling per day pension, and sometime after again recovered his eyesight.'[26,27]

It is appropriate to finish this chapter on the scramble to get out of the valley by following the adventures of the man who was supposedly the last to do so. Nobody can be certain of the first man back as no one has sought to establish this doubtful honour, although there could not have been too many ahead of Cardigan. To be the last man back, however, was certainly something to be proud of, and worthy of countless rounds of drinks into old age. TSM Loy Smith has laid good claim to this distinction.
     Smith, it will be remembered, had rejoined the 11th Hussars after his exploits

trying to capture a cannon. He and his comrades were within a stone's throw of the aqueduct when the order came to turn about and retire. Smith sensed a great reluctance among men who had endured so much, achieved so much, and had got so far beyond the objective, to give up. Many felt they had been badly let down. The 11th Hussars had ridden and fought for over two miles; now, as they looked over their shoulders for support, they saw nothing except the enemy. For a few seconds some thought the Uhlans drawn up across their path were the 17th Lancers. Then came the realization that they were utterly on their own with enemy in front and behind. There was a grim determination to fight their way out, but also the question of where everybody else was. Above all, where was the Heavy Brigade?

According to Smith the Regiment that wheeled about bore little resemblance to the ordered ranks that had pounded past the guns a few minutes before. Now they turned in ones and twos; now all dressing had gone; now officers and men were inextricably mingled, troops and squadrons tangled and depleted. They were a band of flurried horsemen on blown horses. Smith states that at this stage the only officers left were Colonel Douglas, Captain Cook and Lieutenants Dunn and Palmer, making no mention of Assistant-Surgeon Wilkin. Wilkin was there – though unnoticed by Smith. There was a momentary pause before the final effort to raise yet another gallop at the enemy barring the way. Some hasty jostling and pushing helped to produce a more tightly knit, compact group. Hurried attempts were made to get the strongest horses to the front of what had become an oval-shaped formation. It was developing into an 'every man for himself' situation, with the strongest men on the strongest horses acknowledged to have the best chances. As Smith himself admits:

'We assumed an oval shape – the best horses in front. Many men whose horses had been wounded, others whose horses were fagged and could not keep up, were overtaken and killed, for the Russian Hussars were now pursuing and shouting "Bussu, bussu English". We were thus being driven on to the line of lancers in our front.'[28]

Smith, understandably, confines his account to what happened to his Regiment, and to himself. He makes no mention of Paget and the 4th Light Dragoons who were nearby, if not actually combined with, the 11th during the breakout. Afterwards Smith gathered together numerous witnesses to discredit the contention that Paget had assumed command of a combined force of the two Regiments for the final phase. With commendable regimental loyalty Smith and his comrades forever remained adamant that the 11th fought their way out as a regiment under their own officers and NCOs. Colonel Douglas was of the same opinion.

Smith began the retirement at the right rear of the oval. As they moved off he glanced to his rear to assess how closely they were being followed. He was concerned to see that several Russian Hussar officers, with their men close behind, appeared about to overtake him. They were near enough for Smith to hold his sword in the 'right rear guard' position to try to ward off the expected blows from behind. When the 11th were within 200 metres of the Uhlans, however, Smith noticed that the pursuing Hussars were dropping back, as if they were handing over the duty of stopping the fleeing British to their comrades ahead. The line of Russian lancers in front stood still, lances at the carry. There was no forward movement, no attempt to charge. As Smith wrote, 'This was the third time this day that I have seen the Russian cavalry remain halted when they should have charged.'

The 11th closed rapidly with the stationary line. Each rider instinctively bent lower over his horse's neck, shortened his reins and gripped his weapon more firmly as he picked the spot ahead where he would burst through. It was at this moment, to everyone's astonishment, that the right-hand Russian squadron 'went about, wheeled to the right, halted and fronted with the greatest precision . . . their lances still at the carry.' There was a gap to go for.

To Smith, as he galloped across the enemy's front, it seemed for a fleeting instant that 'I had been inspecting them'. At this point Smith's account is at variance with Paget's, but in agreement with the Russians'. Whereas Paget claimed the enemy did little or nothing to prevent an escape, Smith states that he heard an order shouted, saw the lance-heads drop and the Uhlans surge forward to attack the British in their flank. There was hand-to-hand combat as the 11th sought to parry the lance thrusts and ride clear. A number were struck down. Sergeant Bentley fell behind and was rescued by Dunn. The great majority, however, broke through. The Russian charge, if it can be called that, lacked spirit.

It was at this point that the 11th (and the Russians) came under artillery fire again. Bojanov's battery on the Causeway Ridge was firing indiscriminately into the mob of jumbled horsemen below it. Men on both sides suffered. Smith's horse staggered, almost fell. Sergeant Joseph, who was riding near by, yelled out to Smith that his horse had a broken leg. Feeling the limpness of the animal, Smith threw himself clear just before it came crashing down. Springing up instantly, he started to run. It was far from reassuring when soldiers galloped past him from the rear shouting, perhaps with a wicked sense of humour at his dangerous and undignified predicament, 'Come along sergeant-major!' While shouts of verbal encouragement were heard Smith received no practical help. As he put it, 'none could assist me, it being every man for himself.'

Smith veered to his left to avoid being ridden down (see Map 15). But his comrades fast disappeared, leaving him to fend for himself. He had over a mile still to go before he might expect sanctuary. Still gripping his sword he swung further left, so that he was moving along the valley, under the Causeway Ridge and quite close to the Russian troops occupying it. To the enemy infantry he made a tempting target. A number of them had a go at the lumbering figure. Muskets and rifles popped, but the running man did not fall. To the marksmen it was annoying to miss, but at the receiving end Smith was having a decidedly

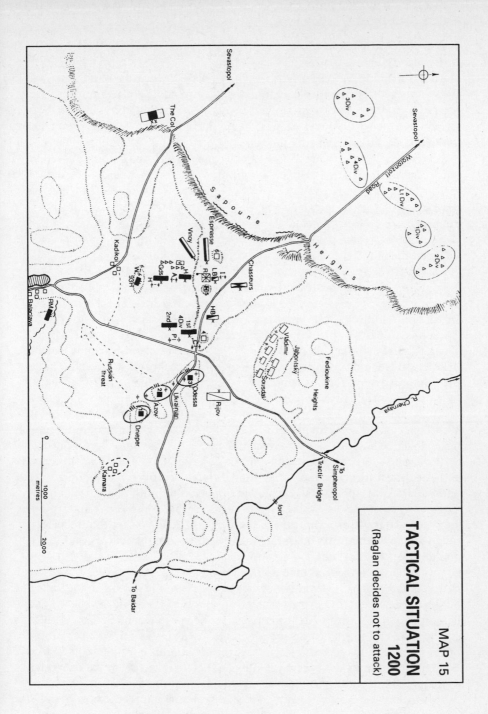

MAP 15

TACTICAL SITUATION
1200
(Raglan decides not to attack)

Sevastopol

Sevastopol

Worontzoff Road

3 Div

4 Div

Lt Div

1 Div

2 Div

The Col

Sapoune

Heights

Vinoy

Esprasse

6

Chasseurs

Kadikoi

W 93H

Gds

A

R

L

H

HB

2nd

4 Div

1st

P

Balaclava

RM

Russian
threat

Odessa

Azov

Ukrainian

Rijov

Dnieper

Kanara

Vladimir

Japonisky

Soustal

Fedioukine
Heights

R Tchernaya

To Simpheropol

Tractir Bridge

To Baidar

Iord

0

1000
metres

2000

211

unpleasant time. Dust was being kicked up at his feet, and with every stride he anticipated the burning thump as a ball hit home.

All this running was infantry work. Smith was slowing, his chest bursting. Then he heard the sound of a galloping horse coming up behind him. He glanced round to see two Russian lancers about 50 metres from him with two dismounted men of his own Regiment about the same distance to his right rear. Smith needed little convincing that they were intent on riding him down but staggered grimly on, vowing to turn at bay when they closed on him. A second quick backward glance, when he thought his pursuers must surely be on him, revealed them wheeling away to attack the other two soldiers. He snatched the chance to stop briefly to regain some wind and to see the fate of his comrades. Smith recorded, 'I saw them circle to the right and meet the two men. When close in front of them, I saw their lances go down.' Smith did not stay to see the outcome but later found out they had both been killed.

This short pause was not enough for the enfeebled Smith to recover his breath. He was almost prostrated with exhaustion, to the extent that he considered just lying down where he was and shamming death. Then the thought struck him that he was not far enough away from the enemy infantry on the Causeway. What if some of them wandered over to loot the corpses scattered along the valley? They were likely to stick a bayonet into a body before handling it. He resolved to continue walking. After a short distance a few more shots came his way. Some riflemen were still practising their skill. Their accuracy was good enough to persuade him to break into a jog again.

After a short time he saw, ahead of him, a man on foot in the uniform of the 17th Lancers. He was moving slowly, stumbling rather than walking. Smith began to catch him up. As he drew near he realized it was an officer wearing a forage cap. As he came up behind him Smith called out, 'This is warm work, sir.' The officer turned to look at Smith. His face was covered in blood and he had 'a very wild appearance'. He made no reply, but Smith recognized him as Captain Morris. Morris was moving towards the left of the valley and would, some time later, flop down near the body of his friend Nolan. Smith decided not to remain with Morris but rather to move more into the centre of the valley and thus lost sight of him.

He had covered the first mile up from near the aqueduct on his horse and then run and walked in a somewhat erratic course to the middle of the valley. He was then about 1,000 metres from the battery they had attacked, and perhaps 750 metres from No. 4 Redoubt. He began to feel more secure, sufficiently so to halt to recover his breath and take a swig of rum from the rubber bottle he kept in his haversack. He began to regain his confidence as well as his breath. Nobody was shooting at him and there were no enemy close enough to threaten him. Smith sheathed his sword and continued.

He was making his way towards the slight, almost imperceptible dip in the ridge between Nos. 3 and 4 Redoubts, inclining to his right (west). He was coming into the area over which the Russian cavalry had retreated from the Heavy Brigade and the shelling from 'C' Troop RHA earlier that morning.

Suddenly, as he came to the top of a slight undulation, he was confronted by three Russians lying on the ground. The one he had almost tripped over was dead but the other two were sitting a few metres away staring at him. Smith halted.

Apart from his sword in its scabbard Smith had no weapon. Either of the Russians could be armed with pistols. A quick shot in the time it would take to draw his sword and advance on them could all too easily ruin his efforts so far. If he rushed one he could get a bullet in the back from the other. Then he saw a carbine lying beside the corpse at his feet. He snatched it up, cocked it and pointed it at first one then the other of the wounded Russians. Both got the message immediately. They raised their hands and gazed at him imploringly. Smith had bluffed them. As he cocked the carbine he had noticed there was no cap on it so it could not be fired. He hastily put on a cap but even then did not know if it had been loaded. Smith went forward to the nearest one. He had been badly injured in the body and was unarmed. The second had lost his left foot just above the ankle. They were artillery casualties and belonged to Rijov's Hussar Brigade.

When Smith examined the one whose foot had been shot off closely he made a strange discovery – the man had the number eleven on his buttons, the same as Smith's. By an extraordinary coincidence all three men were serving in a Regiment designated the 11th Hussars (the Russians being the 11th Kiev Hussars). They wore almost identical brass buttons. Smith bent over the Russian and, 'pointed to the number on his buttons, then to my own, saying as I did so, "English".' The wounded man nodded. Smith decided that so unlikely an occurrence deserved a suitable souvenir. Stooping down he demonstrated to the soldier that he wanted one of his buttons. Anxious to please the Russian tugged desperately at one but did not have the strength to break it off. He unbuttoned his jacket and struggled to get it off. A filthy jacket in exchange for his life was a price well worth paying.

Smith stopped him and, pulling the jacket back over his shoulders said, 'No, poor fellow. I won't take your jacket, most likely you will have to lie here all night.' Still determined to have a button, however, Smith drew his sword to cut one off. The Russian, understandably, thought his end had come; raising his hands he started to gabble what was obviously a last appeal to the Almighty. Smith shook his head, put the sword reassuringly behind his back, before bending over to sever a button. The next day he would replace one of his own buttons with the Russian one. It was to remain on his uniform, polished and cherished, for years to come.

His journey was almost over. He was to finish as he had started, back in a saddle. As he was about to leave the Russians (he seems to have made no attempt to dress their wounds) three loose horses trotted up. Selecting the unwounded one which belonged to the 4th Light Dragoons Smith rode towards No. 4 Redoubt. He was approaching the position of 'C' Troop RHA. In his History of 'C' Battery Colonel Whinyates describes Smith's arrival as witnessed by the gunners as follows:

'While there, a Sergeant-Major of the 11th Hussars came out from the opening between the Redoubts [see Note 29] and made for the Troop. On his arrival at the guns Brandling said to him, "Who on earth gave the order Sergeant-Major?" and the man answered, "Oh sir, God knows! We [just] heard, 'Come on! come on!' called out. My horse was shot when I was at the bottom of the plain; this is a loose horse of another regiment I am on, and this is a Russian loaded carbine I have been defending myself with against some Lancers who attacked and knocked me about." His story was quite true as regards the horse and carbine, and his jacket bore evidence of the struggle he had undergone, for the wadding was all sticking out through the rents made by the lances. He also said that the Russian infantry were hidden in enormous masses at the foot of the hills on the other side of No. 2 Redoubt, and all about there. This was the last man observed by 'C' Troop returning from the Light Cavalry charge, and how he crossed the bottom of the outer plain [the North Valley] amongst the Russians, and got out where he did, is known only to himself.'[29]

Some little distance to the rear was 'I' Troop RHA. As Smith rode slowly through them several called out to congratulate him on his getting back alive. Smith, whose feelings of relief at surviving were giving way to fury at how the Light Brigade had not been supported, could only reply, 'Some one will have to answer for this day's work.' In his diary he was to write that an intelligent private soldier who knew his drill would have done much better. His final sentence on the matter sums up the feelings of many of his companions that day and for long afterwards.

'We had lost our guns, we had taken the enemy's, but had been compelled to leave them behind for want of timely support.'[30]

When Smith rode up to where the remnants of the Brigade were gathered he first dutifully handed over his horse to the 4th Light Dragoons, and then walked over to where the 11th were forming. As he approached the rear of his Regiment they were calling the roll by numbering off from the right of the line. As he slipped into place on the left rear 'someone said, "Number off, sergeant-major", so I called out "Sixty-three" as sixty-two was the last number I heard.'

IV
THE CONSEQUENCES

CHAPTER ELEVEN
The Reckoning

'The order was, "No fires to be lit. No noise to be made." This was indeed a sorry night. Scarcely had any man had more than a little biscuit and a drachm of rum since the day before. It was spent standing in groups talking over the sad misadventures of the day.'
Troop-Sergeant-Major Loy Smith, 11th Hussars

It was five hours after the remnants of the Light Brigade had returned before they were allowed back to their wrecked camp. Five hours of hanging around, five hours of misery and hunger, five hours of waiting in vain for some worthwhile action on the part of the infantry. Balaclava was supposed to have been an infantry battle. They had arrived late, having watched the cavalry attempt to do their work, and now their mounted comrades expected them to finish the job and retake the redoubts. They watched as the infantry scattered skirmishers to their front; they saw the puffs of smoke and heard the popping as individual sharpshooters moved cautiously towards No. 3 Redoubt. Cannons occasionally jumped and banged at distant targets. But the great bulk of the 'red infantry', as Cathcart called his troops, did nothing. Redoubts 1–3 remained firmly in enemy hands. This inaction was incomprehensible to the cavalry.

As time passed the light cavalry became more and more dispirited and disillusioned, convinced that their efforts had been for nothing. Initially, at the roll-call of the Regiments, there was some feeling of elation at having survived, at having achieved an impossibility. When Cardigan rode up to what was left of his Brigade he said, 'Men, it was a hare-brained trick but no fault of mine.'

'Never mind, my lord,' replied an unidentifiable voice, 'we're ready to do it again.'

'No, no, men, you've done enough,' was his lordship's pompous response.

As they sat contemplating their misfortunes, and the Guards' and Highlanders' immobility, morale sank further. If nothing was to happen why

215

could they not return to their camp which was only a few hundred metres away? To have lost so many friends was totally depressing. To watch the farriers pistolling beloved horses to whom so many owed their lives compounded their misery.

Of the horses that survived the charge or the farrier's bullet almost all succumbed to the desperate conditions of the winter that was then but a week or so away. The history of the 11th Hussars records the fact that the Cavalry Division had been reduced from 2,000 to 200 horses by December, 1854. The animals had to be picketed out on the plateau near Sevastopol exposed to glacial winds and where there was 'not a blade of grass for the poor creatures to nibble'. These wretched survivors, led by a trooper, were mostly used to carry biscuits from Balaclava harbour along the awful road up The Col. It was a nightmare journey of seven miles through a quagmire of freezing slush and mud. So covered with sores and so feeble were these horses that they often collapsed to be engulfed by the mud and snow. These carcases lay where they fell until devoured by packs of wild dogs. Fodder became unobtainable as the Commissariat was frequently unable to provide transport from the harbour. On some days a single handful of barley was all the emaciated animals received. Many resorted to eating saddle-flaps, straps or blankets. They gnawed each other's tails to stumps. Those that had been put down had the better deal.

A handful came through it all. Sergeant Avison's mount was an outstanding example. Avison was the farrier-major of the 11th Hussars and had ridden in the charge. His stallion 'Old Bob' had begun his remarkable career in the 15th Hussars, been transferred to the 14th Light Dragoons and later to the 11th. He was ridden by Avison at the Alma and Inkerman, as well as at Balaclava. He was never sick, and eventually returned with the Regiment to England having been renamed 'Crimean Bob'. He eventually died, aged 34 and still in service, in Cahir, Ireland, in 1862. He had a military funeral and a tombstone erected to his memory.

The other animal, last seen sprinting at the heels of the 8th Hussars, was also a survivor and, like 'Crimean Bob', became something of a legend. This was Jemmy the terrier. He recovered from his neck wound and returned to Ireland where Major de Salis had a special collar made for him to which four clasps had been added – Alma, Balaclava, Inkerman and Sevastopol. He wore it on special parade days or on field training. Jemmy was to live on for another four years. He accompanied his Regiment to India and campaigned in the Indian Mutiny for which he was given his fifth clasp – Central India. Regrettably, in September, 1858, when attempting to swim the River Chanbal the current proved too strong and Jemmy was swept away. His body was recovered and his collar kept on prominent display in the officers' mess.

The human loss, though crippling in terms of cavalrymen fit to fight (something not really possible without a horse), was not quite as overwhelming as some accounts suggest. At the first roll call at the head of the valley Kinglake records there were 195 mounted men answering their names. A number of subsequent writers have omitted the word 'mounted' and thus have 195 as the total number of survivors parading. The reality was that some 500 men of all ranks, including wounded, got back. The first count (which included some wounded) by numbering off was only accurate at the time it was taken. A number of wounded (like Morris), some unhorsed soldiers, would not have made it to the first muster. It would be hours, even days, later, when the fate of those missing became known, that the more accurate regimental returns gave the real picture as to how the Brigade had suffered. The 13th Light Dragoons, which initially mustered a mere ten mounted troopers, later submitted a return showing another hundred all ranks had survived (37 were wounded).

It will never be possible to know the exact figures but it is possible to be sufficiently accurate to make some interesting deductions and comments. Cardigan was not far wrong when he wrote in his diary the following day:

'Thursday, 26th October. – The brigade was almost destroyed [true if thinking in terms of an operational force, but not literally] by yesterday's affair. 300 men were killed, wounded and missing. 396 horses were put hors de combat, and 24 officers killed and wounded. . . . In the course of this and the two following days, I had upwards of 30 of the horses of the Light Cavalry Brigade shot, being desperately wounded in the affair of yesterday.'

Cardigan had the regimental returns sent to him by his brigade-major. The present writer has chosen to use those listed in Whinyates' account. They were, 'carefully compiled from the Regimental Records and verified as far as possible by the statements of some of the survivors (prisoners included).' Lined up on the start line had been 658 regimental horsemen, to which should be added Cardigan, Mayow, Maxse, Nolan and the two Sardinian officers (both of whom were wounded and one captured). Therefore a total of some 664 charged. The tally afterwards for all ranks was: killed in action 110 (including seven who died of their wounds); wounded in action 130; prisoners 58. This gives an overall total figure for all types of casualties of 298, or 45 per cent. Whinyates gives the horses killed, or shot later, as about 362, or 54 per cent. A terrible price to pay for nothing, but nowhere near the severity that similar strength battalions suffered sixty years later – and fought on.

Considering what had happened the statistics are more startling when looked at from the point of view of how many returned rather than how many were killed. About 306 came back up the valley with barely a scratch. When the prisoners were released a year later the number of men who had lived to tell the tale was the astounding figure of around 540 (allowing for about ten who died in captivity). Discounting those who later died of wounds or as prisoners only 103, or 15 per cent, were killed on the battlefield. Looked at another way only one man in six of those who rode down the valley, and up it again, died while doing so. Not quite the picture so often painted.

217

Another aspect of the casualty figures that has been neglected, but cries out for a closer look, is the number that actually occurred in the charge itself. Again the great bulk of the literature on the subject indicates that the Brigade was pounded to pieces as it dashed down the valley. If one believes Tennyson the light cavalry were shot at and shelled from three sides as they rode down, and again from all three directions as they rode back. This is, of course, poetic licence and complete nonsense. During the charge they were fired at from three sides, but not from all three simultaneously. Returning, the gunfire was from one flank only.[1]

It is reasonable to assume that about one third of all killed and wounded casualties (discounting prisoners) were inflicted after the Russian guns had been reached. These men would have been hit while fighting through the gunline, attacking the cavalry behind, during the withdrawal up the valley or cutting their way through the Uhlans. This would mean that approximately 156 men were hit during the charge and 77 afterwards. Just over 23 per cent became casualties during the gallop to the guns. Some of these would have been hit but still reached the battery, others would have been unhorsed but not themselves hit. If it is assumed that these roughly cancelled each other out then it follows that nearly 500 men could have reached the guns.

In total 26 Russian guns took the opportunity to fire on the Light Brigade during its charge. The Brigade took some seven and a half minutes to cover the distance. The guns fired approximately 190 rounds during that time and, if the calculations above are roughly right, hit 156 men. They also hit a lot of horses, but a fair number carried gamely on despite their wounds, as a galloping horse takes a lot of stopping.

So why were the Light Brigade not blown away? The Russian gunners were not themselves under fire, they could shoot at their target from three sides and yet nearly three-quarters of the Brigade reached Obolensky's battery. A combination of factors was responsible for this, from the Russian viewpoint, disappointing result. First, the unusually fast pace of the advance. The speed meant that the distance was covered more rapidly than the drill manuals recommended, which meant that the gunners had a reduced time in which to fire. This progressive acceleration meant that the two flanking batteries were shooting at a fast-moving target passing from left to right, or vice versa. After every shot the gunners had to adjust their aim considerably. Firing at the rate of two shots per minute gave the target 30 seconds of movement between firings. In that time the Light Brigade had moved an appreciable distance.

The second difficulty, linked to the rapidity of the attack, was that for the flanking guns the target quickly disappeared out of their arc. The Fedioukine Battery could not swing its cannons to follow the target all the way down the valley. Its firing was limited to about three and a half minutes. Bojanov's battery may have had only two. Add to this the fact that several participants in the charge thought the gunners were firing high and it becomes clear why the accuracy of the shooting was not the best. The instinct of the light cavalry to quicken the pace, despite their brigadier's best efforts to restrain them, had been the right one. It undoubtedly saved lives on this occasion.

*　　*　　*

A large gaggle of horsemen spilled over the Sapoune escarpment and slithered down the slope. Raglan, accompanied by his staff, Canrobert with his, and the war correspondent Russell, were descending into the valley. The British Commander-in-Chief was looking for explanations. His orders had been flouted and his light cavalry seemingly destroyed in an attack that verged on idiocy. According to Captain Portal Raglan was to use the expression, 'The Cavalry has been wantonly sacrificed'. Cardigan was the most likely culprit. It was his Brigade that had gone and he had led them. Russell saw the two meet.

'Now, I was not very far off when Lord Cardigan rode up to the Commander-in-Chief. I saw him rein up and speak to Lord Raglan, who, judging from the way he shook his head as he spoke, and jerked the armless sleeve of his coat, was much moved with anger. But when Lord Cardigan turned to rejoin the fragments of his brigade he seemed in no degree depressed.'[2]

Russell had not been close enough to hear Raglan's furious accusation deftly deflected by Cardigan from himself to his brother-in-law. Raglan angrily voiced his feelings with his opening words:
'What do you mean, sir, by attacking a battery in front contrary to all the usages of war and the customs of the service?'
Cardigan, confident he had the perfect defence and that his conduct that day was irreproachable, replied smugly:
'My lord, I hope you will not blame me, for I received the order to attack from my superior officer in front of the troops.'[3]
That put Lucan firmly 'in the frame'. And there he was destined to remain, at least in his Commander-in-Chief's eyes. When they met, Raglan's barbed denunciation was to leave a wound that would be the source of much pain for years to come.
'You have lost the Light Brigade!' The exact words of Lucan's immediate response are unknown, but they constituted an emphatic denial. As recorded in the final chapter, he had no intention of leaving matters like that. In the meantime rumours quickly spread that, as Russell recorded, 'Lord Raglan had given Lord Cardigan a tremendous wigging [untrue] and had "given it hot" to Lord Lucan too [correct].'[4]
Raglan also met Scarlett within the hearing of the 'C' Troop RHA and, after congratulating him on the performance of the Heavies, he rode off. Whinyates records how:

'He [Raglan] then turned his horse away as if he did not need to hear any more. He seemed very much annoyed and bowed down over the loss of the Light Brigade, and was pointing out with his arm the ground over which he intended the advance should have been made. He and Scarlett being only ten yards in front of 'C' Troop at the time, no divisional staff being present.'[5]

Raglan had more pressing matters to engage his attention than reprimanding his cavalry commanders. He had, in consultation with Canrobert, to make a decision as to what to do next. Was the Light Brigade's retreat to be the end of the battle? Were the infantry finally to be galvanized into serious action, or were the Russians to be left occupying three redoubts and retaining the British guns Raglan had been so anxious to recapture?

It was not only soldiers who felt the situation favourable for offensive action. Russell was later to write of the view just prior to his leaving the escarpment:

'The day was over early . . . there were five good hours of daylight left. The aspect of the plain filled us Ignorantins with expectation. Nothing could be finer than the appearance of the infantry now formed in blocks of scarlet or blue below, and filling the valley with solid-looking masses. Cathcart's division – two fine brigades – were menacing the Russians' hold on the redoubts. . . . The Brigade of Guards were drawn up on the left, near the western slope of the Woronzow Road, and the Highland Brigade (save the 93rd) nearer to Inkerman. The French cavalry on the left – a brigade of infantry in support made a fine show; two of Bosquet's brigades formed on the right covering Balaclava. . . . It was difficult for us who were looking down on the field to believe that our generals were about to accept the situation.'[6]

To do nothing would mean leaving the enemy in possession of most of the battlefield (like the loss of cannons traditionally regarded as proof of a battle lost). More importantly, it would mean the Russians were far closer to Raglan's base at Balaclava than the great bulk of the British Army. They would also control the Woronzoff Road. This would cut the British supply routes from two to one.

From Raglan's perspective at midday, seeing the situation through his eyes, and without the ability to foresee that the enemy would abandon the redoubts in mid-winter, the British position was perilous. Even a tactical stalemate on the Balaclava battlefield would seem to hand yet more strategic advantages to the Russians. The six redoubts had been built as an outer defensive line to the British base. Now the three crucial ones had been lost. The battle had been joined to halt a dangerous, indeed potentially war-winning, thrust at Raglan's communications. Now Liprandi was almost within cannon shot of the entrance to Balaclava gorge. He had the perfect secure start line for a fresh attack to the south-west. An advance of 3,000 metres would put his soldiers in Balaclava. If that happened the siege of Sevastopol could never be sustained. The whole object of the Allied campaign was to take Sevastopol, but that would be impossible without supplies, without secure lines of communications, without a base – without Balaclava.

Raglan had no way of knowing whether the Russians had had enough, whether they would press on, or whether their cavalry reverses had dented their morale sufficiently to outweigh their infantry's successes. If their advance of five hours earlier had been what it seemed, the start of one giant stride to Balaclava, then the Russians were now half-way there. What if the enemy brought up more troops in a few days to reinforce success? Another, perhaps more powerful, advance could not be discounted. Failure to hold it could give the Russians the war. Not much in the way of generalship was required by either side to see both the perils and the opportunities of the situation. Its stark simplicity was clear to any corporal with commonsense. If, however, the British could retake what they had lost and garrison the three key redoubts with 'red' infantry then the chances of the enemy renewing their threat to Balaclava would be greatly reduced.

There was also the question of the Woronzoff Road. If the Russians made no further advance, but just sat in the redoubts it would render the use of the Woronzoff Road impossible for the British. Any troops or wagons moving up to join this road from Balaclava would be under gunfire and threat of attack all the way. It has been suggested by one author that the loss of control over this road was not detrimental to the British in the coming weeks. This contention is made on the basis of a quote from a Captain Montgomery of the 49th Foot. According to him the 'Woronzov Road was never used for the transport of guns . . . a part . . . was used for the transport of ammunition, but we had to cross the mud in the open country to get to this mile of road.'[7]

Ewart, Raglan's ADC with Cathcart, contradicted this view when he wrote:

> 'Despite the glorious conduct of our troops on this occasion, we lost a good deal and gained very little. The eastern portion of an unsupported advanced line of redoubts on the Causeway Heights was captured . . . we were deprived of the use of the Woronzoff Road as a means of communication between our base of operations and the upland, and we had only the unmetalled path which led over the Col to rely on. We shall see that this result of the battle of the 25th was a serious one for the British Army besieging Sevastopol.'[8]

A glance at Map 15 will help in understanding the significance of the roads leading to and from Balaclava. All near-contemporary maps show the basic road system shown on Map 15. The Woronzoff Road carried traffic (people, animals and carts) between Baidar and Sevastopol. Those coming from Baidar who wanted to go to Balaclava turned left at the Causeway crossroads or used some earlier minor tracks. Traffic moving from Simpheropol to Balaclava came along the road that crossed the Chernaya River at the Tractir Bridge, on through the Fedioukine Heights, up onto the Causeway Ridge, crossed the Woronzoff Road and headed straight down the gentle slope to Balaclava. Captain Montgomery called this stretch of ground from the crossroads to the town and harbour 'open country'. He omits to mention the road. It existed in October, 1854. It was the link between Balaclava and the Woronzoff Road which, as Montgomery indicated, may not have been used for guns, but certainly carried other supplies. It followed a natural route to the Tractir Bridge. It was the road the British would have used had they controlled the redoubts to give them an alternative route (from The Col road) to the siege lines. Indeed the Woronzoff route would be slightly shorter than via The Col if wagons were destined for the right of the British positions facing Sevastopol.

The Woronzoff Road was the only metalled one. The others were not properly surfaced and so quickly degenerated into a morass of mud with bad weather and heavy usage. Thousands of men and animals had to be supplied with all the means to live and fight. For the British this entailed a continuous stream of guns, horses, carts and wagons plodding painfully backwards and forwards along The Col route. No road in the Crimea was ever intended to cope with such traffic. The journey of seven miles could take seven hours and leave the men and animals at the point of collapse. Given a choice, two routes are always better than one. Other things being equal, it doubles the speed of movement and halves the congestion.

If Raglan allowed the Russians to remain where they were he would be accepting that they had a stranglehold on his lines of communication. He had got both his infantry divisions assembled in the South Valley before the Light Brigade limped back up the North. At last he was in a position to launch a coordinated attack, supported by artillery. Both the Allied commanders sat side by side; French infantry and the Chasseurs were available to support the British – but nothing happened. Raglan and Canrobert dithered and discussed. Raglan reportedly favoured action, Canrobert advised caution and the British general, ever the diplomat, seldom the soldier, allowed himself to be dissuaded.

The French argument was that Sevastopol was the key to the whole campaign and that if the redoubts were retaken there would be insufficient reliable troops to hold them unless the siege lines were denuded. Every available man and gun was needed to take the city. This was undoubtedly true but presupposed that the Russians would be content merely to watch and do nothing as the daily quota of supplies to feed the guns and the men was hauled painfully up from Balaclava harbour. If that one tenuous track was cut and held attacks on Sevastopol would wither away. To deliberately leave the enemy within long-range cannon shot of the entrance to the British main base was taking an enormous gamble. At 7.30 that morning a Russian assault on the redoubts had been sufficiently worrying to bring two infantry divisions down from the escarpment; eight hours later the Russian capture of the three most important was shrugged off as an inconvenience that could be lived with –

strange logic. In the event the Allies were not made to pay the consequences of this decision due to the inertia and incompetence of the Russians. At the time the decision was made, however, it had all the hallmarks of a classic military blunder. Fortunately, in war mistakes are not confined to one side. It is the side which makes the least that wins.

Without prodding from headquarters Cathcart's efforts to advance had petered out at No. 4 Redoubt. His guns (P Battery) opened fire at extreme range on No. 3, his infantry went to ground. The Duke of Cambridge and Campbell joined him at No. 4 to view the enemy and confer. Their presence was spotted and drew Russian fire. Ewart went so far as to ask Cathcart whether he intended to take No. 3. In that General's view it would be too costly.

With the exception of the Highland Brigade, reunited under Campbell for the close protection of Balaclava, and a French brigade under Vinoy west of Kadikoi, the Allies turned their backs on Balaclava. Cathcart marched his men back up The Col yet again. The next day a despatch from London removed his 'dormant commission' that he had been carrying in his breast pocket in a waterproof bag. From then on Sir George Brown would take over should Raglan die. Cathcart's dilatoriness at Balaclava, which was perhaps the root cause of the way the day's events unfolded, escaped censure. Fate, however, was not as forgiving as Raglan. Eleven days later he was killed by a bullet through the chest at Inkerman. A hill overlooking Sevastopol was named after him. It was the place of his burial and is the site of the British military cemetery and memorial.

Emboldened by the Light Brigade's retreat and the lack of any follow-up the Russians consolidated their hold on the three redoubts with the Azov, Ukrainian, Dnieper and Odessa Regiments. Rijov's demoralized cavalry moved back up the North Valley for the second time that day and formed a link between the Causeway Heights and Jaboritsky's force, which still retained the Fedioukine Heights. Liprandi held a continuous line some 4,000 metres in length from Kamara village in the south-east to the Fedioukines in the north-west. This had been achieved at the cost of some 550 casualties. The seven British cannons captured early in the day were later (Raglan and his staff had been mistaken when they thought they were being removed at 9.30 a.m.) taken to Sevastopol to be put on show in Theatre Square. Bells were rung, and heard in the British lines, to celebrate a victory. Not a clear-cut victory perhaps, but tactically they had been allowed to keep what they had taken in terms of ground and guns. Strategically they were perfectly placed to launch another strike at the most vulnerable part of the British military anatomy – their base. That they failed to do so, and why, is another story.

*   *   *

After numbering off, Smith formed up with about half a dozen dismounted men at the rear of his Regiment. Anxious to obtain spare horses, Colonel

223

Douglas ordered him to take these men back to their camp and bring back any horses that could be found. The little party walked down to the tents. The camp was a shambles. Not only had some fleeing Turks passed through it, grabbing what they could, but a number of Cossacks had ridden into it bent on killing and maiming the tethered horses. Then the Heavy Brigade had advanced over it just prior to their attack on Rijov's cavalry. Paget was to make an amusing comment on the havoc thus caused:

'The attack of the Heavy Brigade was actually in our lines, so we have lost a good deal of property, the answer [by soldiers who had lost kit] about everything being, "Oh, it was knocked over in the attack – I cannot find it;" or "The Turks must have stolen it." There is no doubt that the latter did take a good many things from our tents in their retreat in the morning.'9

Smith could find no uninjured horses. The Cossacks had cruelly speared or gashed every animal they could find. Four horses from Smith's troop had been left in camp; all had received severe sword cuts on their head, back or body. As he could procure no horses Smith elected to remain in camp and thus missed the depressing five-hour wait for permission to stand down that his comrades endured. He went to his tent which, like the great majority of the others, had been collapsed in a crumpled heap of dirty, damp canvas. Nearby he met his friend the Orderly Room Clerk [almost certainly a Sergeant Pickles] whose appointment had meant his being, in more modern parlance, left out of battle. Smith described their meeting:

'He shook hands with me saying, "How glad I am you have escaped, George." I told him I had lost my horse and how fearfully the regiment had been cut up. He then said, "What is this on your busby and jacket?" On picking it off, I found it to be small pieces of flesh that had flown over me when Private Young's arm had been shot off.'10

In an effort to revive Smith's spirits, for he was close to tears over the loss of so many friends and his horse, Pickles resorted to the traditional Army remedy for all ills – a brew of tea. In the early afternoon both of them wandered over the ground in front of the camp where the Heavy Brigade had fought. They came across several dead Russians and a 'young fair-haired Russian officer badly wounded and six or seven of his men, also wounded, sitting huddled together'. They were Hussars, effectively prisoners, but as yet unattended to. At dusk the 11th returned to the camp and the sergeant-major assembled what was left of his troop. Sadly and slowly he called the names. Some thirty-two NCOs and men had paraded early that morning, now under half were present. As far as could be ascertained, of the absentees six were dead, ten wounded and two prisoners. Three of the wounded had lost their right arm and one his left. Two of the wounded died later. Captain Cook, the troop commander, was wounded and Lieutenant Houghton mortally so. As Smith was to write in his diary, 'Thus ended a day of disaster.'

Smith was particularly bitter that the success of the Light Brigade in reaching the guns had, in his opinion, been wasted. Lack of support had turned the gallantry of the charge into a pointless sacrifice. In his diary he pointed his finger at who he regarded as responsible:

'And the world would ask: who was answerable for all this? [lack of support for the charge] The same man that ordered Lord Cardigan with 670 men to charge an army in position and then left them to their fate when he had at his command eight squadrons of Heavy Cavalry and two troops of Horse Artillery, besides a division of infantry with field batteries close in his rear. True, that he advanced two regiments of this brigade a short distance down the valley; but why did he not follow on? What did this avail us, for as soon as he came under fire, he began to retire.'[11]

It was about five o'clock in the afternoon when the uninjured men were at last released to return to their camp. Most of the tents were flattened, belongings looted and still no food was available. Then, just as darkness came, just as they were trying to settle down, trying to salvage some comfort from their dismal circumstances, the order came to move. The remnants of the Brigade stumbled about 400 metres closer to the escarpment. If there was a reason nobody gave it. There they were allowed to rest until midnight, when yet again the order, "On your feet, we're moving." How many times down the centuries have soldiers cursed and groaned when the command has started to play this game of messing (soldiers use a stronger word) the troops about? It always seems to happen at midnight in the pouring rain, although in this case the Light Brigade were spared the wet. They were now on a hillside about a mile in rear of their original position. Somebody passed another order: 'No fires to be lit. No noise to be made.'

If the drawing of how Cardigan spent the night is correct the order forbidding fires did not apply to the staff.[12] It shows his lordship lying asleep wrapped in his cloak close to a fire. At least he was not on his yacht as many writers have suggested. His ADC, Maxse, verified that, far from disappearing off for champagne and a hot meal and bath, Cardigan remained in the field. He had earlier made a point of seeking out the French General D'Allonville to thank him for the charge of the Chasseurs d'Afrique. Although confident he could never be called to account for what had happened, he seemed genuinely upset that his magnificent Brigade had been ravaged. When Lieutenant Earle of the 75th Foot (Cathcart's division) had seen him earlier he thought 'He had never seen a man so grieved'. He was particularly distressed by the loss of his trumpeter, Britten. In this he revealed a more humane, compassionate side to his otherwise unattractive character. Cardigan visited Britten's bedside at Balaclava and arranged to provide funds

225

from his own pocket to improve his comforts on his transfer to hospital at Scutari. Regrettably he died there on 14 February, 1855. Not until 1964 did the bugle he had carried find its way back to his Regiment, then the 17th/21st Lancers.

The Russians captured fifty-eight wounded and unwounded members of the Light Brigade. These were men who were rounded up or caught in the eastern half of the valley. On the 26th Captain Fellows and Trumpet-Major Joy were sent by Raglan down the valley under a flag of truce to make contact with the enemy. They approached the Russians carrying a white table-cloth attached to a lance. Their object was to secure a truce to allow for the burial of the dead. It was not granted. A fat, elderly general, speaking in French, proclaimed that Russians were Christians and as such were quite capable of doing what was necessary. Another visit the next day, however, secured a list of all British prisoners.

For the bulk of the British dead the Russians kept their word but failed to make a decent job of it. The following May 'C' Troop exercised over the North Valley and were distressed at what they saw. Bodies of men and horses had been pushed together into shallow graves from which arm and leg bones protruded. Bits of rotting uniform clung to many of these bones. The white collars of the 17th, the buff of the 13th, the red of the 4th, the cherry-coloured overalls of the 11th and the blue and yellow uniforms of the 8th were all recognized.

The dead were looted, but for the moment the bodies had to be left where they fell. The one notable exception was Nolan. As he was killed so early in the advance his body was easy to find and immediately accessible. Captain Brandling of 'C' Troop arranged his burial, seemingly in the ditch of No. 5 Redoubt. Whinyates recounts how it was done:

'He [Brandling] looked about on the outer side of the [Causeway] ridge for the body. Having found it he came back and took with him Bombardier Ormes (afterwards Riding-Master O. Ormes), and four limber gunners with spades to bury it. The bombardier, on his return, said that the poor fellow's chest had been quite broken away, and that the gold lace and cloth of his jacket very much burnt by the shell which killed him, and which must have burst close by; also that there was only one officer present, who appeared to be a friend, and was much affected, and who took off his watch and sword. The body was then laid in the earth as it was, there was no time to dig a deep grave, as occasional shots were being fired at the Troop. . . . The grave is in the outer plain, and there is a slight bend inwards in the ridge near that place. It would not be visited by the English for some months . . . as the ground was abandoned that night, and for a certain period considered neutral. When Lord Lucan was told that Nolan was buried, his Lordship said, "Ah, I wish we could get poor Charteris' body." The latter officer was his relative.'[13]

226

At a later date, as was the custom, Nolan's personal effects were auctioned off among his fellow officers. Russell was allowed to purchase the cloak Nolan had lent him the night before. He still had it over 40 years later.

There is another version of Lucan's feelings at Nolan's burial. It completely contradicts the letter he was to write to Airey later, in which he claimed total ignorance of Nolan's widespread and repeated criticism of his handling of the cavalry. Lucan wrote:

'Lord Raglan was totally misinformed when he was told that, prior to the action of the 25th October, I entertained any bad feeling whatever towards Captain Nolan. I had never up to that time heard that he had said or done anything to give me annoyance. . . . From the time of our landing in the Crimea, I do not recollect to have exchanged half a dozen words with him . . . but I repeat most positively that prior to the action of the 25th I never had any bad feeling towards Captain Nolan, nor did I know that I had cause for any.'[14]

Perhaps this is correct, but if so Lucan must have been virtually the only man in the cavalry who was unaware of Nolan's forthright and derogatory opinion.

The evidence to the contrary, that in fact Lucan loathed Nolan and was fully aware of what he had been saying, and that he held him responsible for Lucan being accused of losing the Light Brigade has, to the present writer's knowledge, only been published once before. It occurs in an obscure and anonymous manuscript entitled 'Aldershottana' or 'Chinks in my Hut'. The relevant passage tells of a Militia officer who made his own way to the Crimea, reaching Balaclava nine months after the charge. The morning after his arrival he set out to ride to Sevastopol, being accompanied by a Turkish colonel, a parson who claimed to have witnessed the Light Brigade's action and an officer of the Ambulance Service. The visitor expressed a wish to see the famous North Valley. As the party arrived to view the scene the Ambulance Service officer mentioned that he had been standing near Lord Lucan, not far from where Nolan was lying as the survivors came back. The anonymous account continues:

' "It was in this ditch," said this gentleman [the Ambulance Service officer], reining sharply up, "we buried poor Nolan. I knew him," he said: "he was a good fellow, but an ugly man, and made a still uglier corpse; he lay on his back with a deep wound in his left breast. Dr — [this could have been Mouat who helped save Morris who lay close by] had just had his fingers in it, and was wiping them in his handkerchief. Some spoke of carrying the corpse back to our lines. 'No,' said Lord Look-on, 'he met his deserts – a dog's death – and, like a dog, let him be buried in a ditch.' I thought it a hard and cruel speech," said our informant; "but if the poor fellow erred he suffered for it, and there ought to have been nothing of the sort uttered at such a time".'

It is for the reader to decide which account is the more likely.

Private Mitchell's (13th LD) simple experiences that night and the next morning were typical of most soldiers in the Brigade. They well illustrate that the hardships and activities of soldiers, when not involved in, or immediately after, combat, have changed little down the centuries. A Roman legionary in Gaul or a British paratrooper in the Falklands would be familiar with Mitchell's efforts to keep warm, find food and 'win' items of kit from the dead.

'We slept in the open air, and for my part I had a cold night of it, for I had nothing but what I stood upright in, viz, jacket and overalls, for my cloak and blanket were left with my horse . . . and I may say here that from the 25th October I had no cloak through the winter until the end of February. . . . Several men, on unrolling their cloaks after riding back, found grape shot and pieces of shell drop from them, which proves that a well-rolled coat is to a certain extent bullet-proof.

'After a while we were all at rest, if not asleep. I laid down with my back against another man for warmth, but it was a bitter cold night . . . [eventually] I fell asleep, but having no covering, I awoke several times during the night, and early in the morning found there was a sharp white frost.

'I got up and warmed myself at a fire [permission was seemingly given for fires towards dawn] that was burning a little way off. I now bethought myself of a plan to get some breakfast, so taking a camp kettle, I went to a stream and, filling it, placed it on the fire until it boiled; and by that time some of the men were on the move and began to crowd around the fire, so I soon found two or three who had tea, coffee and sugar. . . .

'I went down to the ground whereon the Heavies had charged. . . . I found a Russian Hussar officer lying on his back. He had his long grey coat on, which was open, as well as his jacket under it. His shirt, which was a very fine one, was saturated with gore about the front. On opening it, I saw nearly a dozen sword points, most of them slight, but there was one that had evidently done the work, for it appeared to have nearly gone through him from chest to back. Some one had taken his boots and stockings. Almost close to him a poor fellow was lying with a terrible wound. It must have been a fearful cut [here is evidence that both cut and thrust could be effective], for it had struck him full at the back of the head just above the ear, horizontally, and had gone half way from the back of the head to the face, most likely killing him instantly.

'There were also a few of our Heavies lying here and there. . . . A little further off another man, I think of the Enniskillen Dragoons, was lying. His boots too were gone; but he had on a pair of regimental socks, and, as mine were quite worn out, I thought there would be no harm in ringing the changes, so, after taking his and putting mine on his feet, I did not consider I had robbed the dead. It must be borne in mind that from the day on which we landed – 15th September – until the 25th October, we had not a change of linen, or hardly an opportunity of washing anything.'[15]

Two days later the remains of the Light Brigade exchanged their valley for a camp near a windmill on the windswept Cheronese uplands. There they faced the winter, an infinitely more frightening foe than the Russians. They suffered and survived, but were never again a fighting force to be reckoned with in the Crimea.

CHAPTER TWELVE
The Recriminations

'Alas! alas! it was a sad business, and all without result,
or rather with the result of the destruction of the Light
Brigade. It will be the cause of much ill-blood and
accusation, I promise you.'

Lord George Paget in a letter to his wife dated 26
October, 1854.

'It is wonderful to observe the way that fool the
"British public" kicks a man directly he is down, as in
the instance of unlucky Lucan. I always hated him,
and so did the whole Cavalry Division, but for
heaven's sake let a man have fair play – here is this
unfortunate man catching it over the head and ears,
merely because he obeyed an order given by the thick-
headed Raglan through his still more stupid Q.M.
General Airey, who is about the worst of the whole
headquarters staff.'

Lieutenant Walker Heneage (8th Hussars)[1]

The British love a glorious military failure. The charge of the Light Brigade
was precisely that, and so it has been remembered in the public mind ever since.
It was held (with some justification) to be a sublime example of how British
courage, dash and discipline could conquer impossible odds. It eclipsed
equally brave actions, it outshone far more successful operations as it became
to be regarded as the one redeeming event in an otherwise dismal, depressing
and disease-ridden war. Such a perception was understandable but illogical
and inaccurate. The charge was a costly blunder that should never have
happened. It exposed, for those who cared to look, how incompetent and ill-
prepared for war were many of the senior commanders. Raglan, Airey,
Cathcart, Lucan and Cardigan all contributed to the failure in some way.
Nolan, who so scornfully and in all probability deliberately pointed out the
wrong objective, must take a large portion of the blame for the charge taking
place, but by no means all.

All the generals involved knew full well that there had been an infamous
mistake. Before the day was out the three surviving horsemen of calamity were,

like many men before and since in similar situations, determined to deflect blame from themselves to others. The recriminations and accusations were to fly around for months, indeed years, afterwards. A blunder that the military were anxious to play down and keep 'in house' was eventually trumpeted in the public domain via speeches and debates in the House of Lords, letters and articles in *The Times*, the publication of pamphlets, and a criminal libel action in the Queen's Bench.

Raglan at first blamed Cardigan, but quickly turned on Lucan; Lucan blamed Raglan and Nolan; Cardigan, delighted he had been able to duck any responsibility, was happy to see his brother-in-law found at fault. His conduct in quitting the field so as to be almost the first survivor back, however, later brought public attack from Calthorpe, one of Raglan's ADCs. Nine years afterwards the Battle of Balaclava was refought in the courtroom in the Cardigan v. Calthorpe case.

Possibly on the battlefield, but more probably later that evening, Raglan and Lucan had a more lengthy confrontation as to where the blame lay. The meeting was almost certainly at Lucan's instigation as he was utterly determined to clear his name. It was to become an all-consuming passion involving the newspapers, the Secretary of State, the Commander-in-Chief at the Horse Guards, the House of Lords and even, briefly, the Queen. At that early meeting Raglan found himself at an annoying disadvantage.

Due to the inefficiency of his staff he had no copy of the third and fourth orders he had sent to Lucan. In the rush to get them delivered only the original was written. This meant that Raglan was speaking from memory of what was said in the orders. He did not know or remember the precise wording. Lucan (or more particularly Blunt on his general's instructions) had kept the originals. It may be for this reason that Raglan did not, at this stage, press as to why Lucan had not advanced on the redoubts. Instead he dwelt, according to Lucan, on Lucan's failure to use his discretion, that as a lieutenant-general he had the authority if he did not like an order to decline to carry it out. Lucan recalled Raglan's words as: 'Lord Lucan, you were a Lieutenant-General, and should have therefore exercised your discretion, and, not approving of the charge, should not have caused it to be made.'[2]

The word 'charge' does not appear in the fourth order, nor does the word 'attack'. Raglan's rather pathetic plan envisaged the possibility of the Russians retreating in haste at the sight of a cavalry advance. Lucan was to make the most of this, if it happened, to try to recover the lost cannons. Airey wrote this down in such a way that it was almost incomprehensible to the recipient in his position on the ground. Raglan, at the very last moment and not appreciating the damage it would do, then told Nolan verbally that Lucan was to 'attack immediately'. Giving that verbal message to Nolan of all people was what sealed the fate of the Light Brigade. All the subsequent arguments did not really revolve around the fourth order as it was written. It was certainly confusing. It spoke of following an enemy that was not retiring, it posed queries on how it was to be done, but it did not order a charge or even an attack. Lucan did not understand *how* he was to comply, but, and this is crucial, he fully understood that he had been, as he phrased it in his despatch of the 27th, 'instructed to make a rapid advance to our front, to prevent the enemy carrying the guns lost by the Turkish troops in the morning.'

How he was to do this was the subject of his questioning of Nolan, and of his hesitation in complying. Nolan then introduced Raglan's word 'attack'. Lucan's incredulous retort, 'Attack, sir! Attack what? What guns?' resulted in Nolan pointing to the guns at the end of the North Valley. It was surely this 'attack', this 'charge', that Raglan was referring to when he spoke of Lucan using his discretion. In that meeting Lucan must have vigorously denied responsibility on the grounds that he was complying with the ADC's emphatic, unmistakable *verbal* order to attack that particular enemy battery. Raglan's natural counter to what they both appreciated was something of a suicidal mission was to ask why Lucan had not refused to comply without further confirmation, as it was not reflected in the written message. The present writer is firmly of the view that the controversy, the recriminations afterwards, were not so much about the content of the orders, not about the way they were written, not even about whether they were understood, but rather about whether, in the circumstances, Lucan had any choice but to follow Nolan's verbal order rather than Raglan's written one.

This was what Raglan was so furious about. To charge a battery frontally was always fraught with terrible risks, to do so in the circumstances appertaining at eleven o'clock that morning was unthinkable. Airey's written message, baffling though it was, could in no way be construed as wanting the cavalry to charge down the North Valley. Raglan and Lucan could surely agree on that. But in reprimanding Lucan Raglan forgot his parting words to Nolan. It was Nolan who insisted the cavalry attack as it did. And Nolan was the Commander-in-Chief's ADC. As far as Lucan was concerned Raglan might just as well have been beside him in person. Nolan supposedly knew Raglan's intentions; Nolan was pointing out a specific objective in response to Lucan's questions; Nolan was insolently insisting that Lucan get on with it – at once. Few generals would have demurred again. In Lucan's view he had obeyed Raglan's orders just as Cardigan had obeyed his. Raglan, however, did not (could not) see it that way and, perhaps with some embarrassment, had to ask Lucan for a copy of the fourth order.

Lucan returned to his tent. He was in an agony of frustration. Raglan had been unmoveable, though perhaps when he saw a copy of the actual order Lucan had received he would see sense. Blunt was instructed to copy it out. Portal wrote, 'Lord Lucan is dreadfully cut up about it, but says he can show the order in writing which is his only comfort.'[3] The problem with the order was, as Lucan was yet to appreciate, that it could not be read as telling the cavalry to do what they did. Lucan had done as he was told by Nolan. Nolan's instructions had little, if any, relation to what the Commander-in-Chief wanted, although Lucan could not know that. Now Nolan was dead. Lucan was left as the scapegoat. In his deep depression there was, however, a determination that he would never accept this injustice. He would fight. He summarized his feeling at the time when he wrote:

'I do not intend to bear the smallest particle of responsibility. I gave the order to charge under what I considered a most imperious necessity, and I will not bear one particle of the blame.'[4]

Lucan's obsession, for it was becoming just that, clouded his judgement. He believed that if Cardigan could escape censure because he merely obeyed orders so should he – for the same reason. But if the fault was not his or Cardigan's, and Nolan was dead, then where else could it lie? He was seeking to shift a substantial part of the responsibility onto the shoulders of the Commander-in-Chief, an immense, virtually impossible, undertaking in any army at any time.

One of Lucan's first tasks on the 26th was to write a note to the Quartermaster-General, enclosing the copy of the fourth order. It read:

'Dear General Airey, – I enclose a copy of the order handed me by Captain Nolan yesterday, as desired by Lord Raglan. When his lordship is enabled to give it his attention, I anxiously hope that he will not still think, "I lost the Light Brigade" in that unfortunate affair of yesterday. – believe me, etc.'[5]

When he met Paget he unburdened his feelings with considerable emotion. Paget wrote home that 'Lucan is much cut up; and with tears in his eyes this morning he said how infamous it was to lay the blame on him, and told me what had passed between him and Lord Raglan.'[6]

On the 27th Lucan told Russell:

'He [Lucan] was quite content with *his* charge – he had ordered the Heavy Brigade to charge the Russian horse, and he had nothing to do with Lord Raglan's charge, except to pass on the orders he had received, that the Light Brigade was to charge – that he had not lost them, he had obeyed orders.'[7]

That evening Raglan sent Airey down to Lucan's tent on a mission of conciliation. Having now got a copy of the order Raglan was feeling much more confident. Nothing in the wording indicated a charge or even an attack. Perhaps Nolan had confused things, but it is easy to blame the dead. Anyway, he had no intention of being vindictive; there was no need to pillory Lucan. With a little goodwill and common sense the whole matter could be accepted as one of those regrettable errors of judgement that occur on battlefields. He had earlier that day written his official despatch to London on the battle in which he had suitably glossed over the loss of the Light Brigade and let Lucan down lightly. Airey was to soothe the resentful cavalry commander, get him to accept things and not to keep fanning a fire that would otherwise die down.

As Airey entered Lucan shot at him, 'General Airey, this is a most serious matter.' Indeed it was, but Airey's job was to make it less so. He started off in a light-hearted but less than tactful vein by replying, 'These sort of things will happen in war. It's nothing to Chillianwala.'[8] That it should be linked in any way whatsoever with that awful episode in a battle of five years before in the Sikh Wars was not calculated to calm Lucan. It will be remembered that at that battle a brigade of light cavalry, including the 9th Lancers and 14th Light Dragoons, had fled the battlefield after supposedly mistaking an order of 'Threes right' for one of 'Threes about'. Sikh horsemen had followed up and cut down British gunners and captured British guns. It had cost the British general his job. With the fervour of a man who believes vehemently that he is right Lucan at first refused to be reasoned with. Airey then played his final card. He knew the contents of Raglan's despatch that would go the next day, so reassuringly insisted that it was not a question of blame, and that he was certain Lucan would be more than happy with Raglan's report to London on the affair. This had a suitably mollifying effect. If the official despatch was not to criticize him Lucan seemed content not to press things further.

Raglan's despatch, dated 28 October, was a masterpiece of diplomatic phraseology. No. 1 Redoubt, which the Turks defended against heavy odds for over an hour, fell 'after very little resistance'. The Russian cavalry attack on the 93rd was 'instantly driven back by the vigorous and steady fire of that distinguished regiment'. The advance of the bulk of the enemy cavalry [under Rijov] had 'afforded Brigadier-General Scarlett, under the guidance of Lieutenant-General Lord Lucan, the opportunity of inflicting upon them a most signal defeat. . . . The charge of this brigade was one of the most successful I ever witnessed.' (The reader will recall there was no actual charge and it was artillery fire that forced the Russians to retire.) Then, the crucial comments on the loss of the Light Brigade:

> 'From some misconception of the instruction to advance, the lieutenant-general considered that he was bound to attack at all hazards, and he accordingly ordered Major-General the Earl of Cardigan to move forward with the Light Brigade.
>
> 'This order was obeyed in a most spirited and gallant manner. Lord Cardigan charged with the utmost vigour. . . . The loss they have sustained has, I deeply lament, been very severe in officers, men, and horses, only counterbalanced by the brilliancy of the attack and the gallantry, order, and discipline which distinguished it.'

Even Raglan had difficulty in being complimentary on Cathcart's feeble efforts, but he tried. 'Sir George Cathcart caused one of the redoubts [No. 5] to be reoccupied by the Turks, affording them his support, and he availed himself of the opportunity to assist with his riflemen in silencing two of the enemy's guns.' [There is no evidence of this.]

Lucan was not shown the despatch. Its contents would become public in due course, but hopefully the deliberately pacifying remarks on Lucan's role, and the passage of time, would have taken the heat out of the matter. In addition to the despatch, however, Raglan felt it prudent to send a private letter to the Secretary for War, the Duke of Newcastle. In it he sought to give Balaclava the attributes of a success and, as his conduct would be under scrutiny in London, to ensure that the authorities knew he had in mind Lucan as the man responsible for the loss of the Light Brigade. Lucan, he claimed, 'had made a fatal mistake.' 'The written order sent to him by the Quartermaster-General did not exact that he should attack at all hazards, and contained no expression which could bear that construction.'

In the four weeks that followed, the Battle of Inkerman was fought and won by the British at a cost that made Balaclava look like an insignificant scuffle; Cathcart was killed; the arrival of winter was announced by a tremendous gale that tore apart tents, ripped roofs from buildings, and killed men from exposure; and Lucan and Cardigan resumed their quarrelsome correspondence over reports and returns. Then, in mid-November, newspapers from England arrived.

When Lucan read Raglan's despatch in the newspaper he was outraged. All his fury, that had been partially suppressed by Airey's reassurance that he was not to be blamed officially and publicly, boiled over again. The entire Army, indeed the whole of England, now believed that he had misconstrued an order and had attacked when he should not have done so. In other words he was to blame for the destruction of the Light Brigade, he was professionally incompetent to the extent that he could not even follow orders. Not only that but the despatch spoke in glowing terms of Cardigan's performance. It was insufferable, and it must be retracted at once. He would insist on complete vindication. Lucan picked up his pen.

His plan was to give his version of what happened in a letter to be forwarded to the Duke of Newcastle insisting that it be published as a rebuttal of Raglan's despatch. Such a letter destined for the Secretary of State, in effect complaining about the actions of the Commander-in-Chief must, according to Army regulations, be sent through that officer's headquarters. This was the military way. The chain of command must not be broken. It was not that the officer complained about could alter the letter, but he had to be given the opportunity to comment. Accordingly, Lucan wrote to Raglan on 30 November. By now thoroughly alert to the need for copies of everything Lucan kept a duplicate.

He did not set out his case well. Raglan's view that there had been a 'misconception of the instruction to advance' by Lucan was 'a grave charge and imputation reflecting seriously on my professional character'. He then set out the facts as he saw them. For the first time in writing he attempted to shift much of the blame on to Nolan. He repeated the fourth order and continued:

235

'After careful reading of this order I hesitated, and urged the uselessness of such an attack [no attack was mentioned in the written order], and the dangers attending it; the aide-de-camp, in a most authoritative tone, stated that they were Lord Raglan's orders that the cavalry should attack immediately. I asked him where? and what to do? as neither enemy nor guns were in sight [except those on the Fedioukine and down the North Valley]. He replied in a most disrespectful but significant manner, pointing to the further end of the valley, "There, my lord, is your enemy; there are your guns."

'So distinct in my opinion was your written instruction [it was nothing of the sort, as Lucan well knew], and so positive and urgent were the orders delivered by the aide-de-camp, that I felt it imperative of me to obey, and I informed Lord Cardigan that he was to advance; and to the objections, and in which I entirely agreed, I replied that the order was from your lordship. Having decided against my conviction to make the movement, I did all in my power to render it as little perilous as possible. [Lucan was vulnerable here as he did nothing other than sending some of the Heavies half-way down the valley and then pulling them back.] I formed the brigade in two lines, and led to its support two regiments of heavy cavalry . . .

'. . . doubtless I have discretionary powers; but to take upon myself to disobey an order written by my Commander-in-Chief within a few minutes of its delivery, and given from an elevated position, commanding an entire view of all the batteries and position of the enemy, would have been nothing less than direct disobedience of orders. . . .

'It should also be remembered that the aide-de-camp, well informed of the intentions of his general [Lucan had every right to assume this, and it was in fact correct], and the objects he had in view, after first insisting on an immediate charge, then placed himself in front of one of the leading squadrons. . . .

'I will only ask that your lordship should kindly give the same publicity to this letter that has been given to your report.'[9]

For two weeks Raglan sought to persuade his troublesome cavalry commander to withdraw the letter. Airey was sent on three occasions to remonstrate with him. Raglan knew, as must Lucan had he paused to think clearly, that if this letter was forwarded to London Lucan was finished. The likelihood of the government taking Lucan's side over the Commander-in-Chief was minimal and of publishing his letter zero. Raglan held the trumps and Lucan's letter invited him to play them. Lucan had made much of having to obey a written order to attack immediately when no attack was mentioned – a useless argument. To insist that he could not disobey the written order, muddled and confusing though it was, when that order plainly was not telling him to do what he did, was to make himself look a fool. The strength of Lucan's case was that Nolan was telling him to do something that was *not* in the written message. Lucan missed this crucial point, as have many historians since. He had asked for clarification; he was told to charge down the valley. Lucan's only error was doing what his Commander-in-Chief's ADC told him to. It was as simple as that.

Airey's task was to convince Lucan that to send the letter was not in his best interests. Raglan and Airey both knew Lucan would lose but they had no wish to see him destroy himself needlessly, nor did they want any controversy bandied about in the newspapers at home. Many men had died in the charge and public arguments as to responsibility would be bad for the Army and bad for the individuals involved. Lucan, however, was a man with a mission. He could not be persuaded. Considerable discussion revolved around Nolan's behaviour and actions. In one of the meetings Airey must have mentioned the rumour that bad feeling existed between the Divisional Commander and Nolan, as on 7 December Lucan wrote to Raglan denying any such thing. This letter has been quoted in the previous chapter.

On 16 December Raglan yielded to Lucan's insistence. His letter would go to Newcastle, accompanied by a long one of his own. Raglan ignored Nolan's decisive role and thus found it an easy task to demolish Lucan's contentions. He regretted having to forward the letter. He emphasized that he had tried to get Lucan to

'withdraw the communication considering it would not lead to his advantage in the slightest degree. . . . I have but one course to pursue – that of laying the letter before your Grace, and submitting to you such observations on it as I am bound, in justice to myself, to put you in possession of. . . .

'Not only did the Lieutenant-General misconceive the written instruction . . . but there was nothing in that instruction which called upon him to attack at all hazards. . . .

'[He] was so little informed of the enemy that he asked Nolan, "Where and what he was to attack as neither enemy nor guns were in sight. . . ."

'The result of his inattention to the first [third] order was, that it never occurred to him that the second [fourth] was connected with, and a repetition of [to the recipient this was far from obvious], the first [third]. He viewed it as a positive order to attack. . . . I undoubtedly had no intention he should make such an attack. . . .

'He was told that the horse-artillery might accompany the cavalry. He did not bring it up. He was informed the French cavalry was on his left. He did not invite their cooperation. He had the whole of the Heavy Cavalry at his disposal. He mentions having brought up only two regiments in support.'[10]

Although it is possible to pick holes in this letter it was more than sufficient to damn Lucan. Lucan had failed to highlight the fact that he had complied with Nolan's verbal order rather than the written one. It would cost him his job.

This crucial point was picked up in London, not because it was in the letters, but through the clearer thinking of Colonel Mundy, the Under-Secretary of State. The correspondence had crossed his desk before going to Newcastle and, as was the custom, he attached his own opinions before passing them to the Secretary. In his minute he admits that it was quite possible that Nolan had indicated the battery down the valley, and went on to say:

> 'But even so the Lieutenant-General ought to have acted on the written orders of the Commander-in-Chief and not upon the oral ones of the aide-de-camp. It is evident indeed that he derived his resolution to attack at all hazards, and contrary to his own and Lord Cardigan's expressed opinion, not from Lord Raglan's note, which could by no possibility be construed in that sense, but from the hurried and as the Lieutenant-General says, "most disrespectful" remark of the Staff Officer.'[11]

At last somebody had put on paper the real issue. The main culprit was Nolan, but could, in the circumstances at the time, Lucan be reasonably expected to disregard the ADC's categorical verbal order? Mundy thought he could. So did Newcastle, and Lord Hardinge at the Horse Guards when the matter was shown to him. Lucan had better be recalled.

Perhaps his aristocratic head would placate the growing clamour in the newspapers for an inquiry into the deplorable state of the Army and the incompetence of the high command in the Crimea. After Inkerman Russell had written to *The Times*, 'I am . . . convinced that Lord Raglan is utterly incompetent to lead an army through any arduous task.'[12] The government's response to the vitriolic out-pourings of the press was to make Raglan a field-marshal (as a reward for Alma). But it was not only the newspapers that wanted blood. After the Christmas recess both Houses of Parliament debated the Crimea. In the Lords Newcastle was asked for the text of the order that had caused the Light Brigade to charge on the grounds that two versions existed (which they did). Newcastle side-stepped, stating that correspondence had only just been concluded and that he needed time to prepare a statement. In the Commons the Member for Bristol, Mr Henry Berkeley, gave notice of his intention to demand an inquiry. In these circumstances Lucan, right or wrong, stood no chance. A culprit had to be found – and quickly.

In the Crimea Lucan had, within hours of writing it, convinced himself that his letter would never be published. Abandoning caution and military procedures, he had sent a copy to his solicitor with instructions to find out whether the original would be published and, if not, to ensure the copy was placed in *The Times*. The solicitor, a more prudent professional than his client, did no such thing. He immediately had it shown to the Secretary of State before Raglan's correspondence arrived on his desk and before Mundy penned his minute. Raglan's and Lucan's letters arrived in London on 8 January, 1855. On the 12th Newcastle wrote privately to Raglan to say that he was 'very sorry for the unfortunate course taken by Lord Lucan. I may tell you that I had already seen his letter to you, for he had sent it to this country to be published if you did not send it to me. I presume he is under the impression that I shall publish it with the despatch from you. Of course I shall do nothing of the kind.'[13]

Unknown to him, Lucan's scheme to publicize his position had been thwarted.

Two weeks later, on 26 January, Lord Hardinge wrote from the Horse Guards to the Secretary of State, using all the heavy, pompous terminology of the government of the day, to say that Lucan should be sacked:

'The papers having been referred by your Grace to me, I concur with Lord Raglan that the terms he used in his despatch were appropriate: and as a good understanding between the Field-Marshal commanding the forces in the field and the Lieutenant-General commanding the Cavalry Divisions are conditions especially necessary for advantageously carrying on the public service, I recommend that Lieutenant-General Lord Lucan should be recalled; and if your Grace and her Majesty's Government concur in this view, I will submit my recommendation to her Majesty, and take her Majesty's pleasure on the subject.'[14]

The next day, the Queen's pleasure having been obtained, Newcastle wrote to Raglan informing him of Lucan's recall. The Commander-in-Chief was to

'inform his lordship that he should resign command of the Cavalry Division and return to England. . . .

'I must observe that, apart from any consideration of the merits of the question raised by Lord Lucan, the position in which he has now placed himself towards your lordship renders his withdrawal from the army under your command in all respects advisable.'[15]

In a private note Newcastle added, 'Spare his feelings as much as you can, but I despair of his ever seeing the justice and propriety of this decision.'[16]

How right he was. Lucan was shocked, disgusted and furious that his splendid letter could have backfired in this way. He believed with a burning sincerity that he was being disgracefully treated. He had no intention of giving up. Nevertheless, it had never crossed his mind that he would lose his division and return home disgraced. Among his officers few lamented his going, although many felt he was being unjustly made the scapegoat by his enemies in headquarters. He was given the news in a note from Raglan, who throughout had never liked to meet him face to face, on 13 February. He sailed the following day. He had said goodbye to his staff, to Campbell, and to Canrobert – but not to Raglan. As he stood at the rail of his ship as it sailed under the cliffs of the Cheronese he is said to have looked up towards headquarters and silently shaken his fist.

Lucan arrived in England on the 1st of March. On the 2nd he sent his son, Lord Bingham, who had returned with him, to the Horse Guards with a letter insisting on a court-martial. It was refused. There was quite enough Army dirty linen being washed in public as it was. But Lucan could not be suppressed. It was only when he got home that he was shown Raglan's covering letter to his own. To Lucan, Raglan had saved his own neck by destroying him. He renewed his request for a court-martial. Writing from Hanover Square on 5 March he said:

'Until this day I have been kept uninformed of the letter from Lord Raglan, which appears to have been addressed by his lordship to the Minister of War, when forwarding mine of the 30th of November last. This letter contains entirely new matter, and is replete with new charges, reflecting more seriously than before on my professional judgement and character.'[17]

Again the response was negative. The adjutant-general curtly told him, 'The Commander-in-Chief at Horse Guards cannot recommend that your conduct on the 25th October should be investigated by a court-martial.'

Still unbowed, he wrote to Mr Berkeley, asking that he pursue his demand for an inquiry, and he spoke out in the two places nobody could stop him – the House of Lords and the columns of *The Times*. His efforts in both places were generally well received within the cavalry and in the Crimea. Portal wrote home to his family after the publication in the newspaper of Lucan's defence that it was 'a most excellent letter'. Most soldiers would probably have accepted the remarks of Lieutenant Heneage, quoted at the start of this chapter and who was to win the VC in India, as fair comment – if somewhat colourfully expressed. With this Lucan the industrious, Lucan the meticulous, Lucan the conscientious and Lucan the unlucky had to be content.

*　　*　　*

Finally, the fate of the four horsemen of calamity.

Raglan gave the order. It was a spur of the moment decision made in the erroneous belief that the enemy were taking away the British guns; it called for the cavalry to undertake an operation best suited to infantry; it spoke of following an enemy that was not retiring; it took no account of the fact that the recipient could not see what he could; in its written form it was so confusing as to be almost incomprehensible; at the last minute he told Nolan that the cavalry was to attack immediately, something that was not in the written message. For all this Raglan must share substantial responsibility for the loss of the Light Brigade.

The responsibility of a commander-in-chief for the orders he gives is absolute. Raglan never liked giving direct orders; he preferred to propose, to recommend or make his wishes known. If orders are not understood by the recipient it is invariably the fault of the originator (and also in this case Airey who wrote them down). From the tone of Raglan's letters to Newcastle afterwards, from the way he implied his orders were clear when they were plainly not, and how the fourth order should have been read as an extension of the third (written almost an hour earlier) when there was nothing in it to indicate this, all point to the fact that Raglan knew in his heart where the fault lay.

The appalling winter of 1854/55 crippled the Army. Raglan was blamed for the suffering and death that the weather and maladministration brought. Prolonged physical hardship sapped the morale of officers and men. Senior officers and staff had houses or buildings for shelter, junior officers and soldiers had tents, the horses almost always nothing. The death rate among humans and animals was directly in proportion to their exposure to the elements. Lucan was an exception in this respect, grimly determined to set an example by living like his men. Had he not been recalled he would have seen out the winter in the field. The same cannot be said for Cardigan, Paget or the Duke of Cambridge, all of whom contrived to return to England.

Raglan survived the winter but not the summer. The 18th of June was Waterloo Day, the 39th anniversary of the battle, a day that Raglan had celebrated for most of his life. On this occasion it was commemorated by a 4.00 a.m. attack on Sevastopol. It failed with heavy loss. Ten days later Raglan was dead. Cholera had broken out again. General Estcourt had died of it a few days earlier, and many thought this was what had killed the Commander-in-Chief. It was not so. When Florence Nightingale heard the news she wrote to her parents

'It was impossible not to love him. . . . He had died, so the doctors in the camp had said, without sufficient reason. It was *not* cholera. The diarrhoea was slight but he was so *depressed*. . . . He was not a *very great* general, but he was a *very good* man.'[18]

The first part of the final sentence is probably a fair judgement, the second is more debatable.

His body started the long journey home on 3 July. In the late afternoon sunshine the 9-pounder gun, drawn by eight horses and with a platform fixed across the barrel, that was to carry the coffin entered the farmyard in front of his headquarters. As the coffin was brought out the Grenadier Guards guard of honour presented arms. It was covered with the Union Jack and a black cloak on top of which were his plumed hat, his sword, and a bunch of immortelles placed there by the recently appointed French Commander-in-Chief, General Pélissier. He was the son of a sergeant, and his favourite expression (usually with reference to costly assaults) was, 'You cannot make an omelette without breaking eggs.' Pélissier was extraordinarily ugly. The photographer Roger Fenton thought his face resembled that of a wild boar. He was also fiery, fat and short, so much so that with his podgy little legs he found it exceedingly uncomfortable to ride. For much of the time he went careering about in a trap. To counterbalance a hot temper he had a full measure of Gallic emotion, which he had demonstrated by the flowers and weeping for nearly an hour at Raglan's deathbed.

At a quarter past four the drums rolled and, to the melancholy music of the 'Dead March' played by three combined infantry bands, the cortège began its six-mile journey down to Kazatch Bay where the Commander-in-Chief's ship, the *Caradoc*, waited. As the coffin left the yard a salute of nineteen guns was fired by two batteries of field artillery on a hill opposite. The ceremonial, and the size of the procession and route lining contingents, was impressive. The siege lines around Sevastopol had been denuded for the occasion but the Russian guns were obligingly silent. For the mile that separated the British and French headquarters a double rank of British infantry, consisting of fifty men and three officers from every Regiment in the Crimea, lined the road. From the French headquarters to the Bay units from the Imperial Guard and 1st Army Corps took over.

At the wheels of the gun carriage rode the four Commanders-in-Chief – General Simpson, General della Marmora, General Pélissier (probably with difficulty) and Omar Pasha. Immediately in rear walked Raglan's favourite horse, 'Miss Mary', saddled but riderless. General officers, relatives and personal staff followed. The escort was composed of four squadrons of British, two of Sardinian and eight of French cavalry, plus two troops of French horse artillery and one battery of British field artillery. At intervals along the route the French had posted gun batteries which fired salutes as the coffin passed by. At the wharf the Royal Navy took over. Raglan's body was taken by launch out to the *Caradoc*, which had been painted black for the occasion, while yet two more batteries on shore boomed out their last salute. Within half an hour the *Caradoc* steamed away with the signal 'Farewell' flying at her masthead.

Nolan delivered the order. Like many of those he rode with, Nolan paid for his arrogant impetuosity with his life, a fine talented light cavalryman who allowed his wild excitement of the moment to combine with the contempt he felt for the way the cavalry had been commanded, to blind him to the probable consequences of what he did. Nolan had watched the battle unfold before him; he knew what Raglan was seeking to do when he gave the fourth order; he had probably read the written message; he was quickly briefed by Airey before he left; he heard Raglan's parting shot that the cavalry was to attack at once; he confronted a hesitant and questioning Lucan whom he despised; his fury boiling over, he insolently pointed out the wrong objective. Nothing can be absolutely certain but the present writer believes the balance of probabilities is that Nolan launched the Light Brigade down the North Valley knowing it was not Raglan's intention. He must therefore take the bulk of the blame for its loss.

Lucan received the order. He did not understand its contents; he was uncertain how he should carry it out; he questioned the ADC as to what was required; he was told to attack immediately; he demanded to know what to attack; he was told the battery at the end of the North Valley; he gave orders accordingly. For this he was denounced by Raglan, and a fair number of writers since, as the chief culprit for the charge and its consequences. It is hard to believe that many generals of any age, if forcefully given a direct order to attack an objective at once, would decline. Lucan had to assume Nolan knew what he was saying to be the General's wishes, and that Raglan, due to his elevated position, was better informed than he was. In theory, if he felt the attack to be impossible, he could have insisted on further clarification. But to have instigated a further heated public argument with the Commander-in-Chief's ADC, to decline attack orders on the battlefield, to delay while verification was sought was certainly more than Lucan was prepared to do. For this error of judgement, for with hindsight error it was, Lucan must take a small share of responsibility.

Back in England, after the failure of his attempts to obtain a court-martial, Lucan's popularity was at a low ebb. A pamphlet was published which condemned him for being out of date and incompetent, even suggesting he had failed in his duty by not leading the charge in person. Lucan published a lengthy reply. Perhaps in a belated attempt to placate him Lucan was made KCB in 1855 and shortly afterwards Colonel of the 8th Hussars. Both events led to scathing letters in the newspapers to which he felt obliged to reply. He retreated to Ireland, incensed that his useless brother-in-law had been made inspector-general of cavalry.

Nevertheless he was to outlive Cardigan by twenty years and, although he never commanded troops again, he had the consolation of military honours and promotion. He became a general, was made Gold Stick and Colonel of the 1st Life Guards in 1865. In 1869 he was made GCB and, in 1887 on the occasion of Queen Victoria's Jubilee, he was promoted field-marshal. When he died the following year aged 88 he was the oldest serving soldier in the Army (field-marshals never officially retire). The Chestnut Troop of the Royal Horse Artillery pulled the gun carriage carrying his coffin. It was escorted by a hundred NCOs and men from the Life Guards.

Cardigan executed the order. After Raglan's initial accusation and his telling response nobody seriously blamed him for instigating the charge. He expressed his doubts as to the wisdom of complying, was told there was no changing matters and got on with it. He did at brigade level what Lucan had just done at divisional – queried an apparently dubious order and been told to implement it at once. Cardigan can be criticized for his failure to attack the Russians after the Heavy Brigade action, and for his leaving his brigade after the charge, but not during it.

He rode out the dreadful storm of 14 November on his yacht and five days later wrote to Raglan saying he would be

'obliged shortly to ask you for leave of absence on sick certificate. Were it not for bad health, I assure you I would have no wish to go, for you know you have no keener soldier in your army. [He then went on to try

243

to get his sick certificate] without having to explain my ailments in detail before a Medical Board.'[19]

He felt that the warm climate of Naples best suited to his sickness and he wanted to depart as quickly as possible. He finished with typical Cardigan duplicity, 'but I will follow your wishes and advice, even to the detriment of my health.' For good measure he enclosed two envelopes containing his latest complaints against Lucan. Raglan, however, was unmoved and insisted his lordship go through the indignity of the doctors' examination.

It was another two weeks before the Medical Board certified that Cardigan should not have to face a Crimean winter. On 8 December he left. At Constantinople, still unable to stop carping about his treatment, he picked up his pen again. His living on his yacht had provoked furious rows with Lucan over the need for orderlies to keep plying between Balaclava and his Brigade. He sent off his final protest:

'I cannot leave the country without affording you an opportunity of knowing how the duties of the cavalry command are carried on. . . . [Lucan] had taken the opportunity of commenting on the permission I received from your lordship to live on my yacht. . . . Can it be believed that any other general officers commanding brigades can be so treated in this army except those who have the misfortune to serve in the Cavalry Division?'[20]

Cardigan did not stay in Naples but sailed via Marseilles for Dover where he landed on 13 January, 1855. He was agreeably surprised by his reception. A crowd on the quay gave 'Three cheers for Balaclava'. In contrast to his days as a lieutenant-colonel when he had been mobbed for his extreme un-popularity, now he was cheered as a hero. The press lauded his achievements, his pictures appeared everywhere, and even a woollen jacket similar to the one he wore in the Crimea was copied and sold well. It was called (and still is) a 'cardigan'. Victoria invited him to stay at Windsor for three days. He sat by her side at dinner. While there Prince Albert invited him to give an account of the Charge, in which he blamed Lucan for the direction it took. At the Horse Guards he was overjoyed to be told he was shortly to be appointed inspector-general of cavalry. It would give him un-paralleled scope to indulge his obsession with the details of dress and drill. Cardigan revelled in the veneration he received. He took immense pleasure in the fact that, as he saw it, the Palace and the public were recognizing his achievements.

The adulation went on and on. He had been given a 'halo of heroism' as *The Times* called it. It was unhealthy. On 15 February he arrived at a Mansion House banquet in the uniform he had worn at Balaclava and riding 'Ronald'. The poor animal was continuously having to swish his tail to thwart souvenir hunters as they scrambled to pluck hairs from it. His speech on his actions in the charge was inaccurate and boastful. Bands played 'See the conquering hero comes' when he arrived at railway stations. Leicester gave him an illuminated address, Yorkshire a sword, and his home county of Northamptonshire a forty-yard-long testimonial made of sheets of paper pasted together bearing nearly 5,000 signatures. The Queen made him KCB in the summer of 1855. This was the period when Lucan was struggling to redeem his reputation. How galling and unjust Cardigan's popularity must have seemed to him.

In January, 1856, however, his image began to fade. The findings of an 'Inquiry into the Supplies of the British Army in the Crimea' had been presented to both Houses of Parliament and published. Among those called to account for the misery and death were Airey, Lucan and Cardigan. Both the latter rushed into print to defend themselves and blame each other. Cardigan demanded an inquiry into the inquiry. There was so much publicity, so many accusations that in July a board of inquiry composed of general officers was set up. It became known as the 'Whitewashing Board'. Like so many official inquiries before and since individuals were exonerated and 'the system' blamed.

In April, 1856, the war ended in a victory of sorts for the Allies and the troops came home. In December Raglan's former ADC, Calthorpe, published a book entitled *Letters from Headquarters*. In it he made some highly adverse comments on Cardigan's behaviour at Balaclava. In particular why was Cardigan not around when needed to rally his brigade or give further orders after the charge? It went so far as to say:

'This was the moment [as the first line neared the battery] when a general was most required, but unfortunately Lord Cardigan was not present. On coming up to the battery . . . a gun was fired close to him . . . his horse took fright – swerved round – and galloped off with him to the rear, passing on the way the 4th Light Dragoons and 8th Hussars, before those regiments got up to the battery.'

This was woefully inaccurate but horribly damaging to Cardigan's gilded reputation.

Cardigan demanded a retraction, which he did not get. He then embarked on a campaign to have all future editions of the book stopped, Calthorpe (by then a colonel) dismissed, prevented from exchanging into the 5th Dragoon Guards and court-martialled. He failed in all of them. Not until June, 1863, did he take Calthorpe to court. Lord Chief Justice Cockburn, assisted by three other judges, presided in Westminster Hall. The court established beyond doubt that Cardigan had reached, and gone beyond, the gun position. Nevertheless, his personal withdrawal before the bulk of his Brigade was still left open to unfavourable interpretation. By this time perhaps the feelings of many were summed up best by the diarist Charles Grenville when he wrote:

'The world is weary of Cardigan and his fanfaronades, and of the Crimean accusations and recriminations, and it is time that the subject should be allowed to drop.'[21]

On 26 March, 1868, Cardigan fell from his horse near his home at Deane Park. When he was found, lying near some stones, his face was blue and he was foaming at the mouth. He was carried to his room and attended for two days by doctors from as far away as London. He never said another word and died two days later in his seventy-first year. His death was put down to a seizure or congestion of the brain – a heart attack. Whether it was brought on by his fall or vice versa could not be established.

His body lay in state for two days in the ballroom. Some 2,000 people came in curiosity to file past him. Piers Compton in his book *Cardigan of Balaclava* well describes the scene. Cardigan himself would have loved it.

'The coffin stood on a dais, approached by two steps, in the centre. The shutters were closed, and blinds covered every window along the length of the house. The walls of the room were hung with black. Its only light came from one or two oil lamps whose dimness mingled with the flames of candles that flickered in massive candlesticks standing at each of the corners of the richly draped coffin. His peer's robes, and the regalia of his orders of knighthood, were placed upon the coffin. Also displayed upon the pall were the uniform and busby he had worn, and the sword he had carried, at Balaclava. His coronet and crest, embroidered in gold and silver thread, were shown on silken banners.'[22]

The April wind was chilly on the day he was buried. The procession from the courtyard of Deane Park to St Peter's church was headed by Cardigan's valet, Mr Matthews. The coffin was covered in crimson velvet on which were placed his decorations. The ten pall bearers were all former officers of the 11th Hussars, including Colonel Douglas who had led the Regiment in the charge. There was a band of local players. Behind walked 'Ronald'. Lady Cardigan was later to claim that the elderly animal had to be dosed with laudanum to keep him quiet. Of more questionable veracity was the story that afterwards the horse would not hold up its head to be photographed until a trumpeter had been sent into the nearby bushes to sound the 'charge'. The route was packed with people from his estate and villages from miles around. They had come to catch a last glimpse of a man who had become a legend in his lifetime – and remains so to this day.

'Ronald' survived him by over four years. He died eighteen years after leading one of the most famous cavalry charges in history. His head, tail and one hoof mounted as an ink stand are on display at Deane Park to this day. The location of two other hooves are the guardroom of Windsor Castle and the officers' mess of the 11th Hussars – the fourth is unknown.

246

Epilogue – A Balaclava Banquet

'Honour the charge they made!
Honour the Light Brigade.'

Lord Tennyson

'. . . dead or living, drunk or dry, Soldier I wish you
well.'

A.E. Housman

On 25 October, 1875, the Alexandra Palace was the venue for a special function. Foregathering that day were some 120 non-commissioned officers and
men who had one memorable thing in common – they had charged with the
Light Brigade at Balaclava exactly 21 years before. They were assembling for
the first Balaclava Banquet, which was something of a misnomer as the
organizers had gone to great lengths to ensure that it was exclusively for those
who had actually made that mad dash down an unnamed valley in the
Crimea. The fact that others who were at the Battle of Balaclava (of which
the charge of the Light Brigade was only a part), particularly the Heavy
Brigade, were excluded, caused considerable resentment. It was to be the first
of many annual gatherings of a gradually dwindling band of survivors. Like
the annual Waterloo dinner before it, and the more recent El Alamein
reunion, it was hugely popular with participants, old soldiers with memories
to share, events to relive, stories to retell (and embellish), and friendships to
renew.

Suitably strenuous efforts had been made to prepare the palace. *The
Illustrated London News* of five days later recorded:

'Along the hall . . . was a well arranged museum of relics consisting of
arms and bullet-riddled and sabre-cut helmets and other
portions of uniform. There was also in the collection articles found in
the baggage of Prince Menschikoff, which was abandoned by him on the
field of battle. There was the head of the charger which carried the Earl of
Cardigan while leading the charge. This was sent by the Countess
of Cardigan. But a more remarkable object was a living horse, a high-
caste chestnut Arab, the oldest charger that has survived the Crimean
War, if not the oldest in the British service. . . . As the beautiful little beast

247

stood bridled and saddled at the Alexandra Palace he looked quite young and fit for another campaign.'

Twenty-one years on the Light Brigade's charge was firmly established in the public's mind as the glorious highlight in an otherwise squalid and almost pointless war in which countless men had died, not so much at the hands of the enemy, as from disease and atrocious living conditions. The Army high command had been blamed for incompetence. Official enquiries had been held into why there was totally inadequate transport for supplies and the wounded, desperately primitive medical facilities, shortages of everything from food to blankets and clothing. The press and public wanted some redeeming feature, some event they could look to with pride. The charge of the Light Brigade was that event. It occurred on the same day in October as the Battle of Agincourt and had already been immortalized in verse by Tennyson, as Agincourt had by Shakespeare.

The men climbing the steps of the palace to be welcomed and ushered into the reception area were middle-aged, mostly heavily bearded and, with the exception of a small minority in uniform, mainly wearing waistcoats, bow ties and double-breasted suits. Not all had prospered since leaving the Army, although there was a preponderance of thickening waistlines. Some were in menial jobs, a few were out of work, but all had made a special effort for the occasion. As *The Illustrated London News* reported:

'It was gratifying to see that to a man they were dressed respectably and seemed to be occupying comfortable positions. Their hearty greetings of one another was in itself a sight to see. Some of them who had been companions in the same regiment had never met since they left the Army.'

All were proudly wearing their medals, not neatly mounted as today, but pinned on separately much as the wearer wished. At 53 John Brooks, formerly of the 13th Light Dragoons, was older than most, but then he was one of only a handful of men present wearing the blue and red Sutlej Campaign Medal. He had fought in all the major battles in the Sikh Wars from 1845–49. When he had ridden in the first line he was probably the most experienced soldier of his rank in the Brigade. His wounds, when his horse was blown to pieces by a shell, had been severe. But although they led to a medical discharge he was to live until 1911.

The pale blue ribbon, with its narrow yellow edges, of the Crimean War Medal was on the chest of every guest. A substantial number had four clasps on the ribbon, one for each battle or siege at which the holder had been present – the Alma, Balaclava, Inkerman and Sevastopol. One such man was Richard Brown, the former servant of the commanding officer of the 11th Hussars, described as 'handsome and honest; a model soldier. Never in the defaulters' book'.

A somewhat different character, but instantly recognizable to soldiers the world over, was 'old Jack Penn' – the man who, predictably, had dismounted to acquire a souvenir from the Russian officer he had slain. There have been men like Penn in uniform since armies first marched. The son of a soldier (his father had been a farrier-major in the 14th Light Dragoons), he grew up into a typical hard-drinking, hard-fighting 'old sweat', always in trouble in peacetime but a boon on the battlefield. Experience gained him promotions, and misdemeanours, usually related to the bottle, subsequent demotions. Penn had galloped down the valley as a corporal in the 17th Lancers, was promoted sergeant four months later, and busted to private seven months after that. That day, however, he was one of the most bemedalled and well known veterans; not only did he have the Distinguished Conduct Medal, Turkish Medal and medals for the First Afghan and First and Second Sikh Wars, but his Crimean ribbon was one with all four battle clasps.

The clasps, in the form of elongated oak leaves, had been much criticized as resembling fancy wine or whisky decanter labels, so this design was never used again – Crimean medals now being instantly recognizable because of it. Also being worn with particular pride was a comparatively new decoration for an 'act of valour or self-sacrifice or extreme devotion to duty in the presence of the enemy' – the Victoria Cross.

The Crimean War had brought home the fact that Britain had no suitable way of rewarding exceptional bravery on the battlefield. Queen Victoria, ever one to acknowledge the value of her soldiers and sailors, took a personal interest in remedying this omission; hence the cross that bears her name. It is made from ingots of bronze from melted-down Russian cannons captured at Sevastopol. Eighty-four were awarded for outstanding acts of courage in the Crimea. Of these six were earned in connection with the Light Brigade's charge. Three of these holders were present at Alexandra Palace: former sergeant, but now Lieutenant and Quartermaster, John Berryman of the 17th Lancers who was still serving; Lieutenant and Quartermaster Charles Wooden, also a former sergeant of the 17th Lancers but currently serving in the 104th Bengal Fusiliers; and ex-corporal Joseph Malone who, although now dressed in a civilian suit, had similarly risen to commissioned rank in the 13th Light Dragoons before finishing his career as riding master of the 6th Dragoons.

Berryman had captured three prisoners at the Alma (remarkable, as the cavalry had little to do at this battle) before gaining his Cross for helping to save the life of Captain Webb in the dreadful aftermath of the charge. In 1875 his career was far from over; he was to add the South African Medal (for the Zulu War) to his VC, Crimean Medal (with four clasps), Turkish Medal and Indian Mutiny Medal before becoming an honorary major. Wooden was one of only two men born German ever to win the VC. He had been instrumental in helping to save the life of his commanding officer. Surely he was greeted inside the palace by some of his old comrades as 'Tish me – the devil', the nickname he had acquired when this extraordinary response was his answer, in broken English, to being challenged one night by a sentry. Little did anybody realize that within a year Wooden was to be lying in an unmarked grave in Dover, having taken his own life. Malone had got his award for assisting Berryman in Captain Webb's rescue. A soldier who had been with him at the time, and was present at the banquet, was former Private James Lamb – the man who claimed to have lost the draw for the VC that went to Malone.

The principal hall of the palace, some 200 feet in length, had been divided into a reception and a dining area. By the time all the guests had assembled there was a haze of tobacco smoke almost as thick as the cannon smoke into which they had all ridden that morning long ago. Certainly the roar of conversation made listening to the background music of the 8th Hussars band a virtual impossibility. In regimental groups everybody wanted to catch up, to find out what had happened to missing friends, to reminisce, and to relive those never-to-be-forgotten twenty minutes that had accidentally set them apart from everybody else for ever.

Mingling with the crowd was Thomas Morley who, as a corporal in the 17th Lancers, claimed to have acted with conspicuous gallantry after the charge and to have saved the life of an officer at Inkerman. He appeared to subscribe to the view that the pen was at least as mighty as the sword as he spent much effort between 1857 and 1892 in submitting unsuccessful written claims for the award of the VC; a strange man with a grudge perhaps, but very much the soldier. At the outbreak of the American Civil War he enlisted in the Union Army and rose to the rank of captain, having served throughout the conflict. Afterwards he joined the Ayrshire Yeomanry with the rank of troop sergeant-major. By 1875 he was within a year of finally resigning from service life as a regimental sergeant-major.

There were other soldiers present who had, or would, wield the pen to considerable effect. Albert Mitchell of the 13th Light Dragoons was one of many who stepped into dead men's shoes when, the day following the charge, he was promoted corporal. After leaving the Army as a sergeant in 1862 he wrote one of the best accounts by a participant, not only of the charge, but of the war generally. It would be published under the title *Recollections of One of the Light Brigade*. A man of similar talents with six medals on his suit was ex-TSM George Loy Smith of the 11th Hussars who had the reputation of a strict (some said fearsome) disciplinarian. Smith is depicted, mounted, on the right of Lady Butler's famous painting 'After the Charge'. His extensive diaries of service life in India and the Crimea have been published in book form as recently as 1987 under the title *A Victorian RSM*. Then there was William Pennington, also of the 11th Hussars, who was currently the manager of the Sadler's Wells Theatre. Since leaving the Army he had become a prominent Shakespearean actor who was later to write his autobiography, which included a powerful account of the charge, which he insisted (with some claim to accuracy) on calling an advance.

Among the 4th Light Dragoons were former Sergeant John Howes and Private Thomas Armes. Howes had made strenuous but unsuccessful attempts to bring away a Russian gun from Obolensky's battery. Armes had been promoted corporal the day before the battle but had not been able to put his stripes up, so rode as a private. He had been left for dead after a Cossack put a lance through his leg while lying on the ground, but survived to be nursed by Florence Nightingale at the Scutari hospital. Before leaving for the Crimea Armes had been presented with a New Testament by the 'Ladies of Totnes'. He always carried it in the field but somehow it was lost at the Alma. Incredibly, it was returned to him many years later covered with bloodstains by a soldier who had picked it up on the battlefield.

Enhancing his regimental reputation as a somewhat fanciful raconteur was ex-TSM John Linkon of the 13th Light Dragoons, but now the 'man from the Pru'; he had been an agent for the Prudential Insurance Company for the last five years. Surely he was regaling all who would listen with his hair-raising stories of his experiences as a Russian prisoner. If not that, then of how, when his own horse was killed, he was able to catch the riderless mount of Captain Nolan. Linkon was a big man, over six feet tall, who was to live to be 94. At the time of his death in 1910 he was living in the workhouse but was much respected as a fine, upright old soldier with a long white beard who always wore his medals.

At the appointed moment dinner was announced after the appropriate call by an 8th Hussar trumpeter had stilled the noise. The banqueting half of the hall had been carefully set out with a top table for senior guests and five regimental tables for the veterans. Again the *Illustrated London News* described the scene:

'The latter [banqueting chamber] was very handsomely decorated. Behind the chairman's seat was a trophy, having the Russian flag as a centrepiece, encircled by the English, French, Italian and Turkish [Italy, or more correctly Sardinia, and Turkey were allied with Britain and France in the latter part of the Crimean War] flags. Effigies in armour kept guard on each side of this trophy, and along the side walls were military emblems and mottoes. The tables were rich in ornaments of plate and choice fruit and flowers.'

Although the occasion was an 'other ranks' celebration some officers were invited, and there were also a few special guests sitting down to dine who had not participated. The 'chair' in the centre of the top table was occupied by Colonel Robert White who, as a captain, had led the directing squadron of the 17th Lancers only to have the Brigade Commander's sword placed across his chest when he unwittingly increased the pace too early. Baron de Grancey, military attaché at the French Embassy in London, in the uniform of the Chasseurs d'Afrique, sat on White's right. Next to him was Commandant Canovaro, naval attaché to the Italian Legation, and on his right again Major Sir George Wombwell of the 17th Lancers. Also on this side was Sir Edward Lee and the directors of Alexandra Palace. The seats on White's left were taken by a number of officers such as Lord Tredegar, Sir Godfrey Morgan and, among others, 'Mr J. Malone, 6th Dragoons; Mr J. Wooden, 104th Regiment; [and] Mr J. Berryman'.

The noise level rose again, the band played bravely on, wine flowed and an excellent meal was consumed at a leisurely pace. There was, of course, a toast-master. Mr Samuel Wilson late of the 8th Hussars, not to be confused with William Wilson of the same Regiment who at sixteen had been the youngest participant in the charge, rose to announce the first of numerous toasts – the Queen. This 'was drunk with great cheering' and was succeeded by the national anthem sung by a mixed choir brought in for the occasion. There followed one for the Prince of Wales; after which there was a slight pause as a Mrs Stirling, who had seemingly declined the invitation to dinner, came in and took a seat near Sir Edward Lee. Then it was the turn of the chairman of the Balaclava Commemoration Committee, a Mr Edward Woodham, formerly of the 11th Hussars.

> 'The Chairman, rising amid some commotion, said: Comrades I have now to propose the third toast – (a voice: "Order, please") – "The British Flag". I know not what to say about it. One could say so much that perhaps the best thing would be to say nothing at all. [Needless to say he went on to say a great deal] . . . We will drink to the British flag with all honours (loud cheers).'

Patriotism was undoubtedly the theme for the evening. Sir Edward Lee was the next on his feet. By this time it mattered little what a speaker said or how long he spoke for, which was perhaps just as well for Sir Edward. He rambled on amiably and enthusiastically, but perhaps not too coherently, about 'chivalry', 'deeds of arms', 'the 300 Spartans at Thermopylae', 'shreds of the Union Jack', 'Agincourt', 'Prince Rupert' (alluding to the English Civil War cavalry general), laced with appropriate quotes from Shakespeare's *King Henry V*. Eventually he got to the point and, raising his glass, proposed 'The gallant six hundred!', which triggered a storm of prolonged cheering.

Lord Tredegar, Sir George Wombwell and Colonel Trevelyan had their say next. Then a specially written song entitled 'The Light Brigade' was sung, with the band and choir giving up the struggle to be heard above the bellowing of the chorus. This was followed by Pennington (chosen for his acting talents) rising to recite Tennyson's poem to a totally receptive and responsive audience. The final toast, however, secured silence. The chairman, reminding everybody of the friends left behind for ever in that remote valley, proposed, 'The memory of the dead'. All stood solemnly, if not soberly, while the 8th Hussars bandsmen played the 'Dead March'.

By this time the Balaclava banquet had merged happily and hazily into the Balaclava booze-up. As the celebrations grew ever more alcoholic a number of the officers on the top table did the decent thing and slipped away. Within minutes a befuddled group of old soldiers had invaded the upper end of the room where they waved and shouted for silence. After a minute or so of yelling there was a lull of sufficient duration for one of their number to call for three cheers. An unidentified old sweat had a soldier's sense of humour, and an instinctive feel for the mood of the gathering when he called for, 'Three cheers for Cardigan and Nolan!' Every man responded wholeheartedly three times. As the last yell died away some wag piped up with, 'And three cheers for those who know how to look after themselves!' Three deafening roars rocked the roof.

The Light Brigade's charge is still probably the most remembered cavalry action of all time. It achieved nothing, it should never have happened, and it revealed serious shortcomings among senior commanders. Nevertheless, it came to represent the embodiment of the cavalry spirit, the cavalry *élan*, that anything was possible with enough dash and daring. The fact that almost everything that could have gone wrong went wrong has not diminished its fame, its veneration almost, as the epitome of old fashioned gallantry and glory. The four horsemen responsible would certainly settle for that.

Appendix 1

THE LIGHT BRIGADE[1]

Brigade Commander – Maj.-Gen. The Earl of Cardigan
Brigade Major – Lt-Col. Mayow
Brigade ADC – Lieut. Maxse
Regimental ADC – Capt. Lockwood (8H)
Regimental ADC – Cornet Wombwell (17L)
Brigade Trumpeter – Trumpeter Britten

11th Hussars	17th Lancers	13th Light Dragoons
CO: Lt-Col. Douglas	Capt. Morris	Capt. Oldham
2ic: –	–	–
Adjt: Asst-Surg. Wilkin[2]	Cornet Chadwick	Lieut. Smith
RSM: Sgt-Maj. Bull	RSM Ffennell	Cornet Gardner[3]
Capt. Cook	Capt. Morgan	Capt. Goad
Lieut. Dunn	Capt. Webb	Capt. Jenyns
Lieut. Houghton	Capt. White	Capt. Tremayne
Lieut. Palmer	Capt. Winter	Lieut. Jervis
Lieut. Trevelyan	Lieut. Gordon	Cornet Chamberlayne
ORs: 135	Lieut. Thomson	Cornet Montgomery
	Cornet Cleveland	ORs: 120
	ORs: 136	

4th Light Dragoons	8th Hussars
CO: Col. Paget	Lt-Col. Sherwell
2ic: Maj. Halket	Maj. De Salis
Adjt: Cornet Martyn	Lieut. Seager
RSM: RSM Jennings	–
Capt. Brown	Capt. Tomkinson
Capt. Hutton	Lieut. Clutterbuck
Capt. Low	Lieut. Fitzgibbon
Capt. Portal	Lieut. Heneage
Lieut. Joliffe	Lieut. Phillips
Lieut. Sparke	Cornet Clowes
Cornet King	Cornet Mussenden
Cornet Hunt	ORs: 104
ORs: 115	

[1] Only the numbers who charged are shown. To these must be added Capt. Nolan and two Sardinian officers making an estimated total of 668.
[2] Cornet Yates (adjt) did not charge; Wilkin may have stood in for him.
[3] Although an officer Gardner did the duties of RSM during the charge.

Appendix 2

THE CAVALRY DIVISION

Divisional Commander – Lt-Gen. Lord Lucan
Asst. Adjt-Gen. – Lt-Col. Lord Paulet
Asst. QM Gen. – Maj. McMahon
ADC – Capt. Walker
ADC – Capt. Charteris
ADC – Lieut. Lord Bingham
Interpreter – Mr Blunt

| The Light Brigade | The Heavy Brigade |
| (Maj-Gen. The Earl of Cardigan) | (Brig-Gen. Scarlett) |

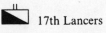 13th Light Dragoons   4th Dragoon Guards

I Troop RHA

17th Lancers   5th Dragoon Guards

11th Hussars   1st (Royal) Dragoons

4th Light Dragoons   2nd Dragoons (Greys)

8th Hussars   6th Dragoons

255

Appendix 3

BRITISH AND FRENCH ARMIES AT BALACLAVA

British	*French*
C-in-C: Gen. Lord Raglan	Gen. Canrobert
1st Lt-Gen. HRH Duke of Cambridge	1st Gen. Bosquet
1st Maj-Gen. Bentinck	1st Brig-Gen. Espinasse

<table>
<tr><td>

3rd Bn. Grenadier Guards
1st Bn. Coldstream Guards
1st Bn. Scots Fusilier Guards

</td><td>

1st Zouaves
7th Regt. de Ligne

</td></tr>
</table>

2nd Col. Cameron 2nd Brig-Gen. Vinoy

42nd Highlanders 20th Regt. de Ligne
79th Highlanders 27th Regt. de Ligne
A and H Batteries RA

4th Lt-Gen. Sir G. Cathcart Cav Brig-Gen. D'Allonville

1st Brig-Gen. Goldie 1st Chasseurs d'Afrique
4th Chasseurs d'Afrique

20th Foot
21st Fusiliers
57th Foot

2nd Brig-Gen. Torrens

46th Foot
63rd Foot
68th Foot
1st Rifle Brigade

P Battery RA

Cav Lt-Gen. Lord Lucan

Lt. Maj-Gen. The Earl of Cardigan

Hy. Brig-Gen. Scarlett

I Troop RHA
Balaclava defences
93rd Highlanders
2 Bns Royal Marines
Invalid Battalion
W Battery RA

Appendix 4

RUSSIAN ORDER OF BATTLE AT BALACLAVA

Divisional/Force Commander – Lt.-Gen. Liprandi
Staff officer – Staff-Major Guersivanov
ADC – Capt. Baron Willebrandt
Artillery Commander – Col. Nemov

Left	*Centre*	*Right*
Maj-Gen. Gribbe	Left Centre Maj-Gen. Semiakin	Col. Skiuderi
24th Dnieper	23rd Azov	24th Odessa
4th Rifle	24th Dnieper	4th Rifle
	4th Rifle	53rd Don Cossack

Composite Uhlan (Lancers)

Right Centre Maj-Gen Levutsky

53rd Don Cos.
6 guns No. 6 Lt. Bty.

23rd Ukrainian (Jaeger)

8 guns No. 7 Lt. Bty

4 guns No. 4 Bty.

4 guns No. 4 Bty.
4 guns No. 7 Lt. Bty.

6 guns No. 6 Bty.
4 guns No. 4 Bty.

6th Hussar Brigade
Maj-Gen. Rijov
11th Kiev Hussars

Flank Guard
Maj-Gen. Jaboritsky
31st Vladimir

12th Ingermanland H.

32nd Sousdal

1st Ural Cossacks

6th Rifle

Black Sea
Foot Cos.

8 guns 12th Lt. Horse Bty.
8 guns No. 3 Bty. Don. Cos.

10 guns No. 1 Bty.
4 guns No. 2 Lt. Bty.

257

Appendix 5

The Last Survivors

Disease, heat and mutineers took their toll of veterans in India during the years immediately following the Crimea. Captain Morris died there, supposedly from the effects of the sun on the silver plate he had in his head as a result of his wounds sustained at Balaclava. The great success of the 21st anniversary celebrations prompted the formation the following year of The Balaclava Commemoration Society. One of its aims was to have an annual commemoration dinner. Initially the rules were too loosely drawn and arguments arose as to whether men who were 'in camp' but did not charge could be members. After considerable discussion and consultation with members the rules were amended to read, 'That no person shall be eligible to become a Member unless he actually rode in the Charge'. In 1879 the Society listed 22 officers and 200 other rank survivors.

In 1899 42 survivors sat down to dinner at St James's Restaurant. Seven years later the number attending had dwindled to 29, while on the last occasion, held on 25 October, 1913, only nine managed to come.

In 1897 Queen Victoria's Diamond Jubilee was celebrated in London in a style befitting the high noon of the Empire. The climax was the Jubilee Procession with its 50,000 troops said to constitute the largest military force ever assembled in London. It marched through the city in two seemingly endless columns, one led by Captain Ames of the Horse Guards, at six foot eight inches the tallest man in the British Army, the other by Field-Marshal Lord Roberts of Kandahar. They were converging on St Pauls. London had never seen its like before and never would again. There were Hussars from Canada, camel troops from Bikaner, Dyak headhunters from Borneo in bright red pillbox hats, Hong Kong Chinese police in conical coolie hats, Hausas from the Niger, Jamaicans in white gaiters and embroidered coats, Cypriots wearing fezzes and dark-skinned Indian lancers in scarlet and gold. Watching it pass beneath them were some 73 elderly gentlemen who had charged with the Light Brigade 43 years earlier.

That they did so was due to the generosity of one man with a house and business offices at 158 Fleet Street. This was a Mr T. Harrison Roberts, a publisher of periodicals. He had written to all survivors that he knew of, inviting them to watch the procession from the windows of his premises. Nothing would be too good for his honoured guests as his letter makes clear:

'In order to avoid all possible fatigue, I propose that you should be at the above address by five o'clock on the evening of Monday the 21st. You will thus avoid all the crowd which is sure to assemble on the day itself . . .

'I provide all accommodation for you, including comfortable sleeping arrangements, food, drinks, etc., etc., on the premises, so that you are put to no trouble or expense of any sort. If you are living out of London, I will . . . send you a free railway ticket . . . and another free ticket home again. . . . I will . . . send a representative to meet you at the London station. . . .

'I want you all to have a real good holiday – to see your Queen, and to let your Queen see you. . . .

'My office is in the centre of Fleet Street; it is a big place and will give you a fine view of the whole Jubilee Procession.'

Roberts received 24 replies of which eight were from men who had not ridden in the charge. Help from the Balaclava Commemoration Society eventually sorted out the genuine from the bogus and 73 duly arrived. But Roberts had been dismayed to discover that so many of the old veterans were destitute, or nearly so. He launched an appeal in his publication the *Illustrated Bits*. The public were generous, and a Roberts Relief Fund was established from which between 7 and 15 shillings was sent weekly to the 40 most deserving survivors. With the money came a postcard which the old soldier had to fill in and return. Until it was received the following week's payment was not made.

The Roberts Fund ceased when the last recipient, Private James Olley, died in September, 1920. At the outbreak of the First World War there were fourteen survivors of the charge still alive, including two officers. They are listed below with the date of their deaths:

Private John Boxall, 4th Light Dragoons – 23 August, 1914
Lieutenant Edward Phillips, 8th Hussars – 18 April, 1915
Private John Whitehead, 4th Light Dragoons – 16 May, 1915
Private James Mustard, 17th Lancers – 4 February, 1916
Private George Gibson, 13th Light Dragoons – 6 June, 1916
Private Thomas Warr, 11th Hussars – 15 June, 1916
Private John Smith Parkinson, 11th Hussars – 12 January, 1917
Captain Percy Shawe Smith, 13th Light Dragoons – 8 February, 1917
Private William S.J. Fulton, 8th Hussars – 29 March, 1918
Private James Olley, 4th Light Dragoons – 4 September, 1920
Corporal Ashley Kilvert, 11th Hussars – 17 October, 1920
Private William Henry Pennington, 11th Hussars – 1 May, 1923
Private Edwin Hughes, 13th Light Dragoons – 18 May, 1927

Appendix 6

A Gun Accident, 1811

'One occurrence I witnessed here [was] almost incredible: a Portugese governor arrived at Colombo, early in the year 1811; on the firing of the salute, Gunner Richard Clark was blown from the mouth of his gun right into the air, and alighted upon a rock at a considerable distance in the harbour, yet escaped without a bone being broken, almost unhurt. It was the most miraculous escape I ever witnessed; he was but an awkward soldier at the best; the gun of which he was No. 1, [at that time the No. 1 was the spongeman] went off by accident. . . . The gun was just loaded when she went off, through the negligence of Clark in not spunging properly. He was not at his proper distance, like the other man, nor yet near enough to receive the whole flash. To the astonishment of everyone, he was seen in the air, the sponge-staff grasped in his right hand, the hammerhead downwards, which first struck the rock as he alighted on his breech [a splendidly quaint way of describing it]. The rock was thickly covered with seaweed. A party was sent down to bring up the body, as all concluded him killed upon the spot; he was brought up only stunned and slightly singed, and was at his duty again in a few days; while No. 5 who served the vent, had his thumb, with which the motion-hole is stopped during loading, so severely burned, it was feared he must have lost it, and it was only saved by the skill of the surgeon. . . .

'If the gun goes off in loading, the thumb is witness whether he [the ventsman] did his duty or not, if it is burned he receives praise, if it is not he is punished. The thumb is sometimes so severly injured that amputation is necessary.'

From *The Life of Alexander Alexander*, edited by John Howell, Edinburgh, 1830.

Sources

Listed below are the books, letters and other documents found of particular value during research.

Books

Adye, Lt-Col. J., *A Review of the Crimean War*, E.P. Publishing, 1973

Anglesey, The Marquess of, *A History of British Cavalry Vol 3*, Leo Cooper, London, 1973

Anitschkof, Captain on Russian Imperial Staff, A Russian Account of the Campaign in the Crimea, *United Services Magazine*, 1860.

Baden-Powell, Captain R.S.S., *Cavalry Instruction*, Harrison and Sons, London, 1885

Barthorp, M., *Heroes of the Crimea*, Blandford, 1991

Barrett, C.R.B., *History of the 13th Hussars Vol 1*, London, 1911

Bidwell, Brig. S., *The Royal Horse Artillery*, Leo Cooper, London, 1973

Boganovitch, M.N., *The Eastern War of 1854–6* (translated from the Russian)

Brett-Smith, R., *The 11th Hussars*, Leo Cooper, London, 1969

Calthorpe, Lieut. The Hon. S.J.D., *Letters from Headquarters*, Murray, London, 1858

Cardigan, The Earl of, *Eight Months on Active Service*, Clowes and Sons, London, 1855

Compton, P., *Cardigan of Balaclava*, Robert Hale, London, 1972

Ewart, Lt-Gen. J.A., *The Story of a Soldier's Life Vol 1*, 1881

ffrench Blake, Lt-Col. R.L.V., *The Crimean War*, Leo Cooper, London, 1971

ffrench Blake, Lt-Col. R.L.V., *The 17th Lancers*, Hamish Hamilton, London, 1968

Fortescue, J.W., *History of the 17th Lancers*, London, 1895

Harris, J., *The Gallant Six Hundred*, Hutchinson and Co. Ltd., London, 1973

Hibbert, C., *The Destruction of Lord Raglan*, Longmans, London, 1961

Higginson, Gen. Sir George, *71 years of a Guardsman's Life*, London, 1916

Horse Guards, The, *Regulations for the Instruction, Formations and Movements of the Cavalry*, London, 1851

Joycelyn, Col. J.R.J., *The Story of the Royal Artillery (Crimean Period)*, Murray, 1911

Kinglake A.W., *The Invasion of the Crimea* (various editions), William Blackwood and Sons, London, 1887–88.

Lummis, Cannon W.M. and Wynn, K.G., *Honour the Light Brigade*, J.B. Hayward and Son, London, 1973

Mollo, J. and B., *Into the Valley of Death*, Windrow and Green, London, 1991

Moyse-Bartlett, Lt-Col. H., *Louis Edward Nolan and his Influence on British Cavalry*, Leo Cooper, London, 1971

Murray, Rev. R.H., *History of the VIII King's Royal Irish Hussars Vol. 2*, Cambridge, 1928

Paget, Gen. Lord George, *The Light Cavalry Brigade in the Crimea*, Murray, London, 1881

Pemberton, W.B., *Battles of the Crimean War*, Batsford, London, 1962

Pennington, W.H., *Sea, Camp and Stage*, Bristol, 1906

Perrett, B., *At All Costs*, Arms and Armour Press, London, 1993

Russell, H.W., *The Great War with Russia*, Routledge and Sons, Manchester, 1895

Russell, H.W., *Despatches from the Crimea*, Andre Deutsch, 1966

Scott-Daniell, D., *Fourth Hussar*, Aldershot, 1959

Seaton, A., *The Crimean War – A Russian Chronicle*, Batsford, 1977

Smith, G.L., *A Victorian RSM*, D.J. Costello Ltd., Tunbridge Wells, 1987

Strawson, J., *Beggars in Red*, Hutchinson, London, 1991

Sweetman, J., *Raglan*, Arms and Armour Press, London, 1993

Thomas, R. and Scollins, R., *The Russian Army of the Crimean War*, Osprey, 1991

Walker, Gen. Sir C.P.B., *Days of a Soldier's Life*, Chapman and Hall, London, 1894

Whinyates, Col. F.A., *From Coruna to Sevastopol*, W.H. Allen and Co., London, 1884

Woodham-Smith, C., *The Reason Why*, Constable, London, 1953

Letters and other documents

Anon., The Charge of the Light Brigade by one who was in it, *United Services Journal*, April, 1856

Blunt, Sir John, Papers, National Army Museum

Cattell, W. MS Autobiography of Asst-Surgeon William Cattell (attached 5th Dragoon Guards), National Army Museum

Farquharson, K.S., Reminiscences of Crimean Campaigning and Russian Imprisonment, Edinburgh, 1883

Forrest, Gen. W.C., Letters from the Crimea 1854–55, National Army Museum

Gowing, T., A Voice from the Ranks, 1884

Grigg, Pte. J., Told from the Ranks, collected by E. Milton Small, 1897

Jolliffe, Lieut. H.D., Letters, 1854, National Army Museum

Mitchell, Sgt. A., Recollections of one of the Light Brigade, Canterbury, 1885

Morley, T., The Man of the Hour, 1892

Nunnerley, Sgt-Maj. J.I., Short Sketch of the 17th Lancers and Life of Sergeant-Major J.I. Nunnerley, Liverpool, 1884

Pennington, W.H., Left of Six Hundred, privately printed London, 1887

Portal, Capt. C., Letters from the Crimea, Winchester, 1900

Seager, Lieut. E., A Letter from the Crimea, 26 October, 1854, National Army Museum

Tremayne, Capt. A., Letters, National Army Museum

Wightman, Pte. J.W., One of the Six Hundred, *Nineteenth Century Magazine*, May, 1892

Notes

Chapter One

1 Paget, Gen. G., *The Light Cavalry Brigade in the Crimea*, Murray, London, 1881.
2 Ibid., p. 170.
3 Ibid., p. 170.
4 The precise number that charged will always be the subject of contention. I have used the figures provided by Colonel Whinyates in his book *From Coruna to Sevastopol* published in 1884. To his total, which was carefully compiled from Regimental returns, of 658 I have added Lord Cardigan, Lt Col Mayow, Capt Nolan, Lt Maxse and the two Sardinian officers.
5 Paget, op. cit., p. 8.
6 Roundshot bounced off the ground, and was still lethal after several such bounces. A range of 600 metres, for example, would mean the shot carried for that distance before first graze, or bounce.
7 Wightman, Pte. J.W., One of the Six Hundred, *The Nineteenth Century Magazine*, May, 1892.
8 Ibid.
9 Capt. William Morris, who until the previous day had been on Lucan's staff, was now acting as the commanding officer of the 17th Lancers due to the sickness of Col. Lawrenson.
10 Moyse-Bartlett, Lt Col. H., *Louis Edward Nolan and his Influence on British Cavalry*, Leo Cooper, London, 1971, p. 144.
11 Smith, G. L., *A Victorian RSM*, D.J. Costello (Publishers) Ltd., Tunbridge Wells, 1987, p. 78.
12 Featherstone, D., *Weapons & Equipment of the Victorian Soldier*, Blandford, Poole, 1978, p. 42.
13 A lancer recruit required a much longer period of training than other cavalrymen. He had to master no fewer than 55 different exercises with his lance – 22 against cavalry, 18 against infantry plus another 15 general ones for good measure.
14 Featherstone, op. cit., p. 43.
15 How familiar it all sounds. Modern British cavalry has been just as badly neglected. On mobilization for the Gulf War all the tank Regiments in Germany were cannibalized to produce a mere two armoured (cavalry) brigades fit to fight. As in 1854 there was a frantic period of preparation with units being milked of men, tanks, vehicles, equipment, spares and ammunition to produce a modest operational force.
16 Smith, op. cit., p. 77.
17 Veigh was a Regimental character, although there is no evidence, apart from his own account, that he charged wielding his poleaxe. In India during the Mutiny he volunteered to be a grave digger in order to increase his drinking money. On a long march from Gwalior to Secunderabad the Regiment was struck by cholera. Veigh died and was buried, along with six others, in a grave he had himself dug.

18 In April, 1854, *Punch* published the following little ditty about them:
>   'Oh, Pantaloons of cherry,
>   Oh, redder than the berry,
>   For men to fight in things so tight
>   It must be trying, very.'

19 Dunn, despite winning a VC, became *persona non grata* for retiring to his estates in Canada with his commanding officer's wife. Colonel Douglas refused to divorce her. Dunn later bought the lieutenant-colonelcy of the 100th Regiment in Canada and then exchanged into the 33rd Regiment. He was then described as, 'a tall, kindly, good-natured dandy, a bad CO and not a good example to young officers – he was very popular but nearly destroyed the Regiment'. He died of gunshot wounds while on a hunting expedition at Senafe in Abyssinia in 1868. The enquiry found death to be accidental, but in 1951 an article in the 33rd's Regimental Magazine claimed he could have been murdered.

20 Smith, op. cit., p. 134.

21 Ibid., p. 93.

22 Ibid., p. 7.

23 Lummis, Cannon W.M. and Wynn, K.G., *Honour the Light Brigade*, J.B. Hayward and Son, London, 1973, p. 164.

24 Smith, op. cit., p. 104.

25 Anon., *United Service Journal*, April 1856, p. 550.

26 Seager, whose promotion to captain had been so long in coming retired as a Lt Gen. in 1881.

27 Paget, op. cit., p. 1

Chapter Two

1 An extreme example of 'progressive advancement' was the promotion to major-general in 1811 of three officers (Scott, Tipping and Trotter) none of whom had done a day's duty since the end of the American War of Independence in 1781 – 30 years previously. When the Peninsular War began one general, and he was not the most senior, had carried his Regimental Colours at Culloden in the '45 Rebellion.

2 Sweetman, J., *Raglan*, Arms and Armour Press, London, 1993, p. 94.

3 Hibbert, C., *The Destruction of Lord Raglan*, Longmans, London, 1961, p. 4.

4 Russell, W.H., *The Great War with Russia*, London, 1895, p. 306.

5 Kinglake, A.W., *Invasion of the Crimea*, 4th Edition, Vol 2, p. 17.

6 Wolseley, F-M, Sir E., *The Story of a Soldier's Life*, 1903, p. 236.

7 Quoted in Moyse-Bartlett, op. cit., p. 100.

8 Ibid., p. 121.

9 Ibid., p. 127.

10 Ibid., p. 141.

11 Ibid., p. 159.

12 Woodham-Smith, C., *The Reason Why*, London, 1953, p. 107.

13 Ibid., p. 133.

14 Ibid., p. 48.

15 Smith, op. cit., p. 44.

16 Ibid., p. 69.

Chapter Three

1 The Cavalry Division initially had two brigades, each of four Regiments of about 300 men. The 4th Light Dragoons (Light Brigade) did not sail until 19 July, and the Greys (Heavy Brigade) later still.
2 Mitchell, Sgt. A, *Recollections of One of the Light Brigade*, Canterbury, 1885, p. 17.
3 Smith, op. cit., p. 80.
4 Anglesey, The Marquess of, *A History of the British Cavalry 1816–1919* Vol. 2, Leo Cooper, London, 1975, p. 34.
5 Calthorpe, Col. S.J.G., *Letters from Headquarters*, Murray, London, 1858, p. 7.
6 Ibid., p. 15.
7 Raglan's despatch, P.R.O. W.O. 1/368. In this despatch Raglan warned of the grave shortage of horses. This led to Horse Guards instructing each of the eleven cavalry Regiments in England to buy 40 extra horses and be prepared to release 40 trained animals for the Army of the East, as the expeditionary force was called.
8 Calthorpe, op. cit., p. 15.
9 Moyse-Bartlett, op. cit., p. 163.
10 Russell, op. cit., p. 118.
11 Cardigan, Earl of, *Eight Months on Active Service*, Clowes and Sons, London, 1855, p. 12.
12 Ibid., p. 13.
13 Ibid., p. 13.
14 Lucan's Divisional Orders and Correspondence, 1858.
15 Quoted in Woodham-Smith, op. cit., p. 149.
16 Ibid., p. 152.
17 Calthorpe, op. cit., p.
18 Quoted in Woodham-Smith, op. cit., p. 153.
19 Ibid., p. 154.
20 Ibid., p. 154.
21 Ibid., p. 155.
22 On this 'sore-back reconnaissance' the horses were expected to carry a rider with full marching order averaging some 20 stone. In addition there was extra ammunition, two blankets, 36lbs of barley, two hay nets with hay, 3lbs of biscuits, 3lbs of meat, plus a keg of 3 pints of water.
23 Forrest, Maj. W.C., Letter dated 27 August, 1854, NAM.
24 Compton, P., *Cardigan of Balaclava*, Robert Hale, London, 1972, p. 141.
25 Cattell, Asst. Surg. W., MS Autobiography, NAM, p. 11.
26 Forrest, Maj. W.C., Letter dated 27 August, 1854, NAM.
27 Paget, op. cit., p. 217.
28 Anglesey, op. cit., p. 37.
29 Portal, Capt. R., *Letters from the Crimea 1854–1855*, Winchester, 1900, NAM.
30 Kinglake, op. cit., p. 117.
31 According to Calthorpe's informant (a French staff officer) the French had lost 7,000 men to cholera and sickness and had, at that time (Sept. 1854), 12,000–15,000 men in hospital. These huge numbers required 4,000 orderlies, guards, and cooks to look after them. The British had by then lost 700 men and had 1,900 in hospital.
32 Quoted by Harris, op. cit., p. 92.
33 Ibid., p. 93.
34 Ibid., p. 94.
35 Calthorpe, op. cit., p. 51.

36 Ibid., p. 16.
37 Quoted by Harris, op. cit., p. 99.
38 Russell, op. cit., p. 17.
39 Ibid., p. 22.
40 Paget, op. cit., p. 18.
41 Smith, op. cit., p. 99.
42 Ibid., p. 99.
43 Ibid., p. 99.
44 Calthorpe, op. cit., p. 62.
45 Brett-Smith, R., *The 11th Hussars*, Leo Cooper, London, 1969, p. 110.
46 Quoted by Hibbert, op. cit., p. 51.
47 At the Bulganek there was a small foretaste of things to come at Balaclava. Raglan sat on higher ground and was thus able to see things his cavalry commanders could not, while Airey and Nolan were involved in transmitting unpalatable orders to Lucan and Cardigan.
48 Raglan had his reasons for not pursuing. Firstly, he could not commit himself without the French, who refused to be drawn forward until they were good and ready. Then he had no way of knowing if his 900 cavalry would defeat the 3,000 enemy horsemen (so far unbloodied) confronting him. He could not know, as we do now, that the Russians were in something of a blind panic to get away across the Katcha River. With hindsight it is not unreasonable to speculate that, had there been a vigorous pursuit after the Alma, the war would have ended there and then.
49 Forrest, Maj. W.C., Letter dated 12 October, 1854, NAM.
50 Russell, op. cit., p. 116.
51 Ibid., p. 90. Nolan was referring to a patrol sent out by Lucan on the day after the Alma battle.
52 Quoted in Woodham-Smith, op. cit., p. 197.
53 Smith, op. cit., p. 112.
54 Russell, op. cit., p. 99.
55 Cardigan, op. cit., p. 81.
56 Portal, *Letters from the Crimea*, Winchester, 1900, NAM.
57 Paget, op. cit., p. 62.
58 Pemberton, W.B., *Battles of the Crimean War*, Batsford, London, 1962, p. 126.
59 Paget, op. cit., p. 64.
60 Ibid., p. 57.

Chapter Four
1 Ewart, J.A., *A Soldier's Life*, Vol 1, London, 1881, p. 104.
2 Raglan's special telescope is on display in the NAM.
3 Pennington, W.H., *Sea, Camp, and Stage*, London, 1906, p. 42.
4 Farquharson, K.S., *Reminiscences of Crimean Campaigning and Russian Imprisonment*, 1883, p. 68.
5 It had been Cathcart's strongly held view that if the Allies had attacked Sevastopol on 28 September it would have fallen. He considered that it could be achieved with the loss of 500 men – a tiny fraction of those who would die in the siege lines and of sickness later. Raglan rejected the suggestion (which would almost certainly have succeeded), and this had peeved Cathcart further.
6 There was also a so-called third line of defence, the corvette HMS *Diamond*. She

had been positioned at the head of the harbour, broadside on to the land. Her task was to fire directly up the gorge in a last ditch effort to halt an attack on the harbour.

7   Russell, op. cit., p. 120.
8   Ibid., p. 134. I believe Russell's estimate of 8 o'clock as being the time of Raglan's arrival on the escarpment is too late by about half an hour. Raglan certainly witnessed the fall of No. 1 Redoubt which occurred around 7.30 am.
9   Time would have been saved if Raglan had ordered the infantry to move as soon as he heard the bombardment starting, or on the arrival of Charteris from Lucan. Instead he waited until after his none too hurried personal reconnaissance. Some writers have suggested he did this, but according to his ADC Calthorpe this is not so. See Calthorpe, op. cit., p. 123.
10 One other suggestion has been made for this seeming tactical error. This was that the French (and Raglan invariably did as his Allies wanted) thought that a move down the Woronzoff road would bring on a major battle, which they wanted to avoid.
11 Russell, op. cit., p. 127.
12 Ibid., p. 127.
13 Ibid., p. 128.
14 Pennington, op. cit., p. 44.
15 Anon., The Charge of the British Cavalry at Balaclava, *United Services Magazine* No. 329, April, 1856, p. 549.
16 Trumpet-Major Joy later sounded the 'Charge' for the Heavy Brigade attack. His bugle is in the NAM.
17 Paget, op. cit., p. 162.
18 Smith, op. cit., p. 125.
19 Paget, op. cit., p. 162.
20 Ibid., p. 163.
21 Ibid., p. 164.
22 Gunner David Jenkins received the French Military Medal and Gunners Jacob McGarry and John Barrett the Sardinian Medal.
23 Selby, J., *The Thin Red Line*, Hamish Hamilton, London, 1970, p. 116.
24 Smith, op. cit., p. 126.
25 Compton, op. cit., p. 164.

Chapter Five
1   Mitchell, op. cit., p. 81.
2   Russell, op. cit., p. 144.
3   The positioning of the Light Brigade at this stage is important. Calthorpe's description is general, whereas those of Paget and Smith are more specific.
4   Smith, op. cit., p. 127.
5   Paget, op. cit., p. 167.
6   I believe that two errors have crept into many accounts of the battle at this point. Firstly, Raglan's reference to 'wavering Turks' did not refer to those defending the redoubts. They had 'wavered' much earlier and were now nearing Balaclava. Secondly, when Raglan gave his second order to the cavalry the North Valley was empty, or nearly so. Had he spotted Rijov's mass of cavalry at 8.30 he would have

warned Lucan, and his order would have been quite different. No contemporary source indicates this.

7   Many accounts have Rijov's force advancing up the North Valley, having crossed the river at the ford well south of the Tractir Bridge. I do not accept this view for three reasons:

a) If they had used this route they would have been in view from Raglan's position earlier, and thus his second order would have been different.

b) It made more military sense to follow the right-hand infantry column over the bridge, and follow the road which led to precisely the area in which they wished to deploy.

c) Seaton, who made extensive use of Russian sources for his book *The Crimean War – A Russian Chronicle*, describes Skiuderi's column as having to, 'dislodge an enemy picket on the Tractir Bridge [it was not there] and move along the main road on to Redoubt No. 3, being followed by Rijov's cavalry force in column of attack.' In addition the French War Ministry maps prepared a year later show Rijov's cavalry using the Tractir Bridge.

8   Kinglake, op. cit., p. 400.
9   Smith, op. cit., p. 130.
10 Kinglake, op. cit., p. 211.
11 Ibid., p. 210.
12 Paget, op. cit., p. 174.
13 Forrest, Letter dated 27 October, 1854, NAM.
14 Whinyates, op. cit., p. 130.
15 Forrest, Letter dated 27 October, 1854, NAM.
16 Whinyates, op. cit., p. 134.
17 Forrest, Letter dated 27 October, 1854, NAM.
18 Whinyates, op. cit., p. 137.
19 Forrest, Letter dated 27 October, 1854, NAM.
20 Russell, op. cit., p. 151.
21 Higginson, Gen. Sir G., *71 Years of a Guardsman's Life*, 1916, p. 185.
22 Harris, op. cit., p. 189.
23 Mitchell, op. cit., p. 82.
24 Woodham-Smith, op. cit., p. 232.

Chapter Six
1   Calthorpe, op. cit., p. 126.
2   Ewart, Lt Gen, J.A., *The Story of a Soldier's Life*, London, 1881, p. 265.
3   Selby, op. cit., p. 167.
4   Ibid., p. 167.
5   Mitchell, op. cit., p. 82.
6   Paget, op. cit., p. 72.
7   Calthorpe, op. cit., p. 129.
8   Anglesey, op. cit., p. 83.
9   Despite a diligent search I was unable to locate Blunt's original letters. This was disappointing as they throw considerable light on what happened at critical times at Lucan's HQ. Harris and Selby have both used his papers and indicated they were kept at the NAM. The Museum cannot trace them. Nevertheless, I have used the important information they contain from Harris' and Selby's books.

10 Quoted in Harris, op. cit., p. 203.
11 Walker, op. cit., p. 135.
12 Paget, op. cit., p. 218.
13 Tremayne, Lt Col. A., quoted in the *Journal of the 13th/18th Hussars*. April, 1964, p. 67.
14 Quoted in Harris, op. cit., p. 203.
15 Kinglake, op. cit., p. 248 and Lucan's despatch to Raglan dated 27th October, 1854.
16 Kinglake, op. cit., p. 249.
17 Ibid., p. 249.
18 Paget, op. cit., p. 247.
19 Kinglake, op. cit., p. 401.
20 Ibid., p. 401.
21 Tremayne, op. cit., p. 67.
22 Mitchell, op. cit., p. 83.

Chapter Seven

1 Whether or not all the guns had to limber up is uncertain. Those that had only a short way to go would undoubtedly have been manhandled. Some guns, however, had 100–150 metres to move and it is likely the teams were needed for this.
2 Seaton, op. cit., p. 149.
3 Apart from Colonel Obolensky, the names I have used for the Russian detachment are fictitious. The actions of the gunners have been carefully reconstructed to portray the procedures followed by a typical detachment facing the Light Brigade charge.
4 For a unique accident of this nature see Appendix 6.
5 The horses of a battery were normally kept well back, away from the guns in action. Horses were highly vulnerable so casualties among the teams could mean the guns could not be moved.
6 As at least two survivors of the charge described seeing a Russian gunner applying a 'spluttering fuze' (portfire) to a vent, I have assumed the Don Battery fired its guns the old way, using linstocks and portfires. The RA, however, had introduced the friction tube with percussion powder just a year earlier.
7 In getting off eleven shots in about four minutes Sulina's detachment had done well. No doubt the urgency of the situation, and omitting to swab the barrel at the end, had helped. In 1848 the *Madras Artillery Manual* gave an example of a light gun in action during a hypothetical charge by cavalry against a gun battery. It assumed the cavalry were a mile away at the start, and that they trotted for the first half mile, galloped the next quarter, and charged the final quarter – the whole taking six minutes. During this time each gun would be expected to fire eleven times (the last two being canister).

Chapter Eight

1 Wightman, op. cit.
2 Kinglake, op. cit., p. 256.
3 Nunnerley, J.I., *Short Sketch of 17th Lancers and Life of Sergeant-Major Nunnerley*, c. 1890, p. 18.
4 Anglesey, op. cit., p. 92.

5   Tremayne, op. cit., p. 67.
6   Cattell, op. cit., p. 14.
7   Paget, op. cit., p. 205.
8   Quoted by McGuffie, T.H., *Rank and File*, St Martins Press, New York, 1966, p. 384.
9   Morley, T., *The Cause of the Charge*, 1892, p. 9.
10   Ibid., p. 10.
11   Mitchell, op. cit., p. 84.
12   Quoted by Barthorp, M., *Heroes of the Crimea*, Blandford, 1991, p. 51.
13   Quoted by McGuffie, op. cit., p. 385.
14   Ibid., p. 385.
15   Cattell, op. cit., p. 14.
16   Quoted by McGuffie, op. cit., p. 385.
17   Morley, op. cit., p. 10.
18   Ibid., p. 10.
19   Quoted by Anglesey, op. cit., p. 95.
20   Cardigan, op. cit., p. 90.
21   Paget, op. cit., p. 178.
22   Ibid., p. 179.
23   Ibid., p. 195.
24   Ibid., p. 180.
25   Smith, op. cit., p. 132.
26   Ibid., p. 132.
27   Anon., op. cit., p. 553.
28   Pennington, op. cit., p. 55.
29   Ibid., p. 56.
30   Seager, Lt., E., Letter from the Crimea dated 26 October, 1854, NAM.
31   Calthorpe, op. cit., p. 129.
32   Cattell, op. cit., p. 11.
33   Kinglake, op. cit., p. 400.
34   Walker, op. cit., p. 136.
35   Anglesey, op. cit., p. 94.
36   Forrest, op. cit., p. 3.
37   Kinglake, op. cit., p. 324.
38   Ibid., p. 400.
39   Smith, op. cit., p. 131.
40   Jocelyn, Col. J.R.J., *History of the Royal Artillery (Crimean Period)*, 1911, p. 213.
41   Whinyates, op. cit., p. 139.
42   Ibid., p. 139.

Chapter Nine
1   Walker, op. cit, p. 138.
2   Kinglake, op. cit., p. 402.
3   Morley, op. cit., p. 10.
4   Seaton, op. cit., p. 150.
5   Ibid., p. 150.
6   Barrett, op. cit., p. 364.
7   Cattell, op. cit., p. 14.

8   Quoted in McGuffie, op. cit., p. 385.
9   Ibid., p. 385.
10 Ibid., p. 385.
11 Morley, op. cit., p. 10.
12 Paget, op. cit., p. 247.
13 Smith, op. cit., p. 133.
14 Paget, op. cit., p. 248.
15 Smith, op. cit., p. 135.
16 Ibid., p. 136.
17 Paget, op. cit., p. 248.
18 Ibid., p. 183.
19 Ibid., p. 185.
20 Anon., op. cit., p. 554.
20 Anon., op. cit., p. 554.
21 Small, E.M., *Told from the Ranks*, 1897, p. 66.
22 Pennington, op. cit., p. 57.
23 Seager, Letter from the Crimea dated 26 October, 1854, NAM.
24 Pennington, op. cit., p. 57.

Chapter Ten
1   Mitchell, op. cit., p. 85.
2   Compton, op. cit., p. 179.
3   Smith, op. cit., p. 152.
4   Whinyates, op. cit., p. 166.
5   Ibid., p. 168.
6   Ibid., p. 168.
7   Ibid., p. 170.
8   Harris, op. cit., p. 249.
9   Ibid., p. 251.
10 Paget, op. cit., p. 248.
11 Ibid., p. 189.
12 Ibid., p. 190.
13 Ibid., p. 191.
14 Ibid., p. 191.
15 Quoted in Barthorp, op. cit, p. 54.
16 Seager, Letter from the Crimea dated 26th October, 1854, NAM.
17 Pennington, op. cit., p. 58.
18 Seaton, op. cit., p. 152. The Russian guns that opened fire would have been Bojanov's battery near No. 3 Redoubt.
19 Ibid., p. 154.
20 Harris, op. cit., p. 245.
21 Kinglake, op. cit., p. 354.
22 Selby, op. cit., p. 168.
23 Cattell, op. cit., p. 13.
24 Mitchell, op. cit., p. 86.
25 Ibid., p. 87.
26 Ibid., p. 87.

27 Rum issues were not confined to the RN. The present author recalls a daily issue being given to his platoon on operations in the Malayan jungle in the late 1950s.
28 Smith, op. cit., p. 138.
29 Whinyates, op. cit., p. 140. While there is no doubt Smith arrived at 'C' Troop, he could not have appeared from between Nos. 2 and 3 as Whinyates suggests. He undoubtedly crossed the Causeway between Nos. 3 and 4.
30 Smith, op. cit., p. 144.

Chapter Eleven

1 The Fedioukine battery did not get back into firing positions before the Light Brigade returned. It could well have redeployed shortly afterwards and opened fire again, frightening away the Turks who had picked up Nolan's body.
2 Russell, op. cit., p. 160.
3 Kinglake, op. cit., p. 402.
4 Russell, op. cit., p. 159.
5 Whinyates, op. cit., p. 143.
6 Russell, op. cit., p. 165.
7 Quoted by ffrench-Blake, R.L.V., *The Crimean War*, Leo Cooper, London, 1971, p. 79.
8 Ewart, op. cit., p. 133.
9 Paget, op. cit., p. 72. Soldiers' grouses and excuses change little over the centuries. 2 Para complained bitterly that their belongings in their rucksacks had been rifled after the battle at Goose Green – they blamed the RM!
10 Smith, op. cit., p. 145.
11 Ibid., p. 147.
12 I have not been able to confirm the name of the artist.
13 Whinyates, op. cit., p. 142.
14 Russell, op. cit., p. 161.
15 Mitchell, op. cit., p. 88.

Chapter Twelve

1 Moyse-Bartlett, op. cit., p. 244. Heneage went on to win the VC as a captain during the Indian Mutiny. He led a successful squadron charge on an enemy camp and two gun batteries, capturing two under heavy fire.
2 Harris, op. cit., p. 262.
3 Portal, Letter to his mother dated 26 October, 1854, NAM.
4 Woodham-Smith, op. cit., p. 264.
5 Kinglake, op. cit., p. 414.
6 Paget, op. cit., p. 73.
7 Russell, op. cit., p. 160.
8 Woodham-Smith, op. cit., p. 267.
9 Kinglake, op. cit., p. 416.
10 Ibid., p. 417.
11 Moyse-Bartlett, op. cit., p. 238.
12 Quoted in Harris, op. cit., p. 274.
13 Woodham-Smith, op. cit., p. 269.
14 Kinglake, op. cit., p. 421.

15 Ibid., p. 420.
16 Woodham-Smith, op. cit., p. 270.
17 Kinglake, op. cit., p. 423.
18 Hibbert, op. cit., p. 296.
19 Woodham-Smith, op. cit., p. 272.
20 Ibid., p. 272.
21 Compton, op. cit., p. 216.
22 Ibid., p. 221.

INDEX

Note: Ranks are abbreviated conventionally; regiments are abbreviated in headings as 4LD, 8H etc.; Bold page numbers refer to maps and sketches; Italic numbers refer to chapter headings; subheadings are generally arranged chronologically.

1st Dragoons (Royals), 63, 105, 108, 170
2nd Dragoons (Scots Greys), 42, 105, 106, 108, 109, 170
4th Dragoon Guards, 105, 108, 109, 110, 171
4th Light Dragoons, 2, 43, 56
   parade strength, 11, 13, 18–19, 71
   in second line of Charge, 162–64
   reach the guns, 164, 178, 182, 183
   among the guns, 189–91
   return from Charge, 197, 200, 201, 209
5th Dragoon Guards, 47, 105, 106, 108
6th Dragoons (Inniskilling), 105, 106, 108
8th Hussars, 42, 44, 57, 243
   parade strength, 11, **13**, 17–18
   in second line, 162–63
   drift off line, 163, 166, 178, 183
   wheel about to attack lancers, 191–92
   return from Charge, 200, 202
   opinion of Cardigan, 198–99
   squadron with Sore-back reconnaissance, 49–50

band plays at Balaclava Banquet, 250, 251, 252
9th Lancers, at Chillianwala, 31
11th Hussars, 7, 10–11, 42–43
   Cardigan as CO, 27, 38–40
   advance to Balaclava, 57, 58, 59, 61–62
   parade strength, 11, **13**, 15–17
   pulled back to second line, 136, 137, 162, 164
   largely miss the guns, 163, 178, 182–83
   pursue Russians towards river, 186–89
   return from Charge, 197, 200–202, 209–10
13th Light Dragoons, 5, 42, 47
   with Sore-back reconnaissance, 49–50
   at Bulganek, 57, 58, 59
   parade strength, 11, 12, **13**, 14
   on right of the line, 136, 152
   among the guns, 182, 183, 185–86
   casualties, 217
16th Lancers, at Aliwal, 5, 31
17th Lancers, 36, 42, 44, 57
   parade strength, 11, 13, 14–15
   to direct line of advance, 19, 136, 152, 159–60
   among the guns, 182, 183–86
   return from Charge, 200, 202
46th Regiment, 56
63rd Regiment, 56
68th Light Infantry, No. 4 Redoubt, 208
93rd Highlanders, 43, 68, 75, 77, 118
   'Thin Red Line', 103–04

Abdelal, Maj, CO 4th Chasseurs, 142–43, 165

ADCs, role of, 28, 29, 67–68, 129

Airey, Brig-Gen Richard, Deputy Quartermaster-General, 3, 67, *230*
  at Bulganek, 58, 59
  at Alma, 60
  writes orders for Raglan, 67, 119, 121, 127
  goes to hurry up 4th Div, 119, 120
  sent to mollify Lucan, 233–34, 236

Alexandra Palace, Balaclava Banquet at, 247–48, 250, 251

Aliwal, Battle of (1846), 5, 31

Alma, Battle of the, 59–61, 87, 267n

Alma, River, 54

Ames, Capt, 258

Aqueduct (North Valley), 179, 181

Armes, Pte Thomas (4LD), 250, 251

Artillery
  arcs of fire, **141**, 142, 145, 159, 218
  case shot (canister), 87, 146, 159
  common shell, 140
  firing drill, 146–48, 150, 260
  howitzers, 87, 140, 146
  range, 75, 86–87
  rate of fire, 87, 142, 149, 150, 151, 218
  roundshot (cannonballs), 85–86, 87, 140, 264n
  Russian, 82, 140, 146–51, 270nn
  shrapnel, 86, 87, 140
  see also small arms

Balaclava, xi
  allied march to, 55, 56–63
  British base at, 63, 220–221

Balaclava Banquet (1875), 247–53

Balaclava, Battle of
  Russian plan of battle, 80–83
  strategic situation, **69**, 70
  tactical situation (0730), **76**
  tactical situation (0800), 91, **92**
  tactical situation (0830), 94, **95**
  tactical situation (0930), **116**, 117–18
  tactical situation (1055), 122
  tactical situation (1200), **211**
  see also under Raglan *and* Light Brigade

Balaclava Commemoration Society, 258

Balaclava Plain, xi–xii, 72, **73**, 74
  see also Causeway Heights; North Valley

Barker, Capt (RA), W Bty. 79–80, 87–88, 103

Barrett, John (W Bty), 268n

Beatson, Col, ADC to Scarlett, 104

Belbec stream, 54, 61–62

Bentley, Sgt (11H), 201–02, 210

Berkeley, Henry, MP, 238, 240

Berryman, Sgt John (17L) VC, 15, 204–05, 249, 252

Bingham, Cornet Lord (Lucan's son), ADC, 80, 84, 113, 240

Bingham, George Charles *see* Lucan, Lt-Gen, Earl of

Blunt, John, interpreter, 80, 84, 88
  Lucan gives 4th order to, 137, 231, 232
  witness to Nolan and Lucan, 130, 132, 134, 135

Bojanov, Capt, CO Russian 7 Light Bty, 143–45, 194, 210

Bosquet, General
  French 2nd Div, 72
  '*C'est magnifique . . .*', 168, **169**

Bowers, Capt (13LD), in Peninsula, 5

Boxall, Pte John (4LD) d.1914, 259

Brandling, Capt (C Troop RHA), 77, 78–79, 87, 101
  tries to support Charge, 173, 174
  after the Charge, 197, 214, 226

British Army
  deficiencies of, 23–24, 56, 245, 248
  incompetence of officers, 29, 31–32, 53
  'progressive advancement' for officers, 25–26, 265n
  purchase of commissions, 35–36
  structure of regiments, 11–12
  inquiry into Crimea, 245, 248
  see also Cavalry

British expeditionary force
  based at Balaclava, 63, 220–21
  conditions in Crimea, 62, 63, 71–72, 228–29
  effect of winter on, 235, 241
  see also Sevastopol

Britten, 'Billy' (17L), brigade trumpeter, 1, 6, 19, 152, 176
  death, 225–26

Brooks, Cpl John (13LD), 14, 248

Brown, Sir George, Commander Light
    Div, 52, 223
Brown, Capt John (4LD), 19
Brown, Trumpeter John (17L), 15
Brown, Richard (11H), 248
Brudenell, James, Lord *see* Cardigan, Gen
    The Earl of
Bugles and trumpets, 1–2, 226
Bulganek, River, skirmish at, 54, 57–59,
    57, 267n
Burghersh, Lord, 49

Calthorpe, Capt, Raglan's ADC and
    nephew, 43, 48, 94, 96, 168
  asserts Nolan given verbal briefing,
    129–30, 158
  sued by Cardigan, 245
Cambridge, HRH Duke of, commander
    1st Div, 43, 68, 223
Campbell, Brig Sir Colin, CO 93rd
    Highlanders, 68, 71, 80, 85, 97
  joins Cathcart at No. 4 Redoubt, 223
  protecting approaches to Balaclava,
    79–80
  and Russian assault on redoubts, 87–88
  and 'Thin Red Line', 103–04
Canovaro, Commandant, at Balaclava
    Banquet, 251
Canrobert, General, C-in-C French Army,
    67, 72, 77, 220, 222
Canrobert's Hill *see* Redoubts, No. 1
Cardigan, Gen The Earl of
  acquaintance with Raglan, 26–27
  career and character, 25, 37–40, 113–14,
    225
  drill methods, 51, 175, 244
  feud with Lucan, 36–37, 54–55, 59, 61,
    63
  Sore-back reconnaissance (June–July
    1854), 45, 49–50
  move to Crimea, 52–53
  at Bulganek, 57, 58–59
  believes he has independent command,
    46–47, 48, 52–53
  permitted to live on yacht, 63, 225, 244
  arrival on battlefield, 89
  inaction during Heavy Bde charge, 90,
    105–06, 112–13, 126–27
  sends ADC to Lucan, 132, 136

  and order to attack, 2, 19–20, 135–37,
    198
  leading the Charge, 152–53, 158–60, 161
  rides through guns, 150, 162, 176
  surrounded by Cossacks, 176–77
  turns back to guns, 195–96
  returns alone, 175, 177, 178, 194–97
  walks at head of 8th Hussars, 198–99
  blames Lucan, 219, 231
  absolved by Raglan, 231, 234
  and death of Britten, 225–26
  applies for sick leave, 243–44
  public adulation of, 244–45
  death, 246
Casualties
  Heavy Brigade, 172
  Light Brigade, 203–08, 216–19
  Russian prisoners, 224–25
  Russian treatment of, 226
Cathcart, Lt-Gen Sir George, Commander
    4th Div, 71–72, 223, 235, 267n
  dilatoriness, 67–68, 78–79, 118, 119–20,
    223
  near No. 4 Redoubt, 196, 223
Cattell, William, surgeon, 50, 168, 184,
    206
Causeway Heights, 3, 72, 156
  Rijov's advance towards, 100–101
  Russian artillery and infantry on, 118,
    125, 143–45
  see also Redoubts
Cavalry
  Nolan's enthusiasm for, 30–31, 32–33
  origins and role of, 10, 60
  theory of the charge, 4–7, 108–9, 158,
    159–60
Cavalry camps, wrecked, 105, 108, 215,
    224, 225
Cavalry Division
  effect of Cardigan-Lucan antagonism, 41,
    46, 48
  mobilization, 10–11
  numerical weakness, 41–43
  pickets and patrols, 71, 83
  Raglan keeps in 'bandbox', 59–61
  at time of Russian attack on redoubts,
    85, 89
  see also Heavy Brigade; Infantry divisions;
    Light Brigade

Chadwick, Cornet John (17L), 15

Chain of Command, 46, 67–68, 235

Chamberlayne, Lt (13LD), 204

Champeron, Col, Chasseurs d'Afrique, 165

Charge, The, 152–74
    first line, 158–62, 182–86
    second line, 162–66, 186–88
    rate of advance, **141**, 142, 144, 145, 147, 148–50
    they reach the guns, 162, 176, 177–78, 182–86
    among the guns, 182–93
    the return, 194–204, 215

Charteris, Capt, Lucan's ADC and nephew, 84, 85, 170, 226

Chernaya, River, 64, 71, 72
    ford, 81
    Tractir Bridge, 74, 99, 222, 269n

Cheronese, Cape, 54

Cheronese uplands, Light Brigade move to, 229

Chetwode, Capt George (8H), 18

Chillianwala, Battle of (1849), 31, 234

Cholera, 2–3, 50, 241, 266n
    among infantry, 56, 71
    on ships, 53

Chorgun village, 72, 80

Clark, Gnr Richard, 260

Clutterbuck, Lt (8H), 166

Col, The, route through Sapoune Heights, 70, 74, 78, 222

Cook, Capt Edwin (11H), 16, 209, 224–25

Cossacks see Russian Army

Crawford, Trumpeter (4LD), 185

Crimea, The, xi–xii, 41, 52

Crimean War
    conditions in, 62, 63, 228–29, 241
    origins of, 21, 23
    supplies, 56

Crimean War Medal, 248, 249

Cruikshank, Mr., commissary officer, 208

Curzon, Cornet, ACD to Raglan, 112

D'Allonville, Gen, Commander French Cavalry Bde, 165, 225

Dashwod, Lt, I Troop RHA, 88

De Grancey, Baron, at Balaclava Banquet, 251

De Ros, Lord, Deputy Quartermaster-General, 46, 47, 53

De Salis, Maj Rodolph (8H), 17, 202, 216

Devna, Light Brigade at, 48

Dickson, Lt (RA), W Bty, 88

Doherty, Lt Col Charles (13LD), 12

Douglas, Lt Col John, CO 11H, 15, 85, 183, 224, 246, 265n
    attacked by Uhlan lancers, 200–2, 209
    order from Lucan, 4, 136, 137
    pursues Russian gunners, 186–87, 189

Doyle, Pte John (8H), 17

Duberley, Capt Henry (8H), 11

Duberley, Mrs, journal, 11, 49

Dudley, Pte (17L), 161

Dunn, Lt Alexander (11H) VC, 16, 201–02, 209, 210, 265n

Earle, Lt (75th Foot), 225

Elliot, Lt, 100, 104, 105

Estcourt, Maj-Gen, Adjutant General, 48, 241

Ewart, Capt, ADC, 77, 103, 196, 205–06, 221
    sent to hurry Cathcart, 67–68, 119–20

Farquharson, Pte (4LD), 71

Farrell, Sgt John (17L), 15

Fedioukine Heights, 3, 72, 77, 78, 91, **169**
    arc of fire from, **141**, 158, 164, 170, 218
    Chasseurs charge Russian guns, 142–43, 164–65, 170
    Jaboritsky deploys on, 82, 117, 125, 126–7, 140, 223
    Russian No. 1 Bty on, 139–40, **141**, 142, 149, 273n

Fellows, Capt, 226

Fenton, Roger, photographer, xiii, 241

Filder, William, Commissary General, 34, 44

Fisher-Rowe, Cornet (4DG), 110

Fitzgibbon, Lt (8H), 166

Flank march around Sevastopol, 61–63

Fletcher, Pte (4LD), 185

Ford, Pte (4LD), 195

Forrest, Maj (4DG), 171
    on Cardigan, 49, 60
    on Heavy Bde attack, 109, 110, 111–12
    on Lucan, 50–51, 60

Fox Strangways, Lt (RA), *90*
France, and origins of Crimean War, 21, 23, 52
Franks, TSM (5DG), 106
French Army, 41–42, 63, 70, 72, 120
  Chasseurs d'Afrique (1st and 4th), 6, 118
    4th charge Fedioukine guns, 142–43, 164–65, 170
    move into North Valley, 120–21, 125, 140
    move towards South Valley, 78
    and Raglan's 4th order, 132
Fulton, Pte William S.J. (8H), 259

Gardner, Cornet George (13LD), RSM, 14
Gibson, Pte George (13LD), 259
Goad, Capt Thomas (13LD), 12
Goad, Cornet George (13LD), 12
Goldie, Brigadier, 1st Infantry Bde, 71
Gore, Maj William (13LD), 12
Govone, Maj Giuseppe (Sardinian), 14, 216
Gowing, Sgt Thomas, *21*
Great Britain, and war with Russia, 23, 52
Grenville, Charles, on Cardigan, 245–46
Gribbe, Maj Gen, Russian Bde Commander, 81, 83, 85, 181
Grigg, Pte (4LD), 191
Guards Brigade (1st Div), 119
Gun spikes, 17

Halkett, Maj John (4LD), 1, 19
Harding, RSM Robert (8H), 18
Hardinge, Capt, 62, 97, 102
Hardinge, Lord, C-in-C British Army, 27–28, 37, 238, 239
Harrison, Sgt Major (8H), 166
Heavy Brigade, 42, 63
  eight squadrons moved to support Campbell, 97, 105
  position (0830), 96, 102
  attack ('charge'), 5, 105–06, **107**, 108–12, 224
  position (0930), 6, 118
  led by Lucan after Light Bde, 142, 167, 169–71
  withdrawn, 171–72, 207
  cheer for returning Light Bde, 198–99
Heneage, Lt Walker, *230*, 240, 273n
Higginson, George (Gren Gds), 112

Highland Brigade (1st Div), 57, 119, 223
Hope, Pte Henry (11H), 16–17
Horse artillery *see* Royal Horse Artillery
Horses
  shortage of, 44, 63, 266n
  transport problems, 42–43, 53
  casualties of Charge, 164, 203–04, 216
  losses, 49, 63, 266n
  maimed by Cossacks, 224
  Nolan's system of training, 32–33
  Black Bess (Pennington), 165
  Crimean Bob (Sgt Avison), 216
  Old Trumpeter (Morris), 15, 183
  Ronald (Cardigan), 20, 153, 244, 246
Houghton, Lt (11H), 199, 225
Howes, Sgt John (4LD), 250–51
Hughes, Pte Edwin (13LD), 14, 259
Hunt, Lt (4LD), 189–90
Hutton, Capt (4LD), 204

Illustrated London News, on Balaclava Banquet, 247–48, 251
India, British Army in, 30, 31–32
Infantry divisions
  1st (Duke of Cambridge), 68
    called forward, 77, 78, 118–19, 126
    see also Highland Brigade
  4th (Cathcart), 71
    Raglan calls forward, 67–68, 77, 78
    slowness of advance, 78–79, 91, 118, 119–20
    see also 68th Light Infantry
Inkerman, Battle of, 223, 235
Ireland, Lucan in, 34–35

Jaboritsky, Maj Gen, Russian Flank Guard, 77, 82
  on Fedioukine Heights, 117, 126–27, 140, 223
Jaenecke, Col-Gen (German 17th Army 1944), xi
'Jemmy', 8H terrier, 18, 216
Jenkins, Gnr David (W Bty), 268n
Jennings, RSM Henry (4LD), 19
Jenyns, Capt Soames (13LD), 12, 49–50, 182, 199, 204
Jeropkine, Col, Uhlan Lancers, 81, 178–79, 181–82, 193
Jervis, Lt (13LD), 178, 185–86

279

Joseph, Sgt (11H), 210
Joy, Trumpet-Major, 84, 106, 130, 152, 226

Kadikoi village, 72, 89, 119–20
Kalamita Bay, Crimea, 54, 56
Kamara, pickets at, 83
Kamiesch, French army based at, 63
Kilvert, Cpl Ashley (11H), 259
Kirdner, Col, 24th Dneiper Regt, 81
Kozhukhov, Lt, Russian light artillery, 203
Krudener, Col, Azov regiment, 85
Kubitovitch, Capt, Uhlan Lancers, 179, 181, 203

Lamb, Pte (17L), 204–05, 250
Lances, 8–9, 100, 264n
Landriani, Lt Giuseppe (Sardinian), 14, 217
Lawrenson, Lt Col John (17L), 14
Lee, Sir Edward, 251, 252
Lee, Pte John (17L), 15, 161
Levett, Pte (11H), 202
Levutsky, Maj Gen, Russian Bde Commander, 81, 87
Light Brigade, 42, 44, 48
  advance to Balaclava, 56–63
  breakfast before Charge, 1–2, 125
  position (0830–0930), 94, 96, 97, 117–18
  does not pursue Rijov, 105–06, 112, 113
  expectation of infantry support, 125, 187, 225
  moved forward (1000), 124–25
  forming up, **13**, 135–36, 137
  survivors, 258–59
  see also 4LD; 8H; 11H; 13LD; 17L; Charge, The
Linkon, TSM John (13LD), 14, 251
Liprandi, Lt Gen, Russian 11th Inf Div, 99, 101, 181
  battle plan, 81, 82, 87, 90, 91
  strategic advantage, 220–21, 223
Lockwood, Capt George (8H) ADC, 19, 176
Low, Capt Alexander (4LD), 19, 83, 200
Lucan, Lt-Gen The Earl of, commanding Cavalry Division
  career and character, 34–37, 63, 241
  drill methods, 50–51

enforcement of discipline, 47, 61–62
feud with Cardigan, 36–37, 54–55, 59, 61, 63
and Cardigan's independence, 46–47, 48–49, 53
and Raglan's refusal to use cavalry, 57, 58–9, 63, 124
nicknamed 'Lord Look-on', 64
on flank march, 62–63
and Campbell, 80, 96
reaction to first Russian attack, 85
reaction to Raglans' 1st order, 91, 94, 96–97
and Heavy Bde charge, 102, 105–06, 113
interpretation of Raglan's 3rd order, 121, 124, 126, 131
omits to use RHA, 172–74
questions Nolan on 4th order, 130–35, 158, 231–32
orders Cardigan to attack, 4, 135–37
leads Heavy Brigade after Light, 142, 171–72
share of responsibility, 134, 137, 225, 242–43
meets Raglan after Charge, 219
disclaims dislike of Nolan, 227
determination to clear name, 231–34, 235–40
recalled, 238–40
ends career as Field Marshal (d.1888), 243

Mackenzie's Farm (Simpheropol Road), 62–63
Malone, Cpl Joseph (13LD) VC, 14, 204–05, 249, 250, 252
Marmora, General della, 242
Marsh, Pte (17L), 161, 185
Martyn, Cornet Fiennes (4LD), 19, 163
Maude, Capt, CO I Troop RHA, 51, 87–88
Maxse, Lt Fritz, 155
Maxse, Lt Henry (21st Foot), Cardigan's Brigade ADC, 19, 136, 176, 184, 225
Mayow, Lt Col George (4DG), Brigade Major, 19, 196
  in the Charge, 176, 184–85, 186
  through the guns, 191, 192
  on way back, 193, 199, 200, 202

McDonald, Capt, ADC, 68
McGarry, Gnr Jacob (W Bty), 268n
McMahon, Maj, Asst QM General, 84–5, 170
Mentschikoff, Prince, 21
Mitchell, Pte Albert (13LD)
  account of war, 42, 228–29, 250
  on retreating Turks, 90–91
  on Heavy Bde charge, 105, 113
  Light Bde moves forward, 124–25
  sees Nolan point to guns, 137
  description of Charge, 160–61
  on Cardigan's return, 195
  returns on foot, 207–08
Monks, Trumpet Major (5DG), 104, 106
Montgomery, Capt (49th Foot), 221, 222
Morale, 51–52, 215–16
Morgan, Capt (17L), 150, 156, 161, 184
Morgan, Sir Godfrey, at Balaclava Banquet, 252
Morley, Cpl Thomas (17L), 15, 139, 155, 156, 178, 250
  description of Charge, 159–60, 162, 185–86
Morris, Capt William, CO 17L, 5, 14–15, 258, 264n
  friendship with Nolan, 15, 32
  asks to pursue Rijov, 112–113
  questions Nolan, 3, 4
  convinced that Nolan ordered direction of Charge, 134
  on Nolan's dash forward, 155, 156
  in the Charge, 159
  behind the guns, 182, 183–84
  returns wounded on foot, 205–07, 212
Morris, General, French Cavalry commander, 142
Mouat, Dr, VC, 206
Mundy, Col, Under Secretary of State, 238
Mustard, Pte James (17L), 185, 259

Napoleon, Prince, French 3rd Division, 43
Newcastle, 5th Duke of, Secretary at War, 33, 52
  Lucan's letter to, 235–36
  Raglan's private letter to, 235
  recalls Lucan 238–39
Nicholas I, Czar of Russia, 21
Nightingale, Florence, 241

Nolan, Capt Louis (15H), ADC to Airey
  career and character, 28–34
  enthusiasm for cavalry, 6, 30–31, 32–33, 135
  friendship with Morris, 15, 32
  reputation of, 33, 43–44, 79, 126
  responsible for remounts, 44, 49
  at Bulganek, 58, 59
  contempt for Cardigan, 63
  public contempt for Lucan, 58, 60–61, 63, 64, 126, 134
  reaction to Heavy Bde action, 112
  understanding of Raglan's order, 129–30, 157–58
  gesture towards guns, 132–33, 134, 135
  passes order to Lucan, 3–4, 130–34, 231–32
  dashes forward, 153, 154–58
  death of, 142, 153, 155, 156, 205–06
  burial, 226–27
  responsibility for attack, 31, 135, 238, 242
North Valley, 72, 74, 117, 179
  view from Light Brigade, 3, 156
Nunnerley, Cpl James (17L), 15, 155, 156

Obolensky, Col Prince, CO 3 Don Bty, 145, 146–47, 149–50
O'Hara, TSM (17L), 186, 206
Old Fort, Kalamita Bay, 54
Oldham, Capt John, CO 13LD, 12, 182
Olley, Pte James (4LD), 259
Omar Pasha, Turkish C-in-C, 242
Orders
  importance of, 115
  and role of messenger (ADC), 129, 133, 232
  and use of discretion, 133, 135, 238
  written/verbal distinction, 132, 232
Orders, Raglan's see Raglan
Ormes, Bombardier O., 226

Paget, Col Lord George, CO 4LD, 1, 18–19, *41*, 63, 183
  on Cardigan–Lucan feud, 54, 64–65
  and Lucan's drill methods, 51
  on Nolan, 126
  position before Charge, 84–85, 86, 96, 109

Paget, Col Lord George, CO 4LD (*cont.*)
  orders from Cardigan, 2
  second line of Charge, 162–63
  among the guns, 183, 186, 189–90
  return from Charge, 200–201, 209, 210
  on Cardigan's behaviour after Charge,
    175, 183, 189, 198
  angry at lack of support, *194*, 197–98
  on recriminations, 133, *230*
  on wrecked lines, 224
Palmer, Lt Roger (11H), 59, 200, 209
Palmerston, Viscount, 52
Parkes, Pte (4LD) (later VC), 185, 189
Parkinson, Pte John Smith (11H), 259
Parliament, debates on Crimea, 238
Paulet, Lord, Asst Adjutant General, 84–85,
    170
Peel, Maj Edmund (11H), 15
Pélissier, General, French C-in-C, 241–42
Peninsular War, 5, 25–26
Penn, Pte John (17L), 15, 185
Pennington, Pte William Henry (11H), 16,
    84, 175, 186, 191, 259
  career as actor, 250, 252
  completes Charge with 8H, 165–66,
    194, 202–03
Perkins, Trumpeter William (11H), 15
Phillips, Lt Edward (8H), 192, 259
Pickets and vedettes, 71, 83, 99
Pickles, Sgt (11H), 224
Pickworth, Sgt (8H), 17
Pilkington, Pte (11H) (Tom Spring), 186
Pipe smokers, 17–18
Pistols *see* small arms
Pollard, Pte (4LD), 195
Portal, Capt (4LD), 43, 51–52, 63, 232,
    240
Priestley, Sgt Joseph (13LD), 59
Prisoners, Russian treatment of, 226

Radziwill, Prince, recognizes Cardigan,
    177
Raglan, Gen Lord, C-in-C
  career, 24–28
  character, 25, 27, 28, 115
  and Cardigan's scandals, 26–27, 39
  inability to give direct order, 59, 127,
    129

and Cardigan–Lucan feud, 41, 46, 48,
    61, 64–65
  at Alma, 60, 267n
  at Bulganek, 57–59, 267n
  keeps cavalry in 'bandbox', 41–42,
    57–61, 94, 118, 124
  on flank march, 62–63
  view from Sapoune Heights, xiii, **66**, **93**,
    98, **123**, **128**
  forbids use of Woronzoff Road, 78
  perception of Russian plans, 91, 118
  1st order to infantry (0730), **66**, 69, 75,
    77–78, 268nn
  1st order to cavalry (0800), 91, **93**, 94
  2nd order to cavalry (0830), 97, **98**,
    268–69n
  visibility of Rijov's force, 101–02
  3rd order to cavalry (0930), 117–18, 121,
    123, 124
  4th order to cavalry (1055), 125–26, 127,
    128, 129–30, 131–34, 158, 231–32
  verbal order to Nolan, 129–30, 132, 231,
    232
  view of the Charge, 167–68
  decides not to follow up after charge,
    220–23
  meeting with Lucan afterwards, 216,
    231–32
  and Lucan's letter to Newcastle, 236–37
  official despatch, 233, 234–35
  private letter to Newcastle, 235
  relations with Cathcart, 72, 234–35
  share of responsibility, *230*, 240–41
  death and funeral, 241–42
Ransom, Sgt Major (17L), 185
Redoubts on Causeway Heights, 72, **73**,
    74–75
  Bojanov's battery fire from near No. 3,
    143–45
  No. 1 (Canrobert's Hill) falls to Russians,
    72, 75, 77–78, 85, 87–89
  No. 4 retaken by 68th Light Infantry,
    208
  No. 5 occupied by Turks, 120, 125, 206,
    234
  Nos. 1–3 held by Russians, 88, 91, 215,
    220, 222–23
  Nos. 5 and 6 unfinished and unoccupied,
    74–75, 94

Raglan's plan to recover, 118, 120–21, 124, 127

Raglans' view of, 91, 94, 96, 97, 117

report of Russians removing guns from, 127, 167

Russian plan of attack, 81–82

true objective of Charge, 127, 131, 133, 156–57

Reilly, Cornet John (4LD), 19

Reynolds, Capt, 'black bottle' affair, 39

Rifles *see* small arms

Rijov, Maj Gen, Russian 6th Hussar Brigade, 77, 82, 99–100, 203, 223

arrives in North Valley, 96, 97, 100–102, 268–69nn

attack on 93rd Highlanders, 103

attack on Heavy Brigade, 103–04, 106, 108, 109–11

with Obolensky, 145–47, 149

sends Cossacks against Light Bde, 178–79, 181, 193

Roads, 74, 221–22

see also Woronzoff Road

Roberts of Kandahar, Field Marshal Lord, 258

Roberts, T. Harrison, veterans' fund, 258–59

Rogers, Pte William (11H), 40

Romania (in Turkish empire), 21, 23

Royal Artillery

E Field Bty, 58

W Bty, 75, 79–80, 87

Royal Horse Artillery, 5, 197

C Troop, 58, 77, 78–79, 87, 101, 110–11

follows Light Bde, 172, 173–74, 213–14

I Troop, 6, 58, 64, 87–88, 89, 214

in support of Cav Div, 105, 131, 172–73, 174, 197

Royal Marines, close protection of Balaclava supply route, 71, 80, 104

Russell, William, *The Times* correspondent, 20, 56, 62, 79

buys Nolan's cloak, 79, 227

on choice of Cardigan, 44, 46, 219

on defence of redoubts, 75, 77

Lucan denies blame, 233

on Raglan, 27, 219, 238

on reaction to Heavy Bde attack, 112

Russia

invades Turkey, 44, **45**, 49

and origins of Crimean War, 21, 23

Russian Army, 36, 41, 52, 62

attack on redoubts, 81–82, 85–89

battle plan, 80–83, 90

centred on Chorgun village, 72, 80

strategic advantage, 71, 220–22, 223

supply route into Sevastopol, **69**, 70, 71

11th Infantry Division (Liprandi), 80–81

6th Hussar Cavalry Brigade (Rijov), 82, 99–100

1st Ural Cossacks, 77, 82

3 Bty of Don Cossacks, 100, 145–51, 179, 270nn

11th Kiev Hussars, 77, 99–100, 213

12th Ingermanland Hussars, 82, 100, 103–04, 108

12th Light Horse Bty, 82, 100, 179

53rd Don Cossacks with, 100, 103

23rd Azov Regiment, 81, 85, 223

23rd Ukrainian Regiment, 81, 82, 223

24th Dnieper Regiment, 81, 223

24th Odessa Jaeger (Rifle) Regiment, 81, 143–45, 223

31st Vladimir Regiment, 82, 140, 143, 165

32nd Sousdal Regiment, 82, 140

53rd Don Cossacks, 77, 81, 83, 100

60th Don Cossacks, 82

Black Sea Foot Cossacks, 82, 140

Cossack Cavalry, 64, 100, 244

at Bulganek, 58, 59

pursuit of Turks, 88, 90–91

riflemen (companies of 4th Rifles), 81, 140

Uhlan Lancers (composite regiment), 81, 83, 174

attack returning survivors, 209–10

move to cut off Light Brigade, 180, 181–82, 191, 193, 200–203

artillery, 82, 85, 86, 87

12th Artillery Bde, 7 Light Bty, 81, 143

16th Artillery Bde, 1 Bty, 139–40, 142

on Causeway Heights, 118, 125

St Arnaud, General, C-in-C French Army, 72

Sapoune Heights, xi, 70, 72, 166–68
   view from, xiii, **66**, **93**, **98**, **123**, **128**

Sardinian Army, officers attached, 14

Scarlett, Brig Gen, commanding Heavy
      Brigade, 99, 102, 104, 171, 199,
      219–20
   Heavy Brigade Charge, 105–06, 108–12,
      234

Scutari, allied armies at, 43–44

Seager, Lt Edward (8H), 18, 51, 166, 192,
      202

Second World War, Crimea in, xi

Semiakin, Maj Gen, Russian Bde
      Commander, 81, 87

Sevastopol, xi, 52, 221, 222
   attack (18 June 1855), 241
   flank march around, 61–63
   siege of, 64, **69**, 70, 267–68nn

Shadwell, Lt, ADC to Campbell, 104

Shakespear, Capt (RHA), 51, 87, 89, 196
   I Troop RHA, 131, 172, 173

Shawe Smith, Capt Percy (13LD), 259

Shegog, Sgt, Scarlett's orderly, 104

Shewell, Lt Col Frederick, CO 8H, 17–18,
      163, 183
   in the Charge, 191–92, 198–99, 202

Sikh Wars (First and Second), 31

Silistria, 44, **45**, 49

Simpheropol, road from, 62, 70, 71

Simpson, General, 242

Sinope (Turkey), bombardment, 23

Skiuderi, Col, Russian Bde Commander,
      81, 82, 88, 99

Small arms
   carbines, 9, 100
   Minié rifle, 82, 86
   pistols and revolvers, 9
   Russian infantry musket, 82, 143
   see also lances; swords

Smith, Lt Percy (13LD), 12, 14, 177, 178,
      204

Smith, Pte George (5DG), 206

Smith, TSM George (13LD), 14

Smith, TSM George Loy (11H), 7, 16–17,
      23, 250
   on Cardigan, 38, 40
   on loading horses, 42–43

Lucan's irrationality, 61–62
   on fall of redoubts, 88
   limited view of North Valley, 96
   the Charge, 164–65, 172
   pursuit towards river, 187–89
   last man back, 208–10, 212–14
   after the Charge, *215*, 223–25
   on Yates, 196

Somerset, Lord Fitroy *see* Raglan, Gen
      Lord

Somosierra, Pass of (1808), 5

Sore-back reconnaissance (June–July 1854),
      45, 49–50, 266n

South Valley, 72, 74

Swords, 7–8, 32, 100, 110–11
   cut or thrust controversy, 7–8, 183–84,
      228

Talbot, Sgt (17L), 161

Taylor, Cpl (8H), 190

Times, The
   Lucan, writes to, 240
   Russell writes to, 238

Tomkinson, Capt Edward (8H), 18, 166

Tractir Bridge (over Chernaya), 74, 99,
      222, 269n

Transport ships, 42, 43, 46, 53, 54, 242

Tredegar, Lord, at Balaclava Banquet, 252

Tremayne, Capt (13LD), 49, 161
   no doubt Nolan ordered direction of
      Charge, 134, 137, 155–56

Trevelyan, Lt Harington (11H), 15, 252

Tunisian troops, in Turkish Army, 74–75

Turkish Army, 43, 71, 103, 104
   in redoubts, 72, 74–75, 77–78, 85
   abandon redoubts, 88–89, 90–91
   retake No. 5 redoubt, 120, 125, 206, 234

Turkish Empire
   and origins of Crimean War, 21, **22**, 23
   Russian invasion of, **45**, 49

Turner, Pte (11H), 164

Uniforms, 11, 12, 14, 15, 17

Varna, move to, 44, 46–48

Veigh, Pte John (17L), 15, 264n

Verkhkitsky, Capt, Uhlan Lancers, 203

Victoria Cross, 249
   Berryman, 204–05, 249

Dunn, 202
Farrell, 204–05
Malone, 204–05, 249, 250
Mouat (Dr), 206
Wooden, 206, 249–50
Victoria, Queen, Diamond Jubilee, 258
Vineyards, xii, 74, 109

Waagman, Prussian engineer, 75
Walker, Capt, ADC to Lucan, 84, 170, 176
  witness to Nolan and Lucan, 130, 133,
    134
Ward, Pte (11H), 164
Warr, Pte Thomas (11H), 259
Wathen, Capt (15H), 37
Webb, Capt (17L), 204–05
Wellington, Duke of
  and Cardigan, 39–40
  and Raglan, 25–26, 27
Weston, TSM John (13LD), 14
Wetherall, Capt, on staff of DOMG, 52,
  62, 91, 94, 119, 120
White, Capt Robert (17L), 15, 159, 161,
  251

Whitehead, Pte John (4LD), 259
Wightman, Pte James (17L), 14, 15, 113
  the Charge, 161, 184–85
  on Nolan, 3, 4, 153–54
Wilkin, Asst-Surg. Henry (11H), 16, 209
Willett, Maj Augustus (17L), 14, 71
Williams, Sgt (8H), 17–18
Williamson, Pte (11H), 59
Wilson, Samuel (8H), 252
Wilson, William (8H), 252
Wombwell, Cornet George (17L) ADC,
    19, 63, 176, 251, 252
Wooden, Sgt Charles (17L) VC, 15, 206,
    249–50, 252
Woodham, Edward (11H), 252
Woronzoff, Prince, 36
Woronzoff Road, 72, 77
  Raglan forbids use of, 67, 78, 268n
  strategic importance of, 220, 221–22

Yates, Cornet John (11H), 11, 16, 196–97
Yorke, Col, CO 1st Dragoons, 170–71
Young, Pte (11H), 164